Shelf Life Assessment of Food

FOOD PRESERVATION TECHNOLOGY SERIES

Series Editor
Gustavo V. Barbosa-Cánovas

Shelf Life Assessment of Food
Editors: Maria Cristina Nicoli, University of Udine, Italy

Cereal Grains: Laboratory Reference and Procedures Manual
Sergio O. Serna-Saldivar

Advances in Fresh-Cut Fruits and Vegetables Processing
Editors: Olga Martín-Belloso and Robert Soliva-Fortuny

Cereal Grains: Properties, Processing, and Nutritional Attributes
Sergio O. Serna-Saldivar

Water Properties of Food, Pharmaceutical, and Biological Materials
Maria del Pilar Buera, Jorge Welti-Chanes, Peter J. Lillford, and Horacio R. Corti

Food Science and Food Biotechnology
Editors: Gustavo F. Gutiérrez-López and Gustavo V. Barbosa-Cánovas

Transport Phenomena in Food Processing
Editors: Jorge Welti-Chanes, Jorge F. Vélez-Ruiz, and Gustavo V. Barbosa-Cánovas

Unit Operations in Food Engineering
Albert Ibarz and Gustavo V. Barbosa-Cánovas

Engineering and Food for the 21st Century
Editors: Jorge Welti-Chanes, Gustavo V. Barbosa-Cánovas, and José Miguel Aguilera

Osmotic Dehydration and Vacuum Impregnation: Applications in Food Industries
Editors: Pedro Fito, Amparo Chiralt, Jose M. Barat, Walter E. L. Spiess, and Diana Behsnilian

Pulsed Electric Fields in Food Processing: Fundamental Aspects and Applications
Editors: Gustavo V. Barbosa-Cánovas and Q. Howard Zhang

Trends in Food Engineering
Editors: Jorge E. Lozano, Cristina Añón, Efrén Parada-Arias, and Gustavo V. Barbosa-Cánovas

Innovations in Food Processing
Editors: Gustavo V. Barbosa-Cánovas and Grahame W. Gould

Shelf Life Assessment of Food

Edited by
Maria Cristina Nicoli

CRC Press is an imprint of the
Taylor & Francis Group, an **informa** business

CRC Press
Taylor & Francis Group
6000 Broken Sound Parkway NW, Suite 300
Boca Raton, FL 33487-2742

First issued in paperback 2016

ISBN 13: 978-1-138-19934-7 (pbk)
ISBN 13: 978-1-4398-4600-1 (hbk)

Library of Congress Cataloging-in-Publication Data

Shelf life assessment of food / [edited by] Maria Cristina Nicoli.
p. cm. -- (Food preservation technology series)
Includes bibliographical references and index.
ISBN 978-1-4398-4600-1
1. Food--Storage. 2. Food--Shelf-life dating. I. Nicoli, Maria Cristina.

TP373.3.S483 2012
641.4'8--dc23 2011042805

Visit the Taylor & Francis Web site at
http://www.taylorandfrancis.com

and the CRC Press Web site at
http://www.crcpress.com

Dedication

To my unforgotten master Carlo Raffaele Lerici, with whom I would have liked to edit this book.
To Maurizio and Laura for their support.

Contents

Preface

The term *shelf life* refers to a complex and fascinating concept whose definition is still far from being exhaustive. According to one of the most widely accepted definitions, shelf life is a finite length of time, after manufacture and packaging, during which the food product retains a required level of quality acceptable for consumption. In this sentence, there are some interesting key words: time, quality, and consumption. They suggest that the food lifetime, namely shelf life, is the result of the relation between product quality and its consumption acceptability. In other terms, the shelf life concept aims to express the complex and sometimes elusive interactions between food and consumers.

This concept implicitly merges a well-defined issue (i.e., the understanding of food quality evolution over storage time) with a more indefinite element (i.e., food acceptability, which depends on regulations, and economic, marketing, and social issues). The latter involves dynamic aspects that may vary greatly from country to country and are liable to change according to societal transformations. It is thus evident that food acceptability, as the result of the combination of choices and considerations accounting for different issues, is very difficult to quantify and give a finite value.

The difficulty to define clearly what food shelf life actually is has had repercussions on the methodologies developed to carry out its assessment. According to the complexity of the shelf life concept, these methodologies should merge scientifically based procedures, mainly addressed to mathematically describe food quality decay during storage, with others. These methodologies are developed to define what food acceptability is and thus transform it into a finite quality value to be used as the acceptability limit. The acceptability limit is the endpoint discriminating acceptable products from ones that are no longer acceptable.

Based on these considerations, it is evident that a shelf life study is a hard job that cannot be confused with a stability study. While the latter is addressed to measure the rate of food quality decay during storage, the aim of a shelf life study is to compute a time value that is the food lifetime, derived by merging the food quality decay rate with the acceptability limit. To reach this goal, a multidisciplinary approach is generally required. This peculiar feature makes the shelf life assessment a complex and challenging task (see Figure 1).

Despite the huge amount of scientific literature in which the term *shelf life* is reported, little effort has been made to identify the food acceptability limits to combine them with rate data. Unfortunately, most shelf life studies are actually stability studies since they fail to determine the acceptability limit. While a variety of mathematical models has been proposed to describe food quality decay, surprisingly little attention has been paid to defining strategies and developing relevant methodologies for the identification of the acceptability limit. The reason probably lies in the indefinite and sometimes "arbitrary" nature of the acceptability limit, which makes this subject scarcely interesting or, more probably, too difficult to be scientifically defined. As a result, despite the availability of a variety of sophisticated mathematical

Figure 1 Joys and sorrows in food shelf life assessment.

equations and statistics techniques to be used to model food quality decay over storage time, food companies hardly ever carry out the identification of the acceptability limit on a rational basis. The contradiction is dreadfully evident: While on one hand accurate and reliable data of quality decay rate can be obtained by applying proper modeling procedures, on the other hand the identification of acceptability limit is still frequently empirically carried out, frustrating any effort into transforming the shelf life issue into a scientifically based task.

These are probably the main reasons that have frequently hindered a unified and coherent view of the process of shelf life assessment. This issue is generally faced by stressing only specific aspects such as kinetic modeling, temperature-based accelerated shelf life testing, description of the main food degradation reactions and shelf life extension issues, and more recently, sensory shelf life assessment.

Working on some topics related to food stability, my coworkers and I were destined to meet the shelf life assessment issue. Some of the less-investigated aspects have been our main research topics for almost a decade. We collected records of a range of cases that represented an invaluable opportunity for rationalizing the number of activities required to estimate or predict food shelf life. In light of these experiences, I developed the idea of the present work on shelf life assessment of food. This book is therefore an attempt to provide an integrated view of the present status of the shelf life assessment issue with the aim of guiding the readers through the possible criteria and current methodologies to be pursued to obtain accurate and reliable shelf life dating. Obviously, the book reflects my view on what a shelf life assessment process should actually be like and how it should be carried out. With that purpose in mind, the book deals with this issue considering a number of sequential steps, each requiring different choices and thus the adoption of proper and dedicated methodologies. The main objective is to assist and support the users with theory and practical examples in planning a shelf life study that best fits their needs.

The main objective was to write a reference book to be used by food industry managers, consultants, and government agency operators dealing with shelf life assessment and quality assurance issues. The organization of the topics was designed in such a way to obtain a comprehensive, scientifically based, and whenever possible, friendly guide aiming to support food industries in the adoption of rational and scientifically based approaches in the shelf life assessment process. The book is even addressed to food scientists who are willing to improve their skills in shelf life assessment to stimulate further debates and research activity in this challenging and open research area. The book could therefore be helpful as a student textbook in advanced master of science and doctoral courses in food science and technology and food engineering.

Since some chapters deal with mathematical modeling, the book requires basic mathematical and statistical knowledge. However, the text is organized into two different parts: The first one allows an understanding of the main concepts beyond the equations, and the second is developed to support those readers who are willing to become more confident with the methodologies under discussion. This more in-depth section can be found in boxes showing examples of experimental data modeling and providing the relevant statistical procedures.

I would like to acknowledge several people who have had a pivotal role in helping me in realizing this project. A warm and grateful thanks go to my coworkers Lara Manzocco, Sonia Calligaris, and Monica Anese for their support, especially for our heated debates, which were helpful in clarifying a number of doubts on the shelf life assessment issue and ultimately creating some new ones. I also take the opportunity to thank Corrado Lagazio, professor in statistics at the University of Udine, for his effort in allowing us, and I hope even the readers, to think *statistically* about shelf life assessment. Furthermore, I would like to thank all the authors for their active participation and willingness to follow my suggestions and indications. I am also grateful for Gustavo Barbosa-Canovas's trust and the challenging opportunity he gave me. A special acknowledgment is given to Steve Zollo for his constant optimistic encouragement and for constant presence when I looked for help.

Maria Cristina Nicoli
University of Udine

List of Contributors

Monica Anese
Dipartimento di Scienze degli Alimenti
University of Udine
Udine, Italy

Sonia Calligaris
Dipartimento di Scienze degli Alimenti
University of Udine
Udine, Italy

Fausto Gardini
Dipartimento di Scienze degli Alimenti
University of Bologna
Bologna, Italy

Lorena Garitta
Instituto Superior Experimental de Tecnologìa Alimentaria
Buenos Aires, Argentina

Guillermo Hough
Instituto Superior Experimental de Tecnologìa Alimentaria
Buenos Aires, Argentina

Rolf Ibald
Department for Logistic Management
European University of Applied Sciences
Brühl, Germany

Judith Kreyenschmidt
Cold-Chain Management Group
University of Bonn
Bonn, Germany

Corrado Lagazio
Dipartimento di Scienze Economiche e Statistiche
University of Udine
Udine, Italy

Rosalba Lanciotti
Dipartimento di Scienze degli Alimenti
University of Bologna
Bologna, Italy

Sara Limbo
Dipartimento di Scienze e Tecnologie Alimentari e Microbiologiche
University of Milan
Milan, Italy

Lara Manzocco
Dipartimento di Scienze degli Alimenti
University of Udine
Udine, Italy

Maria Cristina Nicoli
Dipartimento di Scienze degli Alimenti
University of Udine
Udine, Italy

Luciano Piergiovanni
Dipartimento di Scienze e Tecnologie Alimentari e Microbiologiche
University of Milan
Milan, Italy

CHAPTER 1

An Introduction to Food Shelf Life: Definitions, Basic Concepts, and Regulatory Aspects

Maria Cristina Nicoli

CONTENTS

1.1 INTRODUCTION

Shelf life is an intriguing concept that results from merging scientifically based issues with economic, regulatory, and consumer-related concerns. The latter, which may greatly vary from country to country, are dynamic aspects liable to change according to society transformations, especially in terms of habits and needs. This is the reason there is not a uniform and generally accepted definition of shelf life. In recent decades, its meaning has been subjected to some interesting variations that reflect social and economic evolution. The difficulty to clearly define what shelf life actually is has had repercussions on the methodologies developed to carry out its assessment. According to the complexity of the shelf life concept, these methodologies should merge scientifically based procedures, mainly based on classical kinetic or dedicated statistical techniques, with more "arbitrary" ones that take into account economic and marketing reasons together with considerations of consumer behavior. Several methodologies for shelf life assessment have been developed in different fields of science, especially those dealing with drugs, cosmetics, foods, and electric and electronic devices. Some of these methodologies are quite complex to be easily transferred from one field of science to another as well as from academia to industries, especially to small and medium enterprises. This is particularly true for the food sector; due to the lack of a coherent and unified view of the shelf life issue and of affordable and cost-effective methodologies, companies frequently develop their own shelf life assessment procedures, whose scientific principles are sometimes questionable. It is interesting to note that, despite the fact that shelf life dating is compulsory in many countries for a great variety of food items, a wide and shared agreement on the methodologies and protocols to be used is available in only a few cases.

1.2 PRIMARY AND SECONDARY SHELF LIFE: GENERAL CONCEPTS

Shelf life is an important feature of all foods, including raw materials, ingredients, and semimanufactured products. Any of these products has its own shelf life, and all the subjects involved in the food chain, such as growers, ingredient and packaging suppliers, manufacturers, wholesalers, retailers, and consumers, have a great impact on it and should be aware of it. It is interesting to note that despite the wide literature concerning shelf life of finished products, at present little data on shelf life issues relevant to raw materials, ingredients, and semimanufactured foods are available.

Issues relevant to shelf life assessment, shelf life extension, and shelf life dating are key topics for food scientists working in research and development, processing, and quality assurance and for those involved in national and international control agencies as well as in regulatory bodies.

In general terms, *shelf life* can be defined as a finite length of time after production (in some cases after maturation or aging) and packaging during which the food product retains a required level of quality under well-defined storage conditions. This required quality level allows the product to be acceptable for consumption. Figure 1.1 schematically shows the different life stages of a food product from manufacturing to storage on the shelves.

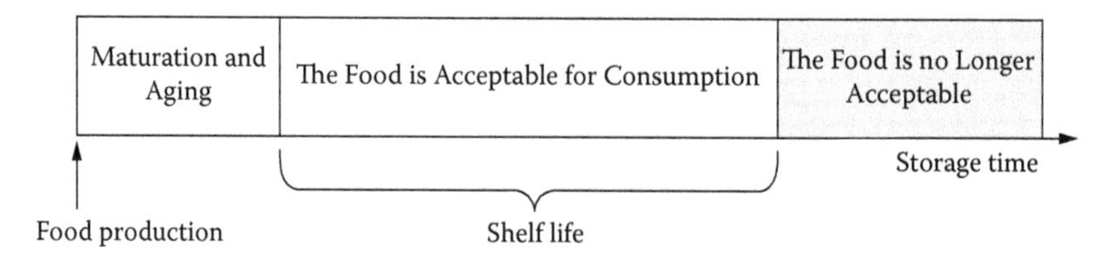

Figure 1.1 Life stages of a food product from manufacturing to maturation and aging, when present, until storage on the shelves.

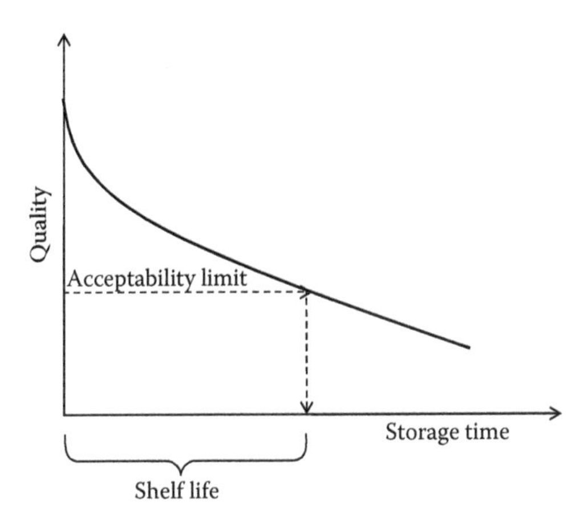

Figure 1.2 Food quality decay versus storage time and relevant primary shelf life value.

This means that any food product is doomed to fail after a certain amount of storage time for a variety of reasons. It is worth underlining that *food quality*, defined as an assemblage of a huge number of properties affecting the degree of food acceptability, is a dynamic state continuously moving to reduced levels (Kramer and Twigg, 1968). It implies that for any kind of food product there should be a defined quality level discriminating products that are still acceptable for consumption from those no longer acceptable. This quality level is generally defined as the *acceptability limit*. The time needed to reach the quality level corresponding to the acceptability limit is the shelf life, namely, primary shelf life, as illustrated in Figure 1.2.

Observing Figure 1.2, it is evident that the shelf life assessment necessarily requires knowledge of the food quality evolution during storage and the quality level corresponding to the acceptability limit. Despite its apparently intuitive meaning, the acceptability limit is difficult to quantify. There are a few examples of compulsory acceptability limits that derive from authority indications. The most complex and common cases are those in which the choice is not supported by any regulatory indication regarding specific constraints to be adopted as acceptability limits. Producers have to identify the acceptability limits of their products according to company policy and quality targets. This is a hard task since potential inaccuracies may be responsible for shelf life overestimations, causing consumer complaints and, in worse cases, product recalls. The identification of the acceptability limit is also hampered by the fact that no clear and widely accepted criteria and relevant methodologies have been defined in the food sector.

Recently, the term *failure time*, usually used in medicine and in engineering, has been increasingly used instead of *shelf life*, even in the food sector. In this case, the positive feature of the term *shelf life* (which implies that the food is still acceptable) is substituted by the negative one (the food is no longer acceptable). In other words, the failure time starts when shelf life ends.

It is interesting to note that the definition of shelf life previously discussed does not mention safety as a possible cause of food failure. Shelf life issues should not be related to safety, as safety is a fundamental requirement for any food product from manufacturing until consumption (European Commission [EC], 2002; Codex Alimentarius Commission [CAC], 2003). Thus, in principle, the food product, properly packed and stored, must remain safe after the expiry date and even after suffering possible temperature abuses during storage and home handling. In other words, although inextricably linked, food shelf life and safe life are not synonyms. As a quality issue, shelf life is expected to be shorter than the corresponding safe life. However, a deeper discussion of the relationship between food shelf life and safety would be greatly advisable. Besides microbiological concerns for which specific regulations and exhaustive guidelines have been developed in the last few decades, potential safety risks may arise in food during storage due to the formation of toxic compounds as a consequence of "silent" degradation reactions. When these events are not responsible for changes in perceivable food quality, they are generally underestimated. This is probably the case of the accumulation into the food matrix of oxidation reaction products or of compounds migrating from the packaging (Kubow, 1990; Ryan et al., 2009; Fiselier et al., 2010).

Packed foods have a primary and sometimes a secondary shelf life (Cappuccio et al., 2001). The latter, even called "pantry" shelf life, is defined as the period after pack opening during which a food product does maintain an acceptable quality level. There is a wide variety of dried and semidried foods, such as roasted coffee, cocoa, dehydrated foods, milk and vegetable powders, pastes and concentrates, flour, pasta, and so on whose domestic consumption can be spread over a significant length of time. Conceptually, the pack opening determines a sudden cascade of changes in the food characteristics (i.e., water activity (a_w), microbial load, etc.) and in the environmental variables affecting the product (i.e., changes in atmospheric composition, higher oxygen or moisture availability, temperature fluctuations, etc.). In all cases, the major consequence is a sudden acceleration of the rate of product quality decay as schematized in Figure 1.3.

In the case of food products with high a_w values, pack opening may introduce some safety issues due to microbial contamination that are negligible as long as the food product is maintained and stored in its original packaging. In these cases, different failure criteria have to be carefully considered, and different indicators and relevant acceptability limits have to be identified. For these reasons, secondary shelf life is of great importance for perishable foods, for which some indications about storage conditions and consumption time limits relevant to the open products are generally reported in the label (Medeiros et al., 2000; Damen and Steenbekkers, 2007; Jevsnik et al., 2008; Lima Tribst et al., 2009). No correspondent information is frequently available in the label for the so-called shelf-stable foods. This is surprising since information on secondary shelf life could contribute to meeting consumer needs and expectations regarding product fidelity. In addition, secondary shelf life is a key issue for semi-manufactured products and ingredients whose use in food industries can be extended over a significant length of time after pack opening. The quality profile of the finished products obviously depends on the quality level of materials and ingredients used.

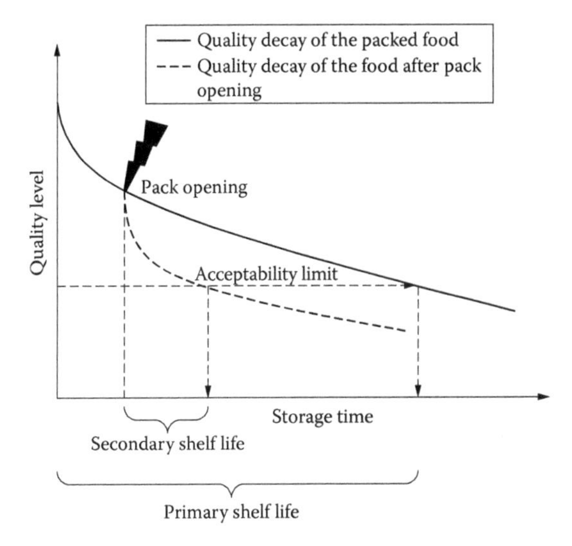

Figure 1.3 Food quality decay versus storage time as affected by pack opening and relevant primary and secondary shelf life values.

Unfortunately, there are few examples in literature of secondary shelf life studies (Cappuccio et al., 2001; Anese et al., 2006; Fu et al., 2009). The reasons are probably the lack of interest by the food industry in this specific issue and the difficulty in planning secondary shelf life studies. In such studies, environmental parameters should change continuously to really simulate use-consumption/storage conditions over time. In addition, the causes of quality decay and relevant descriptors for the open products could be different from those used for the packed ones. However, the application of relatively new devices like time–temperature indicators (TTIs) and radio-frequency identification (RFID) systems could be of great interest even for calculating the remaining secondary shelf life for both finished and semimanufactured products and ingredients (Labuza, 2006; Raab et al., 2008; Abad et al., 2009).

1.3 SHELF LIFE DEFINITIONS FROM A HISTORICAL PERSPECTIVE

Since the 1970s, many definitions of shelf life have been proposed. Interestingly, these definitions reflect the evolution of this concept over time. In 1974, the Institute of Food Technologists (IFT) in the United States defined *shelf life* as "the period between the manufacture and the retail purchase of a food product, during which time the product is in a state of satisfactory quality in terms of nutritional value, taste, texture and appearance." This definition restricts the attention to the time interval of food life from production to purchase, considering the manufacturer and the retailer as the sole subjects involved in affecting product shelf life. This definition does not take into consideration the time interval after food purchase, which includes home storage and handling until preparation and consumption.

The Institute of Food Science and Technology (IFST, 1993) defined *shelf life* as "the period of time during which the food product will: i) remain safe; ii) be certain to retain its desired sensory, chemical, physical, microbiological and functional characteristics; iii) where appropriate, comply with any label declaration of nutrition data, when stored under recommended conditions." This more exhaustive definition stressed the relationship between food shelf life and storage conditions and implicitly enlarged the perspective to the entire time interval from product manufacturing to consumption. On the basis of the IFST definition, safety and quality considerations dictate the end of the food life. However, although food safety and shelf life are inextricably linked issues, as previously mentioned, the food failure time should not be related to the loss of safety. Thus, the shelf life should be mainly intended as the length of time during which the food product is acceptable or, in other terms, able to satisfy specific needs. In this regard, it is interesting to look at the E.U. shelf life definition, which states "the date of minimum durability of a foodstuff shall be the date until which the foodstuff retains its specific properties when properly stored" (Directive 2000/13/EC; EC, 2000). This definition has items in common with quality definition in which terms such as "ability to satisfy given needs" or "fitness for use" can be found. Thus, in principle, shelf life problems are definitely quality problems (Munoz et al., 1992; Hough, 2010). Apart from a few cases in which food acceptability/unacceptability is clearly defined by regulatory bodies, consumers are the main subjects involved in deciding whether a food product is acceptable. As stated by Hough (2010), consumers are the final link of the chain, and at the end of the story, they are the main actors deciding whether the product is able to satisfy their needs.

The role of food–consumer interactions in determining product shelf life was proposed for the first time by Labuza and Schmidl (1988). They defined shelf life as "the duration of period between the packing of the product and the end of consumer quality as determined by the percentage of consumers who are displeased by the product." Thus, the concept of shelf life, initially centered on the product, has progressively moved to a consumer perspective. In this regard, it is worth mentioning the contribution by Gacula (1984), which indicated it is practically impossible to observe food failure times systematically. For this reason, shelf life data exhibit specific statistical features and are governed by probability distribution. These observations anticipated further work addressed to the application of life testing in shelf life studies (Guillet and Rodrigue, 2009; Hough, 2010).

As reported by Hough et al. (2003), from a sensory point of view, food products do not have shelf lives of their own; rather, they will depend on the interaction of the food with consumers, who can accept or reject the food products. It is noteworthy that, on the basis of this interpretation, the shelf life is determined by selecting the number or percentage of consumers the company can tolerate to dissatisfy. On the basis of this point of view, the shelf life concept assumes a different meaning: It cannot be defined as a finite length of time but as a time interval corresponding to a different risk level in terms of percentage of consumers rejecting the product. For an in-depth study, the work by Hough (2010) represents a comprehensive reference book.

1.4 SHELF LIFE VERSUS STABILITY STUDIES

As mentioned, a shelf life study basically aims to compute the time value during which the food product is acceptable for consumption under specified storage conditions. This time interval, that is, the food lifetime, is derived by merging the food quality decay rate with the acceptability limit. As stated, the assessment of the shelf life of a food product necessarily requires knowing both these parameters.

There are many scientific papers claiming to be shelf life studies that do not give shelf life data. The main reason is that they miss identifying the acceptability limit and are thus unable to estimate or predict food shelf life. In most cases, these papers discussed quality decay rates or gave broad descriptions of quality indicator evolutions. These studies dealt with food stability rather than shelf life issues. The discrimination between stability and shelf life studies is of primary importance to avoid basic misunderstanding. In stability experiments, the estimation of the food quality decay rate is of primary concern, while in shelf life experiments it is the estimation/prediction of food failure time. In other words, stability studies are focused on rates, whereas shelf life studies focus on times. Although not suitable to collect shelf life data, stability experiments can be of certain usefulness since they can be performed prior to a shelf life study for a better understanding of behavior of the product on the shelves. In particular, stability tests may be applied to identify the quality descriptor and relevant acceptability limit to be subsequently used in shelf life studies. In addition, stability tests can be conducted instead of shelf life studies when the goal is the identification of technological solutions addressed to extend product stability.

1.5 REASONS TO ASSESS THE SHELF LIFE OF FOODS

A common misunderstanding is the belief that a shelf life study should be applied for problem solving—for instance, when a product shows an unexpected sharp quality decay or suddenly goes bad, probably due to processing problems or improper storage or handling conditions. This is not the case: A shelf life study must be carried out only when foods are correctly processed, packed, and stored, ready to be purchased and consumed. Table 1.1 summarizes typical situations for which a shelf life study is needed.

Table 1.1 Typical Situations for which a Shelf Life Study Needs to Be Carried Out

- New products development
- Changes in formulation or processing conditions
- Changes in packaging materials and atmosphere composition
- Changes in ingredient/packaging suppliers
- Changes in storage conditions (mainly temperature, relative humidity, light exposure)
- During the regular ongoing shelf life surveillance

To maintain acceptable quality when the food is consumed and because dating is compulsory in many countries, a shelf life assessment process is essential in new product development, in monitoring market performances of consolidated productions, as well as when any changes in formulation, processing, packaging, and storage are introduced.

Figure 1.4 outlines the different stages at the basis of the design, development, and market launch of a new food product and relevant shelf life assessment steps. Moving from the first product concept to the early stages of its development, it is necessary to have a rough idea of the expected shelf life that the new product should have when put on the market. Formulation, processing, packaging, and storage conditions will be properly selected with this purpose in mind.

After the first tests, carried out in the lab kitchen or in a pilot plant, and the preliminary safety assessment on the prototype product, a shelf life study needs to be planned. The identification of the main causes responsible for product quality decay at the storage conditions foreseen for the finished product, also considering the

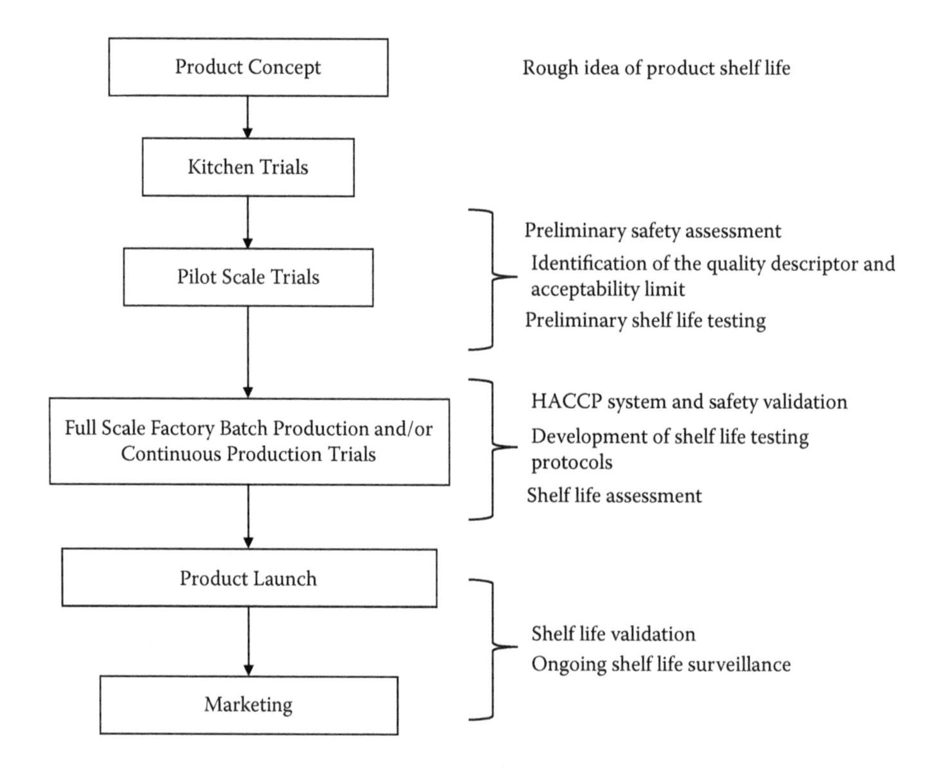

Figure 1.4 Shelf life strategy at different stages of new product development. HACCP, hazard analysis and critical control point. (Modified from Fu, B., and T.P. Labuza. 1997. Shelf life testing: procedures and prediction methods. In *Quality of frozen food*, M.C. Erickson and Y.C. Hung, Eds., 377–415. New York: Chapman & Hall, International Thomson; and Ellis, M.J., and C.M.D. Man. 2000.The methodology of the shelf life determination. In *Shelf-life evaluation of foods*, 2nd ed., D. Man and A. Jones, Eds., 23–33. Gaithersburg, MD: Aspen.)

packaging material that will be adopted, is thus necessary. Afterward, quality decay descriptors have to be selected.

At this stage, stability tests, even performed under accelerated conditions, can be preliminarily carried out to evaluate which, among the quality decay descriptors, is the best critical indicator. Stability test data can also contribute to making easier the choice of a proper acceptability limit for the further shelf life study.

Thus, product shelf life assessment procedures as well as assessment protocols can be definitely defined. Finally, a preliminary shelf life test can be carried out. Moving from the pilot plant to the first production trials, being sure that the product will not suffer safety problems during storage, a first exhaustive shelf life assessment process should be carried out and a dating identified.

The duration of a shelf life study should generally exceed the expected lifetime of the product. In some cases, this may be a year or more. This is particularly true for shelf-stable and frozen foods. Obviously, such a long time necessary for a shelf life study does not coincide with the 6-month development cycle typically required for food products. In some cases, even the duration of accelerated shelf life tests exceeds the time given for the product development. Therefore, in such situations, shelf life dating should be carefully validated after product launch on the market.

Any changes in formulation, processing, packaging, and storage conditions addressed to optimize new or regular productions need to be checked in terms of shelf life, using previously defined protocols. Ongoing monitoring of the regular productions in the market is also necessary so any changes in food storage performance can be noted and appropriate actions taken, if necessary. Furthermore, efficient customer complaint and product quality management systems would be useful for providing early warning signals concerning product shelf life problems (Ellis and Man, 2000). By analyzing historical data, including complaint and recall data, useful information about the "real" performance of the product on the shelves and potential differences between real and lab-assessed shelf life can be obtained.

It is evident that an efficient and ongoing shelf life assessment activity is an expensive process requiring appreciable human and economic resources. However, it is certainly cheaper than product recall and will contribute to maintaining a company's reputation.

Shelf life assessment activities are generally integrated in the quality management systems that oversee both shelf life determination during product development and ongoing monitoring of regular production.

The shelf life issue is also an important requirement of international quality standards; for instance, in the Food Safety Management Systems (UNI EN ISO 22000, 2005), although shelf life testing procedures are not specifically mentioned, the assessment of the failure time of raw materials, ingredients, and finished products is required. International standards for companies supplying retailer-branded foods, such as the International Food Standard (Hauptverband des Deutschen Einzelhandels [HDE], 2007) and the British Retail Consortium (BRC) Global Standard for Food Safety (2008), ask for shelf life protocols to be applied for new product development and ongoing shelf life assessment.

Most of these retail consortia together with producer associations and national and international food agencies have developed specific and helpful guidelines to support producers in the shelf life assessment process, especially in the case of perishable products. These booklets generally provide some basic information on shelf life assessment procedures with practical examples, food safety criteria limits, and a source of information for further readings. In this regard, see the exhaustive guidelines prepared by the Chilled Food Association (2006, 2010) and by the New Zealand Food Safety Authority (2005) and BRC (2006). For instance, in the last, shelf life assessment procedures are divided considering different food groups on the basis of their quality and safety shelf life related issues.

1.6 SHELF LIFE COMMUNICATION

Although dating of food products has been known to exist in the United States in the dairy industry since 1917 (Labuza and Szybist, 1999), at present shelf life dating is still not used worldwide. When applied, there are two different kinds of dating: open and closed dating. The first is a calendar date printed in the food package indicating the time limit suggested for consuming the product. Closed dating is coded numbers printed in the packaging (Labuza, 1982; Labuza and Szybist, 2001). While open dates are thought to communicate shelf life to consumers and to assist them in food purchase, coded dates are for manufacturers to allow product tracking during distribution and retail and to rotate stocks. Open dating refers to the maintenance of a food's perceived freshness, sensory characteristics, or functional properties. As previously mentioned, shelf life and hence open dating should not be related to safety issues.

Open dating is compulsory in the European Union, in many South American and Arabic countries, in Israel, and in Taiwan (Robertson, 2009). In the European Union, the E.U. Directive 2000/13 mandated the use of "best before" and "use by" dates (EC, 2000). The first is used for a stable product, the latter for highly perishable foods. In the United States, dating is not generally required by federal regulations, with the exception of infant formula and baby foods. However, at present, more than 20 countries require dating of some food categories. Among these countries, no uniform and universally accepted dating system is used. Different types of open dates corresponding to different kinds of information can be used: "sell by," "best before," and "use by" (Labuza and Szybist, 1999; U.S. Department of Agriculture [USDA], 2007). In addition to open dates, in the United States closed or coded dates indicating expiring time might be used for shelf-stable products. Table 1.2 summarizes the different types of food dating within the European Union and the United States.

In the countries where an open dating system has been introduced, it affected consumer purchasing habits and their awareness of how to handle and store food products. Concerning this aspect, it is interesting to mention the example of a milk-based product recently launched in the Italian market. This is a shelf-stable product that has to be consumed after freezing. To assist consumers to handle the product correctly, the company decided to put in the label two expiring dates: the first one indicating the suggested time limit before freezing and the second one indicating the

Table 1.2 Open Dating and Closed Dating in the European Union and United States

Open Dating		Closed Dating
European Union	**United States**	**United States**
Mandatory (EU Directive 2000/13/EC)	Not mandatory (except for infant formula and baby foods)	Not mandatory
"Best before": must indicate day/month/year. It is recommended for products that will not keep more than 3 months. "Best before end": must indicate month/year for products that will keep more than 3 months but not more than 18 months. For foodstuffs keeping more than 18 months, it must contain only the year. "Use by": it must contain day/month and possibly the year in encoded form.	"Sell by" date gives information how long the product is for sale "Best before" "Use by"	Packing codes used for shelf-stable foods (i.e., canned foods)

Source: From European Commission (EC). 2000. Directive No 2000/13/EC on the approximation of the laws of the Member States relating to the labelling, presentation and advertising of foodstuffs. *Official Journal of the European Communities*, L109/29, 06.05.2000, pp. 29–42.; USDA Food Safety and Inspection Service. 2007. Food product dating. http://www.fsis.usda.gov/factsheets/food_product_dating/index.asp.

suggested time limit for consumption after freezing. Furthermore, the introduction of the open dating system also heightened awareness along the distribution chain, forcing distributors and retailers toward more precise storage practices (Labuza and Szybist, 1999).

However, it has been observed that in industrialized countries a considerable amount of high-quality packed food products (~30%) is lost because of exceeding the expiry dates. It mainly happens in retail and food service establishments and at home (Floros et al, 2010). This is attributable to several factors, including the relatively low costs of food as well as the difficulty for the majority of consumers to understand exactly the meaning of "best before" and "use by." This situation suggests that there is the need to introduce specific incentives addressed to minimize food wastes, to encourage the development of parallel marketing strategies (i.e., last-minute marketing) (http://www.lastminutemarket.it), as well as to reconsider the food dating systems that should give more detailed information to the different actors in the food chain, such as retailers, food service operators, and consumers.

Dating is generally established by the food industry, and it should be identified as the result of a shelf life study. However, dating and shelf life are not necessarily synonyms. The shelf life value provides scientific bases for open date setting. Depending on the type of product, food dating could include a margin with a magnitude that depends on the kind of product, especially taking into consideration the variability of its initial status (i.e., microbial load) and other needs, such as (a) ensuring safety for highly perishable foods even if improperly stored or handled; (b) keeping as low as possible the percentage of consumers who are displeased by the product at the failure time; and (c) allowing higher product turnover on the shelves. Unfortunately, no shared criteria are available to support producers in estimating the earliness of dating with respect to the shelf life value assessed or predicted on a rational basis.

Open dating does not guarantee consumers that a food product is not spoiled as this could happen if the food was improperly processed or suffered improper practices during distribution and domestic handling. In such conditions, an open date is meaningless and could represent false information. To overcome these problems, TTIs and more recently RFID systems were introduced to signal the premature end of shelf life of food products (Labuza, 2006; Kreyenschmidt, 2008; Kreyenschmidt et al., 2010). TTI and RFID information is thought to integrate open dates to ensure correct shelf life communication to consumers and to allow aware food purchasing and consumption. Although available for almost a decade, TTI use is not popular among food manufacturers, especially in Europe.

1.7 BOOK ORGANIZATION

This book aims to provide an overall view of the present status of the shelf life assessment issue. To this purpose, the assessment process has been approached considering a number of sequential steps, each requiring different choices and, thus, the adoption of proper and dedicated methodologies. Chapter 2 outlines the logic behind the following chapters and gives an overview of the possible approaches to be pursued to design a shelf life study. Obviously, each strategy requires proper testing plans and modeling procedures that are time and cost effective. For this reason, a rational choice of the strategy that best fits a company's needs is absolutely necessary before starting a shelf life study.

Chapter 3 deals with the identification of the acceptability limit, which is probably the most difficult parameter to be defined when developing a shelf life test since it may depend on regulatory, economic, marketing, and social constraints. The chapter discusses criteria and present methodologies to be used for the identification of the acceptability limit as the result of a variety of considerations coming from different decision makers. Chapter 4 discusses the strategies at the basis of the choice of the shelf life indicator, namely, the critical indicator. Conceptually, the choice of the critical indicator should precede the identification of the acceptability limit. However, in most cases the choice of the critical indicator is forced by previous decisions on the nature of the acceptability limit. This is why in this book the identification of an acceptability limit precedes the selection of the critical indicator.

Chapters 5, 6, and 8 deal with kinetic modeling illustrated from different points of view. Chapter 5 is an overview of the classical kinetic approach to be applied in shelf life studies when chemical, physical, or sensorial indicators are used. The chapter critically analyzes possible computation uncertainties and pitfalls and gives some working protocols to support readers in correctly facing a shelf life study based on kinetic modeling. Chapter 6 deals with kinetic modeling when microbial indicators are used. The chapter contains a brief overview of predictive food modeling and the most commonly used mathematical models. The following parts are designed to illustrate how shelf life studies based on microbial indicators should be conducted and how to adapt the models to realistic scenarios. Chapter 8 illustrates how to approach a shelf life study based on kinetic modeling from the packaging point of view. In particular, this chapter is designed to analyze critical situations in which classical accelerating shelf life testing cannot be applied because of packaging performances changing by varying the magnitude of the accelerating factors.

Chapter 7 describes an emerging approach to assess food shelf life based on survival analysis. Although mainly developed to assess shelf life by means of consumer acceptability, in principle this approach can be used to assess food shelf life using any kind of critical indicator.

Last, Chapter 9 presents some case studies relevant to frozen, chilled, and ambient food products. The chapter organization was designed to illustrate how to face different shelf life situations on the basis of the criteria and the methodologies described in the previous chapters. Some practical indications on how to organize the experimental plans are also given.

The overall organization of the chapters was designed in such a way to obtain a comprehensive scientifically based guide aiming to support the readers in the adoption of rational and scientifically based approaches in the shelf life assessment process. To this purpose, links and recalls among the different chapters are introduced to allow a better understanding of the concepts and methodologies and give to the readers an integrated view of the complex and fascinating shelf life issue.

1.8 REFERENCES

Abad, E., Palacio, F., Nuin, M., Gonzales de Zarate, A., Juarros, A., Gomez, J.M., and S. Marco. 2009. RFID smart tag for traceability and cold chain monitoring of foods: demonstration in an intercontinental fresh fish logistic chain. *Journal of Food Engineering* 93: 394–399.

Anese, M., Manzocco, L., and M.C. Nicoli. 2006. Modeling the secondary shelf-life of ground roasted coffee. *Journal of Agriculture Food Chemistry* 54: 5571–5576.

British Retail Consortium. 2006. *BRC Global standard guidelines: shelf life determination.* London: TSO.

British Retail Consortium. 2008. *BRC Global standard for food safety* (English): Issue 5. London: TSO.

Cappuccio, R., Full, G., Lonzarich, V., and O. Savonitti. 2001. Staling of roasted and ground coffee at different temperature. Combining sensory and GC analysis. In *Proceeding of the 19th International Scientific Colloquium on Coffee* (CD-ROM), Trieste, Italy.

Chilled Food Association. 2006. *Guidance on the practical implementation of the EC regulation on microbiological criteria for food staffs.* Kettering, UK: Chilled Food Association.

Chilled Food Association, British Retail Consortium, Food Standard Agency. 2010. *Shelf life of ready to eat food in relation to* L. monocytogenes*—guidance for food business operators.* Kettering, UK: Chilled Food Association.

Codex Alimentarius Commission (CAC). 2003. *Recommended international code of practice. General principles of food hygiene*. FAO/WHO Food Standards, Codex Alimentarius Commission (CACRCP I-1969) Rev. 4-2003, FAO, Rome, Italy. 31.

Damen, F.W.M., and L.P.A. Steenbekkers, 2007. Consumer behavior and knowledge related to freezing and defrosting meat at home: an exploratory study. *British Food Journal* 9(7): 511–518.

Ellis, M.J., and C.M.D. Man. 2000.The methodology of the shelf life determination. In *Shelf-life evaluation of foods,* 2nd ed., ed. D. Man and A. Jones, 23–33. Gaithersburg, MD: Aspen.

European Commission (EC). 2000. Directive No 2000/13/EC on the approximation of the laws of the Member States relating to the labelling, presentation and advertising of foodstuffs. *Official Journal of the European Communities* L109/29, 06.05.2000, pp. 29–42.

European Commission (EC). 2002. Regulation No. 178/2002 of the European Parliament and of the Council of 28 January 2002. REGULATION (EC) No 178/2002 laying down the general principles and requirements of food law, establishing the European Food SafetyAuthority and laying down procedures in matters of food safety. *Official Journal of the European Communities* L31/1, 01/02/2002, pp. 1–34.

Fiselier, K., Rutschmann, E., McCombie, G., and K. Grob. 2010. Migration of di(2-ethylhexyl) maleate from cardboard boxes into foods. *European Food Research and Technology* 230(4): 619–626.

Floros, J.D., Newsome, R., Fischer, W., et al. 2010. Feeding the world: today and tomorrow: the importance of food science and technology. *Comprehensive Reviews in Food Science and Food Safety* 9: 572–599.

Fu, B., and T.P. Labuza. 1997. Shelf life testing: procedures and prediction methods. In *Quality of frozen food*, ed. M.C. Erickson and Y.C. Hung, 377–415. New York: Chapman & Hall, International Thomson.

Fu, Y., Lim, L.T., and P.D. McNicholas. 2009. Changes in enological parameters of white wine packaged in bag-in-box during secondary shelf life. *Journal of Food Science* 74(8): 608–618.

Gacula, M., Jr. 1984. *Statistical methods in food consumer research*. Orlando, FL: Academic Press.

Guillet, M., and N. Rodrigue. 2009. Shelf life testing methodology and data analysis. In *Food packaging and shelf life. A practical guide,* ed. G.L. Roberson, 31–53. Boca Raton, FL: Taylor & Francis.

Hauptverband des Deutschen Einzelhandels (HDE). 2007. *Standard for auditing retailer and whole branded food products*. Version 5, August 2007. Berlin: HDE/Paris: FCD.

Hough, G. 2010. *Sensory shelf life estimation of food products*. Boca Raton, FL: Taylor & Francis.

Hough, G, Langohr, K., Gomez, G., and A. Curia. 2003. Survival analysis applied to sensory shelf life of foods. *Journal of Food Science* 68: 359–362.

Institute of Food Science and Technology (IFST). 1993. *Shelf life of foods: guideline for its determination and prediction*. London: Institute of Food Science and Technology.

Institute of Food Technologists (IFT). 1974. *Shelf life of foods*. Report by Institute of Food Technologists. Expert Panel on Food Safety and Nutrition and the Committee on Public Information. Chicago: Institute of Food Technologists.

ISO 22000. 2005. *ISO 22000, food safety management systems—requirements for any organization in the food chain*. ISO Geneva, Switzerland.

Jevsnik, M., Hlebec, V., and R. Raspor. 2008. Consumers' awareness of food safety from shopping to eating. *Food Control* 19(8): 737–745.

Kramer, A., and B. Twigg. 1968. Measure of frozen food quality and quality changes. In *The freezing preservation of foods*, 4th ed., Vol. 2, ed. D.K. Tressler. Westport, CT: AVI.

Kreyenschmidt, J. 2008. Cold chain-management. *Proceedings of the 3rd International Workshop Cold-Chain-Management, Bonn,* June 2–3.

Kreyenschmidt, J., Christiansen, H., Hübner, A., Raab, V., and B. Petersen. 2010. A novel time-temperature-indicator (TTI) system to support cold chain management. *Journal of Food Science Technology* 208(45): 208–215.

Kubow, S. 1990. Toxicity of dietary lipids peroxidation products. *Trends in Food Science and Technology* 1(3): 67–70.

Labuza, T.P. 1982. *Shelf life dating of food.* Westport, CT: Food and Nutrition Press.

Labuza, T.P. 2006. Time-temperature integrators and the cold chain: What is next? *Proceedings of the 2nd International Workshop Cold-Chain-Management*, May 8–9, 43–51.

Labuza, T.P., and M.K. Schmidl. 1988. Use of sensory data in the shelf life testing of foods: principles and graphical methods for evaluation. *Cereal Foods World* 33: 193–205.

Labuza, T.P., and L.M. Szybist. 1999. Playing the open dating game. *Food Technology* 53(7): 70–85.

Labuza, T.P., and L.M. Szybist. 2001. *Open dating of foods.* Trumbull, CT: Food and Nutrition Press.

Lima Tribst, A.A., de Souza Sant'Ana, A., and P.R. de Massaguer. 2009. Review: microbiological quality and safety of fruit juices—past, present and future perspectives. *Critical Reviews in Microbiology* 35(4): 310–339.

Medeiros, L.C., Sanik, M.M., Miller, E.H., McCombs, K., and C. Miller. 2000. Performance and microbiological growth in ground meat associated with the use of thawing trays. *Journal of Food Quality* 23(4): 409–419.

Munoz, A., Civille, C.V., and B.T. Carr. 1992. *Sensory evolution of quality control.* Chapter 1. NewYork: Van Nostrand Reinhold.

New Zealand Food Safety Authority. 2005. *A guide to calculating the shelf life of foods.* Wellington, NZ: New Zealand Food Safety Authority.

Raab, V., Bruckner, S., Beierle, E., Kampmann, Y., Petersen, B., and J. Kreyenschmidt. 2008. Generic model for the prediction of remaining shelf life in support of cold chain management in pork and poultry supply chains. *Journal of Chain and Network Science* 8(1): 59–73.

Robertson, G.L. 2009. Food packaging and shelf life. In *Food packaging and shelf life. A practical guide*, ed. G.L. Robertson, 1–16. Boca Raton, FL: Taylor & Francis Group.

Ryan, E., McCarthy, F.O., Maguire, A.R., and N.M. O'Brien. 2009. Phytosterol oxidation products: their formation, occurrence, and biological effects. *Food Reviews International* 25(2): 157–174,

USDA Food Safety and Inspection Service. 2007. Food product dating. http://www.fsis.usda.gov/factsheets/food_product_dating/index.asp.

CHAPTER 2

The Shelf Life Assessment Process

Maria Cristina Nicoli

CONTENTS

2.1 SHELF LIFE ASSESSMENT: AN INTEGRATED APPROACH

It was mentioned in Chapter 1 that food quality is a dynamic state continuously moving to reduced levels over time after production (see Figure 1.2 in Chapter 1). By selecting a proper quality indicator, namely, the critical descriptor, food quality decay can be monitored as a function of storage time. For instance, quality decay can be described by monitoring the loss of a peculiar food property or a target nutrient or by measuring the development of undesired changes such as those regarding food color and appearance, oxidation status, microbial load, and so on. Figure 2.1 shows some examples of possible evolutions of a generic quality indicator I over time.

The mathematical description of the evolution of a generic quality indicator I versus storage time can be expressed as follows:

$$I = g(t, \vartheta) \tag{2.1}$$

This mathematical model involves the independent variable time t, the dependent variable quality indicator I, and the parameter ϑ (Van Boekel, 2009; Brandao and

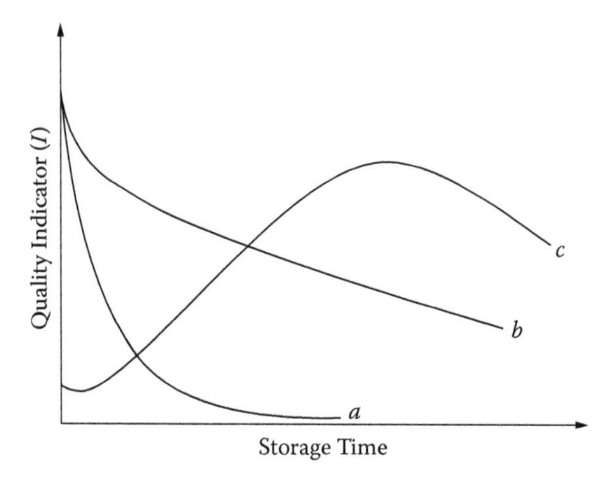

Figure 2.1 Examples of food quality decay (a, b, c) versus storage time described by a generic quality indicator *I*.

Silva, 2009). The term g symbolizes the function describing the evolution of I over time. In practical terms, g models the shape of the curves shown in Figure 2.1, while the term ϑ symbolizes the characteristic parameters of the function g. Readers with previous experience in food kinetics should certainly understand that the term ϑ accounts for kinetic parameters (i.e., rate constant) of the event under observation. For those who do not have any kinetic expertise, no further in-depth knowledge is required for the moment.

Although Equation 2.1 allows us to express, in mathematical terms, the evolution of the quality indicator I over storage time, it does not represent the proper function for calculating product shelf life. As defined in Chapter 1, shelf life is a finite length of time after production and packaging during which the food product retains a required level of quality. This definition implies that a proper tool for calculating shelf life should be a mathematical equation in which the time t is the dependent variable, while the quality indicator I is the independent one. For this reason, the inverse of Equation 2.1 should be considered:

$$t = f(I, \lambda) \tag{2.2}$$

where f is the function describing the inverse of g, and λ is the relevant parameters. The letters g and f used in Equations 2.1 and 2.2, respectively, imply that the correspondent mathematical functions are obviously different.

Shelf life is the value assumed by the variable t corresponding to the acceptability limit I_{lim}. The latter is the I value that discriminates acceptable food products from ones no longer acceptable. Thus, the shelf life SL can be expressed as follows:

$$SL = f(I_{\text{lim}}, \lambda) \tag{2.3}$$

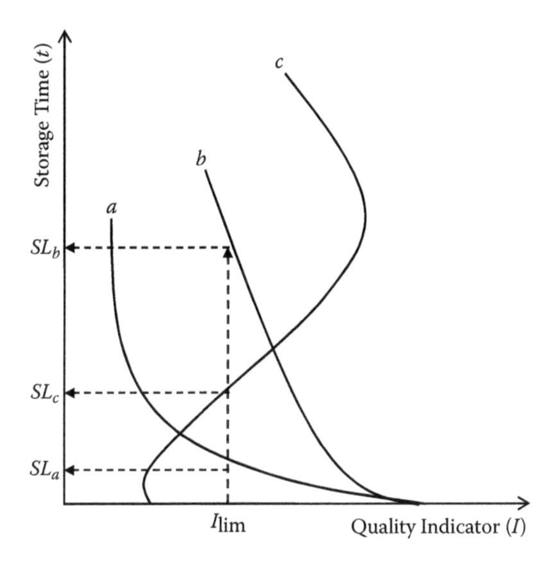

Figure 2.2 Mathematical model $t = f(I,\lambda)$ of the a, b, and c quality decay curves shown in Figure 2.1 and relevant shelf life values (SL_a, SL_b, and SL_c, respectively) assumed for a given acceptability limit I_{lim}.

Figure 2.2 shows the function $t = f(I, \lambda)$ of the quality decay curves reported in Figure 2.1, the I value considered as the acceptability limit I_{lim}, and the corresponding shelf life values SL_a, SL_b, and SL_c.

Let us consider a food product showing the evolution of the critical indicator I over storage time described by curve b in Figure 2.1. Suppose we store aliquots of such product in different environmental conditions (E_1, E_2, E_3) (i.e., different temperatures) and monitor quality changes over time. The evolution of the quality indicator I as a function of storage time is represented in Figure 2.3.

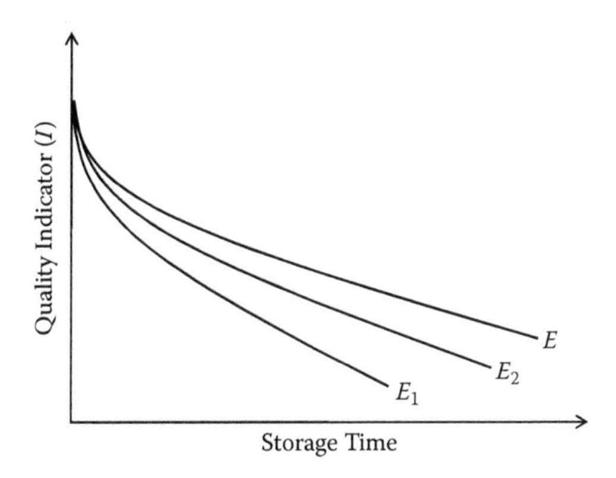

Figure 2.3 Evolution of the quality indicator I of a food product versus storage time under different environmental conditions (E_1, E_2, and E_3).

Observing this figure, it can be noted that the curves show the same shape but proceed at different rates depending on the environmental conditions adopted ($E_1 > E_2 > E_3$). From a mathematical point of view, such curves can be described by the same function g but have different values of parameter ϑ. In practical terms, these parameters assume a kinetic meaning as they allow us to distinguish the three curves mathematically. Such differences are definitely rate differences.

As stated by Labuza in his pioneering papers (Waletzko and Labuza, 1976; Labuza, 1982) and further contributions (Labuza, 1984; Labuza and Schmidl 1985, 1988; Labuza and Taoukis, 1990; Labuza et al., 1992; Labuza and Fu, 1993; Fu and Labuza, 1993, 1997; Nelson and Labuza, 1994; Labuza and Szybist, 2001), the rate at which the food quality moves to reduced levels over time results from the integrated effects of formulation, processing, packaging, and storage conditions. In fact, for a given formulation, processing allows the food product to be obtained in its final form with peculiar sensory and nutritional characteristics and specified safety/stability properties. The maintenance of such characteristics over time depends on the packaging and storage conditions adopted. These interactive effects can be broken down in factors accounting for the overall intrinsic product characteristics C_i, packaging properties P_i, and environmental conditions E_i. Based on the aforementioned considerations, it is quite intuitive to consider ϑ, which are the characteristic kinetic parameters of Equation 2.1 related to C_i, P_i, and E_i factors as follows:

$$\vartheta = h(C_i, P_i, E_i, \alpha) \quad (2.4)$$

The term h indicates the mathematical function, and α is the relevant parameters.

Each factor (C_i, P_i and E_i,) includes a number of different variables, some of them are summarized in Table 2.1.

While the C_i and P_i factors specifically refer to the finished product itself (food + packaging), E_i factors refer to the environmental conditions suffered by the product during distribution, retail, and home storage.

Table 2.1 Some Product Characteristics, Packaging Properties, and Environmental Conditions Expected to Affect Kinetic Parameters ϑ

Product		Shelf
Food Characteristics (C_i)	**Packaging-Related Factors(P_i)**	**Environmental Factors (E_i)**
Food components concentration and their physical state	Packaging barrier properties to moisture, gases, and electromagnetic waves	Temperature
Food physical state	Food-packaging interactions	Light
Catalysts	Temperature dependence of packaging barrier properties	Relative Humidity (RH%)
Inhibitors	Temperature dependence of food-packaging interactions	
Microorganism load		
pH		
Water activity (a_w)		

The generic term C_i includes a wide number of intrinsic variables, including biochemical, chemical, physical, and physicochemical properties, which can also depend on technological operations applied during food manufacture. Although not properly considered an intrinsic variable, the product microbial load can be accounted for simplicity in the C_i factor. The P_i term accounts for the barrier properties of the packaging material adopted, which in turn may affect food composition (i.e., food moisture content, oxygen-sensitive compound concentration, microbial load, etc.) and food sensitivity to the environmental variables over time. The P_i term considers also potential food-packaging interactions deriving from the use of active packaging able to remove undesired compounds from the headspace or release desired ones. Finally, the term E_i includes a number of extrinsic variables, namely, those affecting the food quality decay rate during distribution and retail storage until consumer purchase, storage, and handling.

Chapters 5–7 are dedicated to analyzing the "weight" of C_i, E_i, and P_i, their role in quality decay rates, and their possible exploitation to speed up a shelf life study.

To face a shelf life study successfully, an integrated approach that takes into consideration C_i, E_i, and P_i factors and their interactive effects is needed. A common mistake in planning a shelf life study, especially when accelerated shelf life tests are required, is to consider only some aspects (i.e., some E_i or P_i variables) and ignore the interactive effects of a single or a group of variables on the other ones. For instance, when using temperature to speed up a shelf life study, it is quite common to underestimate temperature-related changes in food composition or physical state and/or structure and/or packaging permeability. For instance, changes in temperature can greatly affect oxygen solubility into the food matrix, affecting the extent of oxidation reactions. Temperature can also cause first- or second-order phase transitions responsible for changes in molecular mobility and reactivity. Similarly, it is erroneous thinking that in a shelf life study only packaging and related variables should be considered.

An integrated approach in a shelf life study is obviously a hard task considering the complexity of Equation 2.4, and some simplifications need to be introduced. To this purpose, it is useful to remember that a shelf life study is addressed to finished packed products; it means that the food products must be correctly processed and packed in a well-defined packaging material previously identified during the product development stage. Thus, in principle, P_i factors should be considered constant. In addition, a further simplification in Equation 2.4 derives from the consideration that, in principle, the composition variables can also be considered constant.

Obviously, the aforementioned considerations do not pertain to when tests are required to assess how new ingredients or packaging materials may affect the shelf behavior of the product. In such cases, composition and packaging variables cannot be considered constant. However, these cases generally require stability rather than shelf life studies since the goal is to know how much the quality decay rate decreases/increases depending on the applied formulation/packaging changes.

A further complication may arise considering the special feature of P_i factors, which show a binary behavior: When the environmental conditions (i.e., temperature) change, the packaging barrier properties may remain constant or vary. On the basis of these considerations, when P_i factors do not show any sensitivity to environmental changes applied during food storage, Equation 2.4 can be simplified as follows:

$$\vartheta = h(E_i, \alpha) \tag{2.5}$$

Equation 2.5 indicates that when C_i and P_i factors are constant, parameters are a function of the sole environmental factors. In other words, since parameters allow expression of how fast the critical indicator I changes over time, it can be inferred that, keeping C_i and P_i factors constant, the rate of food quality decay depends on the sole environmental conditions applied (i.e., temperature, relative humidity, light exposure, etc.).

When P_i factors vary by changing the environmental conditions, a mathematical model describing the effect of the independent variables E_i on the dependent ones P_i should exist. Such a model can be expressed as follows:

$$P_i = d(E_i, \beta) \tag{2.6}$$

where d indicates the mathematical function describing the relationship among E_i and P_i variables, and β is the relevant parameter. Thus, keeping the C_i factors constant, Equation 2. 4 can be expressed as follows:

$$\vartheta = h(d(E_i, \beta), E_i, \alpha) \tag{2.7}$$

In practical terms, Equation 2.7 indicates that when the packaging properties show a sensitivity to environmental variables, such variables perform both a direct and an indirect effect on parameters. For instance, an increase in storage temperature is expected to increase the food quality decay rate. The extent of such an increase is the result of the temperature sensitivity of the event responsible for quality decay and temperature-induced changes of the packaging barrier properties. The latter in turn may have a significant role in accelerating or slowing the degradation event.

Based on these considerations, there is sometimes the habit to approach the shelf life issue from two different perspectives: When Equation 2.5 is satisfied, the shelf life is mainly considered product dependent, while in those cases described by Equation 2.7, the shelf life is assumed to be mainly packaging dependent. When accelerated shelf life tests need to be applied, cases described by Equation 2.7 require a careful design of the shelf life experiments and more complex modeling procedures. Chapter 9 extensively discusses how to approach these situations.

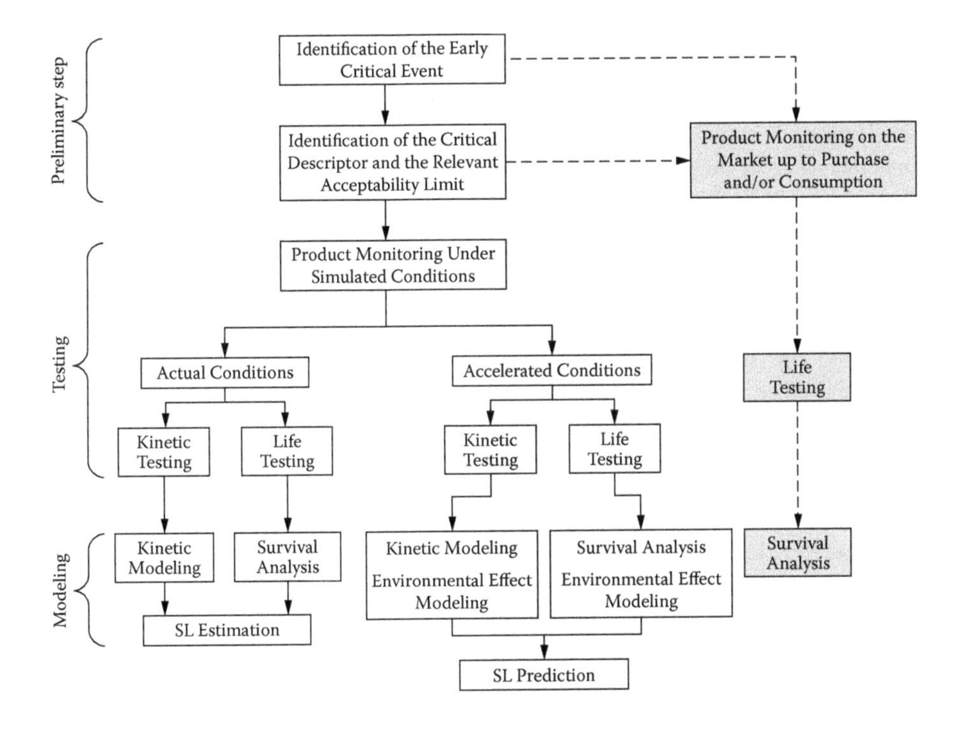

Figure 2.4 General overview of the shelf life assessment process.

2.2 GETTING A SHELF LIFE ASSESSMENT PROCESS STARTED

A shelf life study is designed to provide an objective measure of the length of time, after manufacture, the food product is acceptable for consumption. It is a multistage process that can be pursued following different approaches, each requiring a number of choices. In general terms, a shelf life study can be divided into three steps as shown in Figure 2.4.

The preliminary step is designed to identify the tools necessary to carry out a shelf life study. These are the critical indicator I and the relevant acceptability limit I_{lim}. The second step is focused on sample testing, which must be carried out under specified storage conditions. During testing, experimental data are generated and recorded. Finally, the third step deals with experimental data modeling and shelf life computation.

2.2.1 The Preliminary Step

As previously introduced, food quality decay is not a function of time alone; rather, it is a complex function of composition, environmental, and packaging-related factors (Equation 2.4).

The first step of the shelf life assessment process implies the identification of the event judged to have the most critical impact on food quality at the storage conditions foreseen for the packed product. Foods might fail for a variety of microbiological, enzymatic, chemical, and physical causes that can take place simultaneously or consecutively. Among them, it is generally possible to identify those that are shelf life limiting because they are responsible for early quality changes. Then, a good understanding of these events is imperative in starting a shelf life study. This book does not describe the main biological, chemical, and physical reactions at the basis of food degradation. For more extended coverage of these topics, refer to works by Robertson (2010); Steele (2004); Kilcast and Subramaniam (2011); Man and Jones (2000); Eskin and Robinson (2000); Damodaran et al. (2007); Taub and Singh (1998); and Labuza (1982, 1984).

Based on the common assumption that perishable foods generally undergo microbial or enzymatic spoilage and preserved foods undergo chemical or physical degradation reactions, there is sometimes the erroneous habit to distinguish a microbiological shelf life from a chemical, physical one. Although the mathematical functions to be applied for modeling may differ, both cases need to be faced using the same integrated approach and following the scheme reported in Figure 2.4.

After the identification of the early degradation event affecting food acceptability, a suitable quality decay descriptor (i.e., the critical indicator) should be identified. As mentioned in Chapter 1, at this stage stability tests can be preliminarily carried out to become confident of the product's behavior and to select the critical indicator I that best fits the use. Depending on the main degradation event responsible for food quality decay, chemical, physical, and microbiological indicators can be used. When food staling cannot be properly described by an instrumental index, sensory analysis, carried out by trained panelists or consumers, is frequently used. Although effective in describing perceived quality, sensory analyses may be expensive and time consuming. For this reason, when possible, it would be useful to relate the evolution of sensorial data versus storage time with that obtained using instrumental quality decay descriptors. This approach gives results that are particularly useful for *routine* shelf life assessments that must rely on cheap and easy-to-use quality indicators, which the instrumental ones generally are (see also Chapter 4).

The search for the critical indicator is inextricably linked to the identification of the acceptability limit I_{lim}. In fact, the acceptability limit is the amount of quality change, described by the selected critical indicator, that discriminates acceptable products from ones no longer acceptable. Although the acceptability limit sounds intuitive, it is the most difficult parameter to define when developing a shelf life study. These difficulties are made worse by the lack of dedicated literature or guidelines to support its identification. Basically, the choice of the acceptability limit may derive from legal or voluntary requirements. There are a few examples of compulsory acceptability limits deriving from authority indications. In the majority of cases, the acceptability limit is voluntary, and its choice is that of the producers, who are free to identify this value according to their own policy and quality targets. At present, the acceptability limit is frequently identified empirically, but more rational approaches for its identification are possible. Chapter 3 analyzes the concept of the acceptability limit and discusses possible strategies and current methodologies to be pursued for its rational identification. To my knowledge, it is the first attempt to analyze and discuss the concept of food acceptability limit and to review strategies and present methodologies for its quantification. Chapter 4 illustrates how to identify the food quality descriptor, whose choice is strictly dependent on the nature and characteristics of the acceptability limit.

2.2.2 Testing

After the identification of the critical indicator together with the acceptability limit, shelf life testing is ready to start. Testing is the step that deals with what actually happens to the product and experimental data gathering. Specifically, changes of the critical indicator over time are measured under well-defined storage conditions. This is a labor-intensive and time-consuming activity, and for this reason, it should be carefully planned. Chapters 5–7 give some basic indications of how to design shelf life experiments. In addition, Chapter 9, which illustrates and discusses some shelf life case studies, provides basic details on the relevant experimental designs and equipment needed.

Basically, food monitoring can be performed under actual or accelerated conditions. The simplest approach implies the monitoring of the quality decay of the packed food product under conditions simulating those actually experienced by the product on the shelves. This implies that experiments have to be carried out under environmental conditions (i.e., temperature, light exposure, relative humidity) carefully chosen to simulate the food behavior on market shelves realistically. Such environmental conditions E_i should be kept constant until the end of the shelf life experiments. Obviously, even C_i and P_i factors are constant since product formulation and packaging material have already been selected. In practical terms, experimental data gathered during testing allow the monitoring of food quality decay versus storage time. This easy approach, which leads to direct shelf life estimation, can only be used when quality decay occurs quickly (i.e., for perishable foods) and when changes in the environmental conditions during storage are considered negligible. There are no universal protocols for the direct determination of shelf life. A number of experimental designs have been put forward (Kilcast and Subramaniam, 2011; Ellis and Man, 2000); however, to avoid general mistakes, their application should be critically evaluated, taking into consideration the overall characteristics of the food product to be tested.

When food quality depletion proceeds fairly slowly under actual storage conditions (i.e., for shelf-stable foods), it is convenient to try to accelerate shelf life testing by monitoring the product behavior under environmental conditions able to speed up the quality decay. This procedure allows the extrapolation of the rate of food quality decay in milder conditions usually experienced by the product (Mizrahi, 2010). This approach is generally called accelerated shelf life testing (ASLT). The basic premise in performing ASLT is that the C_i and P_i variables must be considered constant. Even the environmental factors E_i, with the exception of the one selected for the test E_1, must be kept constant. The choice of the environmental variable to be applied in ASLT must fulfill the following requirements: (a) The environmental variable is able to accelerate the event responsible for food quality decay significantly;

(b) a mathematical function describing the dependence of the ϑ parameters on the selected environmental variable exists. Based on these assumptions, Equation 2.5 can be simplified as follows:

$$\vartheta = h(E_1, \alpha) \tag{2.8}$$

where E_1 is the sole environmental variable voluntarily changed to speed up the food quality decay.

The different environmental factors potentially applicable in ASLT are temperature T, relative humidity RH%, and light intensity L. Those with previous experience in food kinetics should have guessed that Equation 2.8 represents the well-known Arrhenius equation where E_1 is the temperature and α is the relevant characteristic coefficient (the activation energy and the frequency factor). Other equations able to describe the dependence of ϑ parameters on the changes in whatever environmental variable can be synthetically expressed in the form of Equation 2.8.

As can be observed in Figure 2.4, shelf life testing, under both actual and accelerating conditions, can be carried out following a kinetic or life testing approach. Additional details on life testing are given in this chapter. In both cases, experimental data are collected under different food storage times at a well-defined environmental condition. Although conceptually similar, the experimental designs required for these approaches differ greatly in terms of amount of experimental data to be collected. Life testing requires wider experimental plans than those adopted in kinetic testing. The main reason lies in the different features of the experimental data obtained: While in kinetic testing the mathematical variable adopted to monitor the evolution of food quality decay versus storage time is continuous, in life testing such a variable is discontinuous and described by binary behavior. Since in the latter case data are governed by probability distribution (Guillet and Rodrigue, 2009), the testing plan must be designed to generate enough data to be correctly processed in the further modeling step. Based on these considerations, it is clear that the two testing approaches are conceptually different and require tailored experimental plans followed by dedicated modeling procedures. In other words, at the end of the preliminary step, the operator must have a clear idea of the testing methodology to be used to develop a proper experimental plan and select suitable software tools for data modeling and shelf life computation.

2.2.3 Modeling

After experimental data have been gathered, data analysis is the next stage. This activity must be performed keeping in mind Equations 2.1 to 2.3.

$$I = g(t, \vartheta) \tag{2.1}$$

$$t = f(I, \lambda) \tag{2.2}$$

$$SL = f(I_{\text{lim}}, \lambda) \quad (2.3)$$

At the beginning, data analysis should lead to the choice of a proper mathematical "skeleton" whose relevant parameters can be estimated by fitting the model to the experimental points (Brandao and Silva, 2009). Data elaboration will allow for finding a suitable g function and to compute the ϑ parameters of Equation 2.1. Afterward, the λ parameters of the inverse function f (Equation 2.2) can be calculated. Finally, by substituting I with I_{lim}, shelf life can be computed.

When ASLT is applied, as discussed in the previous section, a function describing the dependence of the ϑ parameters on the environmental variable chosen to speed up quality decay should be established to satisfy Equation 2.8:

$$\vartheta = h(E_1, \alpha) \quad (2.8)$$

In practical terms, experimental data collected by applying ASLT allow the finding of the h function and relevant α parameters from which ϑ kinetic parameters, referring to the environmental conditions of interest of the g function (Equation 2.1), can be easily extrapolated. Afterward, the λ parameters of the inverse function f (Equation 2.2) can be calculated. Finally, by substituting I with I_{lim}, a predicted shelf life value referring to the environmental conditions of the shelf can be obtained (see Chapters 5 and 6).

The mathematical model should predict the response variable accurately. In this case, the ultimate variable is the shelf life, which can be estimated or predicted depending on whether testing is carried out on actual or accelerated conditions, respectively. A number of different models have been developed to estimate or predict food shelf life. However, it must be pointed out that such models are actually shelf life models only when they include the function operator f, the appropriate parameters λ, and the acceptability limit I_{lim} (Equation 2.3). The first criterion to consider in choosing the correct modeling strategy is to take into consideration the nature of the mathematical variable (continuous or discontinuous) adopted to monitor the evolution of food quality decay versus storage time.

Kinetic modeling is based on the assumption that food quality continuously decreases when the food moves from the manufacturer to consumers. The basic premise is that the critical indicator is a continuous variable whose evolution over storage time can be modeled using classical kinetic or empirical equations (Mizrahi, 2010). By contrast, experimental data gathered in life testing (frequently also called failure, survival, or lifetime data) have a discontinuous nature and require a specific statistical tool called survival analysis. Basically, survival analysis originates from the consideration that it is practically impossible to observe food failure times systematically. This is an emerging approach in the food area that has been extensively applied in other fields of science, such as medicine (to estimate survival times of patients treated with different drugs) or engineering (to test the reliability of components under different types of stress) (Guillet and Rodrigue, 2009).

The fundamental idea is that samples do not fail at the exact same time. This phenomenon is technically referred as "censoring." Based on this assumption, by

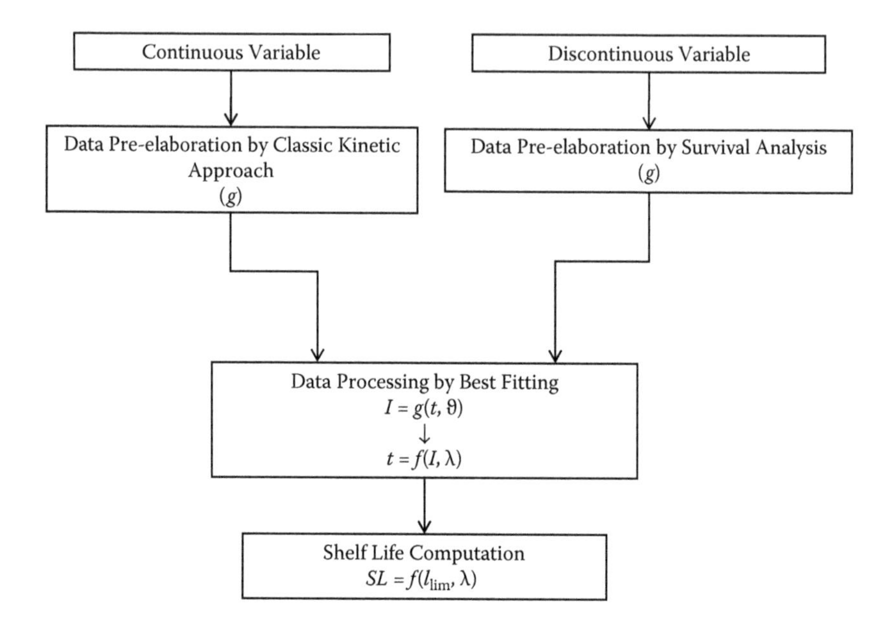

Figure 2.5 Steps of the shelf life modeling process depending on the nature (continuous or discontinuous) of the critical indicator when testing is carried out at actual storage conditions.

analyzing the performance of the food product on the shelves, two different possible results can be collected: (a) The product has failed, that is, it has exceeded the acceptability limit; or (b) the product has not failed, that is, it is still acceptable for consumption. The main feature of this approach is that the state of the sample is described by a binary variable, so shelf life data are governed by probability distributions (Guillet and Rodrigue, 2009).

Figures 2.5 and 2.6 summarize the different steps of the shelf life modeling process depending on the nature of the critical indicator (continuous or binary) when testing is carried out under actual or accelerated storage conditions, respectively.

In can be observed that the two schemes reported in Figures 2.5 and 2.6 only differ for data processing: When ASLT is applied, the ϑ parameters of interest (i.e., those referring to the environmental conditions experienced by the food product on the shelves) have to be estimated using Equation 2.8. When the critical indicator is a continuous variable, the classic kinetic approach has to be applied, whereas survival analysis should be used in the case of a discontinuous variable. Data are then processed by the best-fitting procedure. Finally, the corresponding inverse function is derived, and shelf life data are calculated.

The modeling procedure based on survival analysis is quite complex and can be basically divided into two steps: In the first step, based on a given judgment criterion (sensory, chemical, physical, or microbiological attribute level), acceptable food products are discriminated from the ones no longer acceptable. This is why experimental data with binary behavior are generated. By increasing the storage

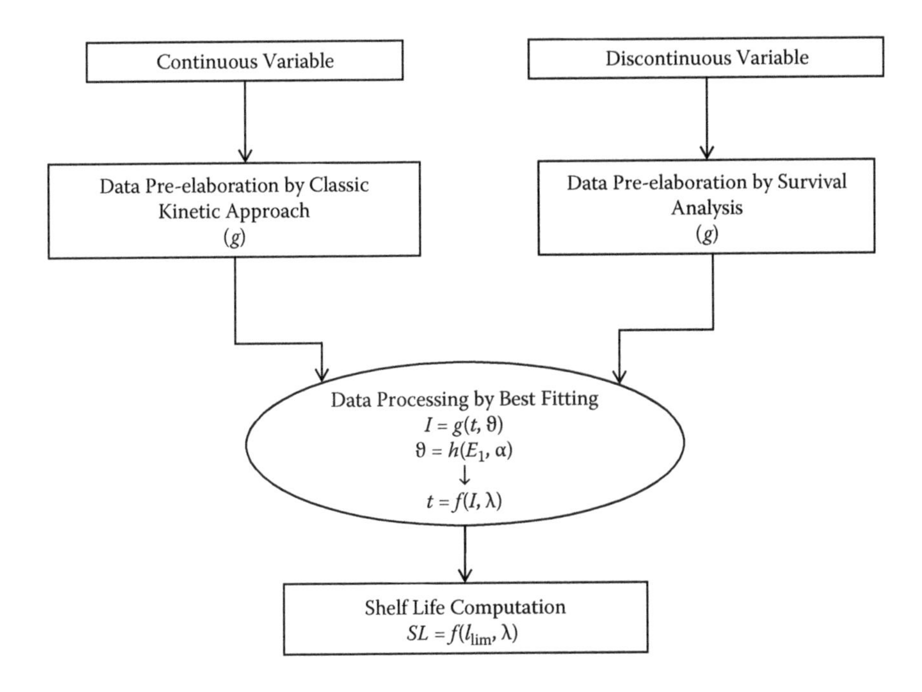

Figure 2.6 Steps of the shelf life modeling process depending on the nature (continuous or discontinuous) of the critical indicator when testing is carried out at accelerated conditions.

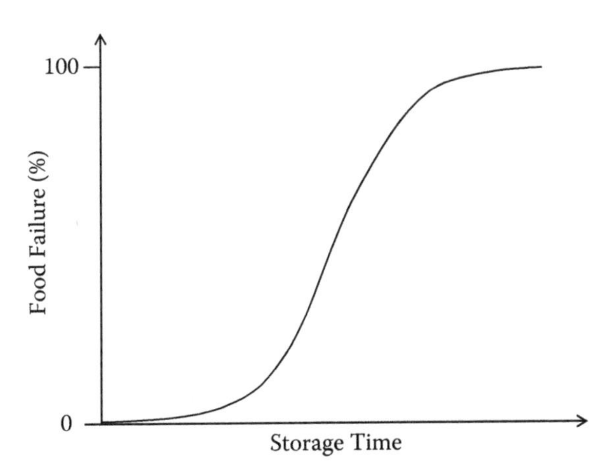

Figure 2.7 Distribution curve of the percentage of food failure versus storage time.

time, the number of unacceptable products is expected to increase to the detriment of the acceptable ones. In the second step, data elaboration allows finding a probability function of food failure versus storage time. In Figure 2.7, an example of the output obtained by data elaboration based on survival modeling is shown. The figure shows the percentage of food failure over storage time. The shelf life value is computed using the inverse function *f* by selecting a suitable percentage of failure the company decides to tolerate I_{lim}.

It is interesting to note that, in the case of the survival analysis, at the beginning of the modeling process, a quality indicator level, namely, the judgment criterion, is used to discriminate products that are acceptable from those no longer acceptable. In the following stage of the process, the ultimate critical indicator I is the probability of food failure, and the acceptability limit I_{lim} is a given risk level of product unacceptability.

Although survival analysis has been mainly applied to assess sensory shelf life by using the consumer acceptability response as the judgment criterion (Hough et al., 2003; Hough 2010), this methodology can even be applied using a physical, chemical, microbiological, or sensory attribute cutoff as the judgment criterion. Unfortunately, at present there are few applications of survival analysis based on judgment criteria other than consumer acceptability.

In summary, survival analysis represents an alternative but powerful approach compared to kinetic modeling. Chapters 5, 6, and 8 describe how to perform shelf life estimation or prediction by means of a kinetic approach. Chapter 7 illustrates how to face a survival analysis methodology and some of its powerful applications. Finally, Chapter 9 illustrates and critically discusses some examples of real shelf life studies in which the aforementioned methodologies are applied.

2.3 SHELF LIFE MONITORING ON THE MARKET

The estimated or predicted shelf life data obtained in simulated conditions hold only when the food product, in its "real" life, actually experiences the same environmental conditions provided for shelf life experiments. Obviously, "simulated" shelf life data cannot take into consideration any temperature fluctuation or any other change in environmental storage conditions during distribution, storage, and home handling. To better understand whether any discrepancies between assessed and real shelf life data exist, monitoring of the shelf life performance of the food product in real conditions and consequent data elaboration would be of great importance to take the necessary actions.

It cannot be considered a classical shelf life study, but it represents what companies did and still do by analyzing historical data concerning the real performance of the product on the shelves, including consumers' complaints and recall data and other information, for instance, those coming from time–temperature indicator (TTI) and radio-frequency identification (RFID) tags. Obviously, this is not the case for new products, for which no historical data are available.

This slow process allows the comparison of simulated with real shelf life data. The latter take into consideration "any moving away" from the storage conditions

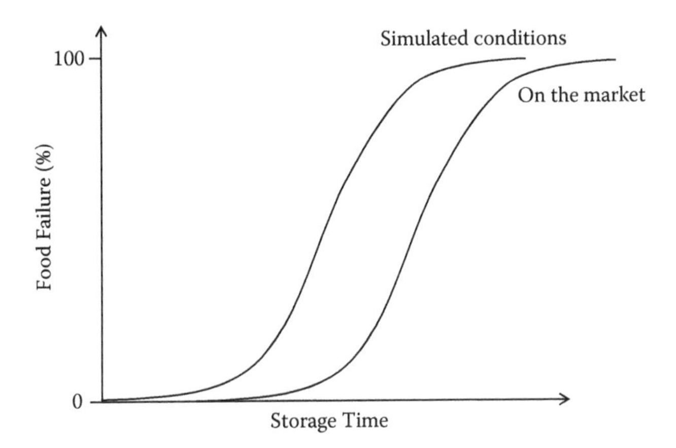

Figure 2.8 Comparison of the failure distribution curve obtained in the lab (simulated conditions) for a given product with that obtained by monitoring its actual performance on the market.

previously considered in shelf life testing. Although originally developed to assess the product life end empirically, this approach, if carried out using proper statistical tools, can represent an additional strategy able to complete information previously obtained from a classic shelf life assessment process. For instance, processing historical data can allow confirming or correcting dating previously estimated by ASLT.

Again, survival analysis methodologies can be successfully applied to compare the percentage of food failure versus storage time under simulated and real storage conditions, as shown in the example depicted in Figure 2.8.

It must be pointed out that the comparison between the two distribution curves allows the company to know how different the two conditions are and to compute the risk of failure of dating obtained from data collected in the lab.

2.4 KINETIC DATA VERSUS LIFETIME DATA

As discussed in Section 2.2.3, survival analysis requires processing more experimental data than necessary in kinetic modeling. For this reason, a shelf life study based on survival analysis is generally more time consuming and costly. Based on this consideration, a question arises: Why should life testing and further dedicated modeling procedures be preferred on the classical kinetic approach? Are there any particular advantages for a food company that chooses life testing instead of kinetic testing? The answer to these questions lies in the different meaning given by the two approaches to the quality level discriminating acceptable products from ones no longer acceptable. As a consequence, the outputs obtained by kinetic and survival analysis modeling procedures show completely different features. The basic premise for both methodologies is that a quality level discriminating acceptable products from ones no longer acceptable must exist. To avoid any misunderstanding, such a quality level is called the cutoff from this point. The cutoff can be objective, and for this reason, it can be expressed by quantitative data (i.e., sensorial score, value of peroxide, viscosity, color, nutrient concentration, microbial growth level, etc.); however, sometimes it can be subjective and rather conceptual, such as that present in the mind of consumers who are asked to judge whether a food product is acceptable or not. In the latter case, the cutoff has a binary feature (yes/no), and it cannot be converted in whatever score or quantitative value.

In the case of the kinetic approach, it is assumed that all food products will fail when they approach the cutoff. It implies that all food products are supposed to have exactly the same quality decay rate. This is because the cutoff value is assumed to be the acceptability limit, and the shelf life is computed accordingly.

In the case of the survival analysis approach, it is assumed that food products do not fail at the same time. This may happen for a variety of reasons: (a) Products belonging to different batches can show reasonably different performance on the shelf; (b) the same product may be judged acceptable by some consumers and unacceptable by others. This is because such a cutoff has a subjective nature, depending on the consumer's opinion. Since it cannot be converted to a quantitative value, it gives only qualitative information. According to this point of view, food products are expected to fail at different times since they are supposed to have different quality decay rates. This is because the cutoff cannot be considered the acceptability limit and why a function describing the evolution of the percentage of food failure versus storage time must be developed. The acceptability limit is then chosen by selecting the percentage of food failure considered tolerable, and the shelf life is computed accordingly. The shelf life is a time limit at which a given percentage of food products is expected to fail. It means that the remaining percentage of products is still supposed to be acceptable. Table 2.2 summarizes the different features and meanings assumed by the terms *cutoff*, *acceptability limit*, and *shelf life* obtained following the kinetic or survival analysis approaches.

In practical terms, the kinetic study gives a perfect picture of the quality decay rate of the tested samples. This does not mean that this approach allows for the finding of a perfect picture of reality. In the case of the survival analysis, the output is a failure function that allows the measure of how the risk of food unacceptability increases over storage time. In other terms, survival analysis allows obtaining a more

Table 2.2 Different Features and Meaning Assumed by the Terms Cutoff, Acceptability Limit, and Shelf Life Obtained Following Kinetic or Survival Analysis Approaches

	Cutoff	Acceptability Limit	Shelf Life
Kinetic approach	Quantitative	Equal to cutoff	Time limit at which 100% of food products are expected to fail
Survival analysis approach	Quantitative or qualitative	Percentage of food failure	Time limit at which a given percentage of food product is expected to fail

complete picture of product behavior on the shelves, giving the go-ahead to the food companies to compute the product shelf life on the basis of the level of failure risk they are willing to run.

Based on the aforementioned considerations, when should the survival analysis approach be preferred to the kinetic one? The first good reason is when the cutoff used to discriminate acceptable food products from unacceptable ones is subjective. This is when consumers are asked to judge whether a food product is acceptable. In such a situation, there is no possibility to follow a kinetic approach. The second good reason is when the understanding of the product failure risk over time is essential for the company. It is certainly essential when a low or very low failure risk is admitted because the cutoff derives from specific legal constraints (i.e., voluntary label claims) or when the food is a niche product that is expected to maintain a high quality level to the end of its life. The knowledge of the product failure risk function is also useful for those companies aiming to develop a tool that allows better management of the position of the product on the market, especially when they have no historical data on food performance on the shelves. Survival analysis is also the preferred tool for shelf life assessment when high production variability has been detected from batch to batch, and the company needs to avoid excessive food failure risk on the market. By contrast, the kinetic approach should be preferred for shelf life studies of regular productions showing low or negligible batch variability and for *routine* shelf life monitoring, when historical data on food performance on the shelves are available. It is certainly the favorite tool when a first shelf life study is required during new product development or in the following stages after product launch.

Table 2.3 rationalizes situations in which the survival analysis approach should be preferred to the kinetic one.

In practical terms, the survival analysis is required when the cutoff has special features (subjective/qualitative) or it is the need, for a variety of reasons, for strict control of the food failure risk. Although essential to face some specific requirements or needs, survival analysis is not a suitable tool to perform routine shelf life

Table 2.3 When the Survival Analysis Approach Should Be Preferred to the Kinetic One

Cutoff Has a Qualitative Feature	Company Has Collected Food Acceptability Data from Consumers
Existence of legal constraints for cutoff determination	The company tolerates low or very low risk of the product cutoff overcoming to avoid legal troubles.
Existence of voluntary constraints for cutoff determination	The company tolerates low or very low risk of the product cutoff overcoming since the product has to maintain a high quality level up to the end of its life.
Need for tools to better manage the position of the product on the market when no historical data are available	The company aims to choose the acceptability limit by merging rational and marketing considerations.
Existence of high production variability	The company aims to minimize an excessive risk of food failure on the market.

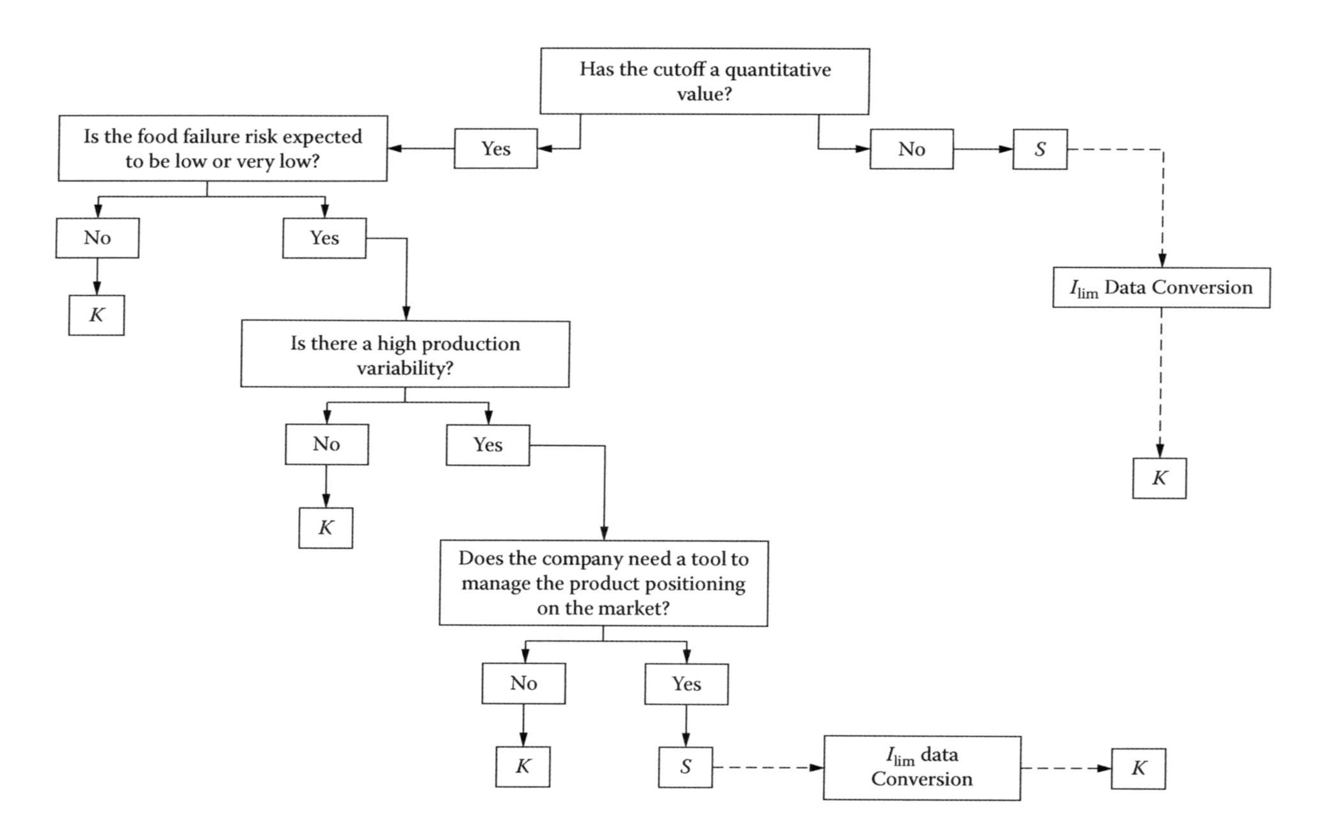

Figure 2.9 Tentative decision tree to support the choice of the survival analysis approach (S) or the kinetic approach (K).

studies. However, in these cases, it is conceptually possible to convert acceptability limits, expressed as the percentage of food failure, into kinetic acceptability limits. In this way, routine kinetic testing and further modeling procedures can be rationally performed based on the entire information acquired by the survival analysis. An exhaustive description of how to merge these different approaches is given in Chapter 4.

Figure 2.9 schematically shows a possible decision tree that should help in choosing the right modeling approach based on specific needs.

2.5 REFERENCES

Brandao, T.R.S., and C.L.M. Silva. 2009. Introduction to integrated predictive modeling. In *Predictive modeling and risk assessment*, R. Costa and K. Kristbersson, Eds., 3–18. New York: Springer Science.

Damodaran, S., Parkin, K.L., and O.R. Fennema. 2007. *Fennema's food chemistry*, 4th ed. Boca Raton, FL: Taylor & Francis.

Ellis, M.J., and C.M.D. Man. 2000. The methodology of the shelf life determination. In *Shelf-life evaluation of foods,* 2nd ed., D. Man and A. Jones, Eds., 23–33. Gaithersburg, MD: Aspen.

Eskin, N.A.M., and D.S. Robinson. 2000. *Food shelf life stability: chemical, biochemical and microbiological changes.* Boca Raton, FL: CRC Press, Taylor & Francis.

Fu, B., and T.P. Labuza. 1993. Shelf life prediction: theory and application. *Food Control* 4(3): 125–133.

Fu, B., and T.P. Labuza. 1997. Shelf life testing: procedures and prediction methods. In *Quality of frozen food,* M.C. Erickson and Y.C. Hung, Eds., 377–415. New York: Chapman and Hall, International Thomson.

Guillet, M., and N. Rodrigue, N. 2009. Shelf life testing methodology and data analysis. In *Food packaging and shelf life. A practical guide*, G.L. Roberson, Ed., 31–53. Boca Raton, FL: Taylor & Francis.

Hough, G. 2010. *Sensory shelf life estimation of food products.* Boca Raton, FL: Taylor & Francis.

Hough, G., Langohr, K., Gomez, G., and A. Curia. 2003. Survival analysis applied to sensory shelf life of foods. *Journal of Food Science* 68: 359–362.

Kilcast, D., and P. Subramaniam. 2011. *Food and beverage stability and shelf life.* Cambridge, UK: Woodhead.

Labuza, T.P., 1982. *Shelf life dating of food.* Westport, CT: Food and Nutrition Press.

Labuza, T.P. 1984. Application of chemical kinetics to deterioration of foods. *Journal Chemical Education* 61(4): 348–358.

Labuza, T.P., and Fu, B. 1993. Growth kinetic for shelf life prediction. Theory and practice. *Journal Industrial Microbiology* 12(3–5): 309–323.

Labuza, T.P., Fu, B., and P.S. Taoukis. 1992. Prediction for shelf life and safety of minimally processed CAP, MAP chilled foods. A review. *Journal of Food Protection* 55(9): 741–750.

Labuza, T.P., and M.K. Schmidl. 1985. Accelerating shelf life testing in foods. *Food Technology* 39(9): 57–64.

Labuza, T.P., and M.K. Schmidl. 1988. Use of sensory data in the shelf life testing of foods: principles and graphical methods for evaluation. *Cereal Foods World* 33: 193–205.

Labuza, T.P., and L.M. Szybist. 2001. *Open dating of foods.* Trumbull, CT: Food and Nutrition Press.

Labuza, T.P., and P.S. Taoukis. 1990. The relationship between processing and shelf life. In *Foods for the '90s*, ed. G.G. Birch, G. Campbell-Platt, and M.G. Lindley, Eds., 73–106. London: Elsevier Applied Science.

Man, D., and A. Jones. 2000. *Shelf-life evaluation of foods,* 2nd ed. Gaithersburg, MD: Aspen.

Mizrahi, S. 2010. Accelerated shelf life testing of foods. In *Food and beverage stability and shelf life*, D. Kilkast and P. Subramaniam, Eds., 482–503. Cambridge, UK: Woodhead.

Nelson, K.A., and T.P. Labuza. 1994. Water activity and food polymer science: implications of the state on Arrhenius and WLF models in predicting shelf life. *Journal Food Engineering* 22: 271–289.

Robertson, G.L. 2009. *Food packaging and shelf life. A practical guide.* Boca Raton, FL: Taylor & Francis.

Singh, R.P. 2000. Scientific principles of shelf life evaluation. In *Shelf-life evaluation of foods,* 2nd ed.. D. Man and A. Jones, Eds., 3–20. Gaithersburg, MD: Aspen.

Steele, R. 2004. *Understanding and measuring shelf life.* Cambridge, UK: Woodhead.

Taub, I.A., and R.P. Singh. 1998. *Food storage stability.* Boca Raton, FL: CRC Press, Taylor & Francis.

Van Boekel, M.A.J.S. 2009. *Kinetic modeling of reactions in foods.* Boca Raton, FL: CRC Press, Taylor & Francis.

Waletzko, P., and T.P. Labuza. 1976. Accelerated shelf life testing of an intermediate moisture food in air and in an oxygen free atmosphere. *Journal Food Science*, 41(6): 1338–1344.

CHAPTER 3

The Acceptability Limit

Lara Manzocco

CONTENTS

3.1 DEFINITION

Shelf life has been defined in a number of different ways, but implicit in each definition is the fact that the end of shelf life is reached in correspondence to a quality level that distinguishes products still acceptable for consumption from those that are no longer acceptable. This shadowed, vague, and somewhat mysterious boundary is referred to as the *acceptability limit* (Figure 3.1).

As reported in Chapter 2, given an adequate kinetic model of quality depletion f and relevant parameters λ, the acceptability limit of the critical indicator I_{lim} is required to compute the shelf life value:

$$SL = f\left(I_{\text{lim}}, \lambda\right) \tag{3.1}$$

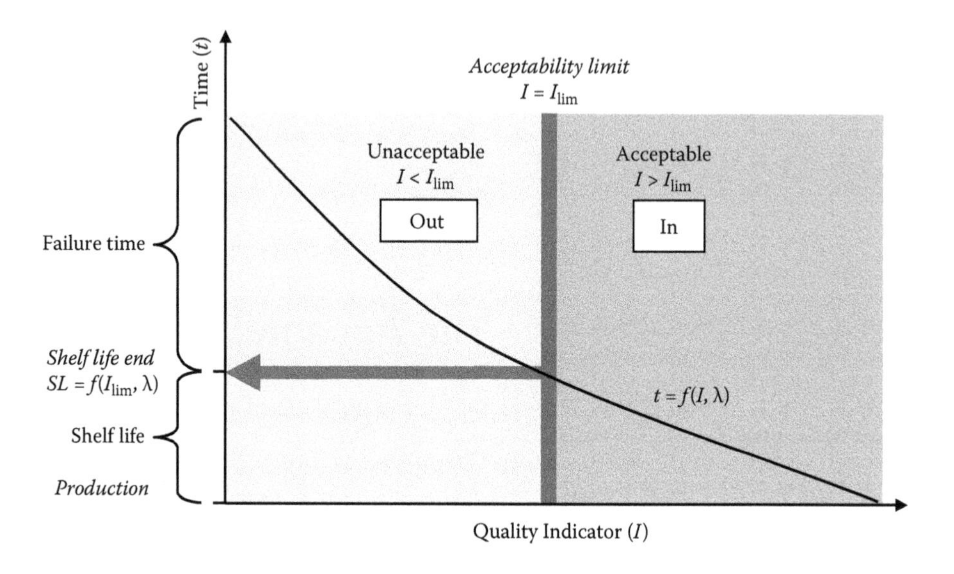

Figure 3.1 Acceptability limit discriminating shelf-life in from shelf-life out food

The acceptability limit is thus the finite value of the critical indicator that has to be inserted in Equation 3.1 to obtain the relevant solution, represented by a finite time value, which is the shelf life.

In principle, the concept of acceptability limit is simple and intuitive. It is the quality level discriminating between shelf life-in and shelf life-out products. While shelf life-in food still meets the requirements for consumption, shelf life-out food is no longer adequate because it does not fulfill the consumption standards. Despite its apparently intuitive nature, the acceptability limit is probably the most difficult parameter to be defined when developing a shelf life test. Although it dramatically affects the final shelf life value (Guerra et al., 2008), it is surprising that discussion of its nature has been largely neglected in the literature, and only rare indications about the possible methodologies for its determination are available with reference to specific food cases.

3.2 THE ACCEPTABILITY LIMIT: END OF SHELF LIFE BUT NOT OF SAFE LIFE

Basically, there are two possibilities for a product to become unacceptable during storage. The first one is mainly relevant to the occurrence of safety issues potentially leading to a risk for consumer health. The second one is the result of quality issues that could beget a risk for consumer dissatisfaction due to poor-quality appearance, low sensory or nutritional quality of the product. Table 3.1 shows these possibilities.

Unacceptability due to food intake under unsafe conditions is critical and can be the result of the growth of pathogenic microbes eventually contaminating food

Table 3.1 Safety and Quality Issues Leading to Food Unacceptability

Issue	Risk	Phenomena Involved	Example
Food safety	Consumer health	Growth of food poisoning microorganisms	Microbial counts exceeding the limits set by the regulation
		Migration of contaminants from packaging	Concentration of inks or plasticizers exceeding the limits set by the regulation
		Formation of toxic compounds	Concentration of histamine or peroxides exceeding the limits set by the regulation
Food quality	Consumer dissatisfaction	Development of biological, chemical, or physical phenomena negatively affecting food sensory properties	Off-color, off-odors, off-flavors, off-texture
		Degradation of beneficial compounds claimed on the label	Concentration of bioactive compound lower than that declared on the label

during its preparation. It can also derive from the presence of contaminants migrating from the packaging into food and of toxic or potentially toxic compounds whose formation occurs during storage. A number of different limit values for commercialization and consumption of food under safe conditions are indicated by regulatory bodies. To this regard, Table 3.2 shows some regulatory indications concerning the presence of pathogens that must be taken into account for commercialization of different food categories.

Similarly, the limit values for overall migration of plastic constituents from packaging materials into foodstuffs are reported by EEC Regulation No.72/2002 and its following amendments. The European Commission also provides an indication of the limit concentration of some toxic or potentially toxic molecules into specific food categories (Table 3.3).

When approaching the definition of the acceptability limit for shelf life assessment of food, a basic question arises: Should the limit values for safe food consumption, indicated by regulatory bodies, be regarded as acceptability limits to estimate shelf life? The only answer to this question is, "No." Shelf life definition should not be

Table 3.2 Limits of the Presence of Pathogens in Some Food Categories (EEC Regulation No. 2073/2005 on Microbiological Criteria for Foodstuff)

Food Category	Microorganism	Limit Value
Ready-to-eat food other than those intended for infants and for special medical purpose	*L. monocytogenes*	100 cfu/g
Live bivalve mollusks	*E. coli*	230 MPN 100 g of mollusks

Table 3.3 Limits of the Presence of Histamine and Peroxides in Some Food Categories

Food Category	Compound	Limit Value	Regulation
Fishery products	Histamine	100–200 mg/kg	EEC No. 2073/2005
Extra virgin and virgin olive oil	Peroxides	20 mEq O_2/kg	EEC No. 2568/1991
Olive-pomace oil	Peroxides	15 mEq O_2/kg	EEC No. 2568/1991

related to safety issues. During storage, any product overcoming the safety limit values reported by the regulation has to be regarded as outside basic standards required for consumption because it has reached the end of its "safe life." Two possible periods can be identified during storage of a perishable food: (a) a length of time during which the product is safe; (b) a length of time during which the product is unsafe. It would be a mistake to consider the limit that discriminates the safe condition from the unsafe one as the acceptability limit indicating the end of shelf life. Shelf life is a quality issue not related to safety. Thus, shelf life is a length of time included in the safety time interval during which the product retains acceptable quality characteristics (Figure 3.2).

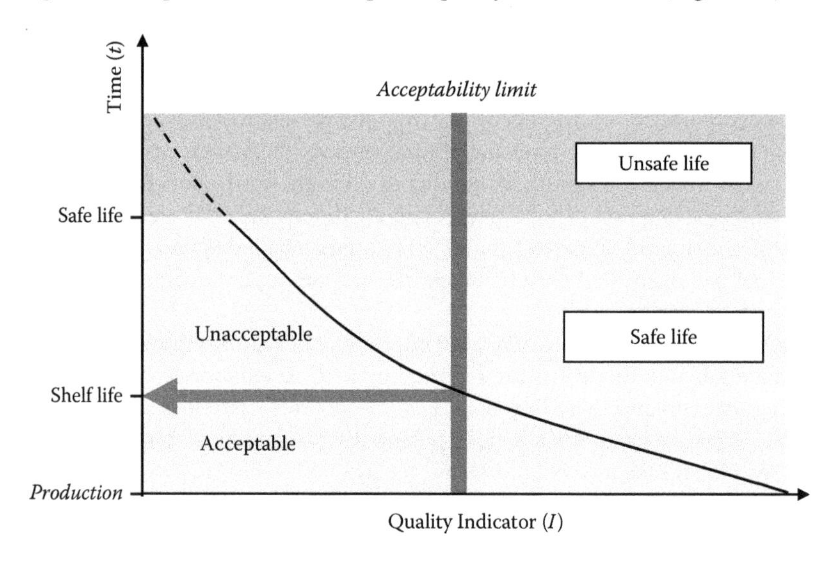

Figure 3.2 Positioning of shelf life within safe life of food.

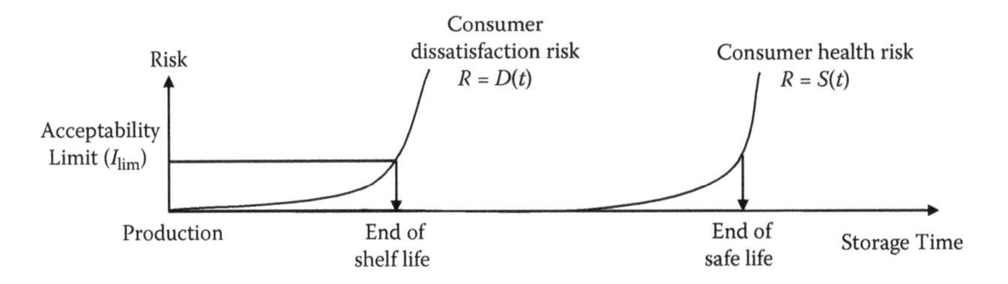

Figure 3.3 Evolution during food storage of unacceptability risk due to consumer dissatisfaction and consumer health risk. The end of shelf life and safe life are also shown.

The shelf life of the product must end largely before any risk for consumer health arises. In other words, shelf life and safe life are not only conceptually but also quantitatively different, with the shelf life much shorter than the safe life of the food:

$$\text{Shelf life} \ll \text{Safe life} \tag{3.2}$$

Conceptually, both safe life and shelf life can be estimated by analyzing the evolution of their risk functions versus storage time (Figure 3.3).

Because

$$R = S(t) \tag{3.3}$$

is the function describing the consumer safety risk as a function of storage time t and

$$R = D(t) \tag{3.4}$$

is the function describing the risk for consumer dissatisfaction due to low sensory or nutritional properties, the acceptability limit should be identified among the following conditions:

$$\begin{cases} D(t) \gg S(t) & (3.5) \\ \lim_{t \to SL} \left[D(t) + S(t) \right] = D(t) & (3.6) \end{cases}$$

If food is correctly designed and the production process properly carried out, unacceptable sensory or nutritional properties are reasonably expected to appear much earlier than any possible safety risk (Equation 3.5). The acceptability limit should be thus identified to estimate a shelf life value within the storage period in which only a sustainable risk for consumer dissatisfaction exists, while absolutely no risk for consumer health is present (Equation 3.6).

Table 3.4 Decision Makers and Characteristics of Acceptability Limits Corresponding to the End of Shelf Life

Decision Maker	Cause	Consumer Awareness	Acceptability Limit Legal Status	Acceptability Limit
Producer	Excessive consumer sensory dissatisfaction	Consumer is aware	Volunteer	Level of consumer sensory dissatisfaction
Producer	Label claim default	Consumer may not be aware	Compulsory	Concentration of bioactive factors voluntarily claimed by the producer

The decision maker of the acceptability limit is the producer itself; the producer is completely responsible for consumer satisfaction and adherence to legal requirements. The producer thus has to choose the acceptability limit by defining a set of requirements for the commercialization of its own product.

Actually, the acceptability limits decided by the producers can have two different natures depending whether consumer dissatisfaction, leading to the end of shelf life, involves sensory or nutritional defaults (Table 3.4).

There is a substantial difference in the capability of consumers to be aware of these defaults. In fact, based on their sensory perception, consumers are generally well aware of their level of sensory dissatisfaction when consuming a food. They make their own evaluation and decide whether the producer should be "punished" or "awarded" with their fidelity. In fact, when sensory dissatisfaction of consumers is the prevalent risk at the end of shelf life, the producer voluntary defines the tolerable level of such sensory dissatisfaction. This choice is the result of considerations merging the different needs of consumers, production, delivery structure, and business. Alternatively, compulsory acceptability limits can derive from volunteer label claims. In fact, the regulation allows producers to claim on the label the presence of compounds potentially exerting beneficial effects on consumer health. In most cases, bioactive compounds or health-effective microorganisms are added to increase food functionality. In this case, although consumers may be satisfied with the sensory quality of the product, they are seldom aware of the eventual decrease of the bioactive compounds to values lower than those claimed on the label. For this reason, according to the regulation, if producers voluntarily decide to make a claim, it is compulsory that they guarantee the conformity of the product to that specific claim over its shelf life.

3.3 THE ACCEPTABILITY LIMIT AS A CONSUMER- AND MARKET-ORIENTED ISSUE

As mentioned, the shelf life value is obtained by merging data describing the quality decay of the food product during storage with the relevant acceptability limit

Table 3.5 Properties of Quality Depletion and Acceptability Limit for a Given Food

Feature	Quality Depletion	Acceptability Limit
Origin	Product	Consumer and market
Character	Univocal	Discretional
Assessment method	Computed from experimental data elaborated by kinetic modeling or survival analysis	Defined by rational choice
Affecting variables	Formulation Packaging Environment (e.g., temperature, light, pressure)	Consumer sensory dissatisfaction Label claim default Social and marketing variables Production, delivery system, and business constraints

value. Quality decay data and acceptability limits have different features and generally require different approaches for their assessment/definition.

Table 3.5 schematically compares nature, methods of assessment, and variables affecting quality decay and acceptability limit.

Quality decay represents an intrinsic characteristic of the product and can be unequivocally defined by applying kinetic modeling or survival analysis of experimental data collected during food storage under specific environmental conditions (Chapters 5–7). For this reason, quality depletion can be regarded as a product-oriented issue.

In general terms, food quality progressively decreases during storage, making it difficult to derive an acceptability limit looking for peculiar mathematical points (e.g., abrupt changes) in the evolution of the critical indicator describing quality depletion. This is why little help to define food acceptability can be found in the analysis of the quality depletion kinetics of foods. The choice of the acceptability limit is a problem of rational evaluation of potential negative consequences deriving from the risk of consumer dissatisfaction. For this reason, the acceptability limit has a complex and fluctuating nature deriving from the fact that its definition is a consumer- and market-oriented issue (Table 3.5). Both the risk for sensory or nutritional default R and other factors, including marketing constraints and variables M, can play a critical role on the final choice of the acceptability limit:

$$\text{Acceptability limit} = f(R,M) \tag{3.7}$$

The acceptability limit is thus a quantitative value that must be inevitably chosen based on discretional, not superficial, considerations, taking into account (a) the risk of sensory or nutritional default; and (b) the effect of aspects relevant to production, delivery system, and business, which may critically condition the risk the producer is willing to tolerate.

Marketing variables and constraints M_i can actually exert a critical role on the final decision of the acceptability limit. Their consideration is often necessary to meet the basic producer requirement to increase/maintain market shares of historical products or to develop a novel market for innovative ones. Sometimes, production, delivery system, and business aspects are so critical they become the dominant factor of the final choice. For instance, during storage frozen food may quickly undergo a risk for consumer sensory dissatisfaction due to the development of oxidative reactions. Producers are well aware of this risk but are forced to choose less-conservative acceptability limits because several large retailing companies require a shelf life longer than 9 months for these products. In such a case, the peculiar marketing situation tends to increase the level of consumer dissatisfaction risk the company can tolerate at the end of shelf life. By contrast, the storage time required for canned food to develop a risk for consumer sensory dissatisfaction is generally much longer than the actual shelf life. The level of consumer dissatisfaction risk the company can tolerate at the end of shelf life is forced to zero, not to satisfy consumers, but to increase product turnover on the shelves.

In general, the acceptability limit identification process ends up in a reasonable compromise among the different needs to

- minimize sensory dissatisfaction as required by the target consumer
- minimize product recalls as required by the producer
- maximize product turnover on the shelves as required by the producer
- maximize shelf life as required by the distribution system

3.4 DISCRIMINATING ACCEPTABLE FROM UNACCEPTABLE FOOD: FROM THE CUTOFF TO THE ACCEPTABILITY LIMIT

As discussed in Chapter 2, a judgment criterion is needed to discriminate acceptable from unacceptable food. This criterion is generally the overcoming of a given attribute level and is defined as the cutoff. The cutoff can be a finite and countable value that may be adopted as an acceptability limit based on the drastic assumption that all product items overcome the cutoff at the same time, showing identical quality decay during storage (Figure 3.4).

In light of these considerations, the concentration of a nutrient claimed on the label by the producer or the intensity in a peculiar flavor could be regarded as possible acceptability limits to estimate shelf life. This is obviously a radical simplification because under real conditions, each product is expected to exceed the cutoff at a different time, reflecting its own variability. In the case of consumer dissatisfaction resulting from undesired sensory properties, additional variability must be considered because the same product can be judged acceptable by some consumers despite being recognized as unacceptable by others. It must be noted that if a strongly altered product is offered to a person, he or she will be definitely and unequivocally able to tell this product is unacceptable. However, when an acceptability limit to be applied in shelf life studies is looked for, only early and minor changes in the sensory properties should be taken into account because no strongly degraded product should reasonably reach the consumer. In these conditions, the uncertainty in discriminating

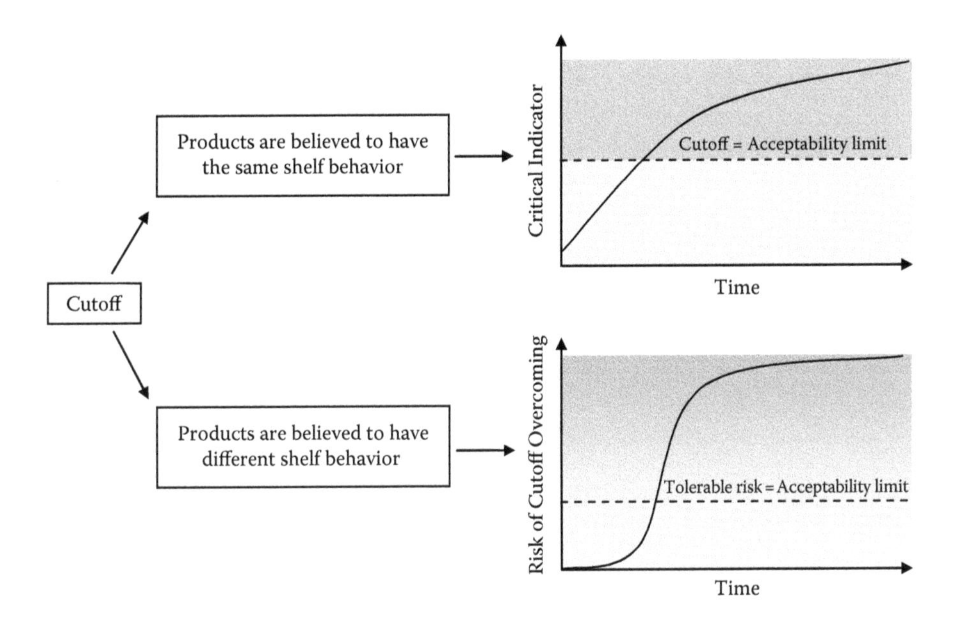

Figure 3.4 Discrimination between acceptable and unacceptable food.

acceptable from unacceptable food increases, and unacceptability appears as a wide boundary zone in which an increasing risk of overcoming the selected cutoff is observed (Figure 3.4). In other words, under real conditions, there is no unequivocal quality level discriminating acceptable from unacceptable food, but the acceptability limit is set corresponding to a tolerable risk of exceeding the cutoff discriminating acceptable from unacceptable foods.

3.5 RISK FOR CONSUMER SENSORY DISSATISFACTION

3.5.1 Variables Affecting Consumer Sensory Dissatisfaction

The existence of a wide sensory boundary discriminating acceptable from unacceptable food is due to the contribution of a high number of different variables affecting the consumer decision regarding food acceptability or unacceptability. In fact, based on the interaction between sensory stimuli from one side and affective, cognitive, and behavioral reactions from the other, the consumer makes the decision whether food is acceptable (Figure 3.5).

As is well known, the perception of sensory stimuli by itself is strongly individual. In addition, the joint elaboration in the human brain of sensory perception with information coming from the surrounding context and previous experiences can dramatically influence the response of acceptance or rejection (Costell et al., 2010). In this regard, it should be noted that affective, emotional, cognitive, and

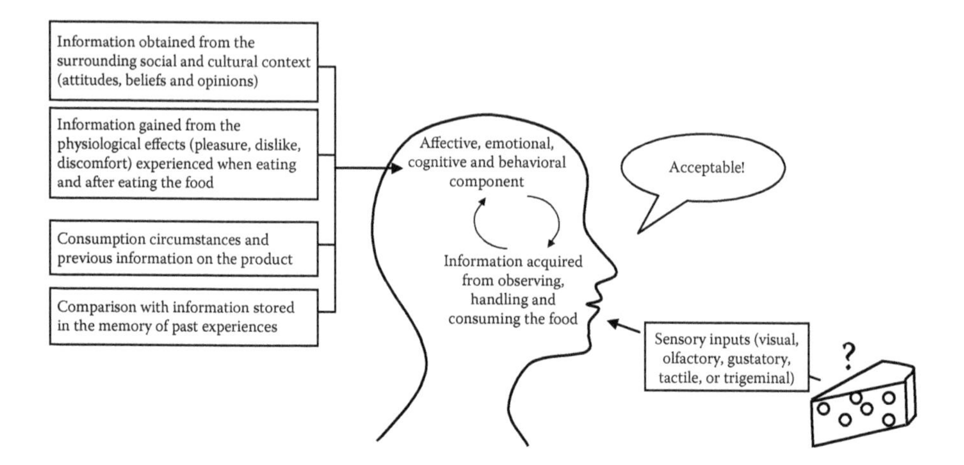

Figure 3.5 Acceptance or rejection of a given food by joint processing of different information in the human brain.

behavioral components of the process by which humans accept or reject food vary from location to location, from country to country, and from culture to culture since people from different locations, countries, and cultures may have different reactions to the same sensory stimuli. In addition, for the same subject, the decision whether the product is acceptable can vary due to the different circumstances of the sensory stimuli as well as in relation to the emotional status of the subject. A man could easily find acceptable a food offered by his new girlfriend in a relaxed situation while finding it unacceptable when prepared by his ex-wife. Similarly, marketing and advertisement strategies are expected to play a critical role on the threshold of food acceptability.

It has been supposed that food can also influence the feelings and mood of consumers (King and Meiselman, 2009). For this reason, it could be hypothesized that food stored for increasing time may induce a different emotional status in the consumers to whom it is offered. Figure 3.6 shows the changes of the main emotional terms consumers used to describe how they felt when presented a fruit salad stored for increasing time.

It is noteworthy that, compared to the just-prepared fruit salad (0-day storage time), the sample stored for 6 days only lost its fresh appearance without evident spoilage. The fruit salad stored for 10 days presented evident alteration, as also indicated by the presence of white yeast colonies growing on the fruit. Negative emotions were generally used with higher frequency by the consumers to describe how they felt when offered fruit salads stored for increasing time. By contrast, frequency of positive emotions decreased with storage time. However, only dramatically altered samples, which were evidently out of shelf life, were able to change the emotional state of consumers. Therefore, the response of acceptability or unacceptability by consumers seems to be mainly driven by rational considerations on the quality level of the product rather than by the emotional reactions induced by its consumption.

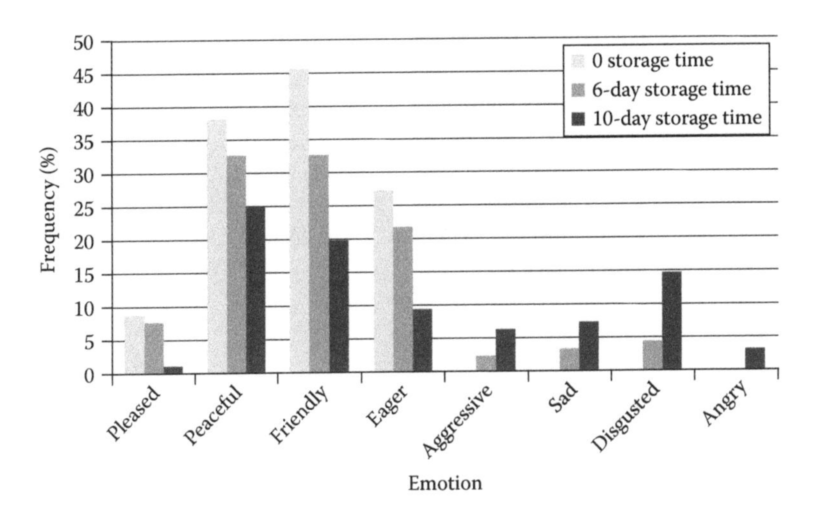

Figure 3.6 Emotional profiles related to fruit salad stored for 0, 6, and 10 days at 4°C

It must be considered that, when an unacceptable food is consumed, not all consumers will have the same physiological or emotional reaction. In fact, the sensitivity of the consumers to minor changes in sensory properties (e.g., off-flavor, off-firmness, off-color) following the development of a specific critical event during food storage could be significantly different. For these reasons, if not strongly spoiled, a product is never simply acceptable or unacceptable to all consumers. It will be unacceptable to a certain percentage of consumers, while other consumers will still find it acceptable. Any time a population of consumers eats an unacceptable food, a risk of sensory dissatisfaction arises. The high subjectivity of the physiological or emotional reactions of consumers to food consumption makes the acceptability limit a matter of probability.

In this regard, Box 3.1 shows an example of computation of the hazard rate of food unacceptability during storage.

3.5.2 Methodologies for the Assessment of Consumer Sensory Dissatisfaction

The assessment of consumer sensory dissatisfaction requires the application of sensory analysis. In some companies, descriptive sensory analysis carried out with panels of trained assessors is performed to describe the evolution of sensory attributes potentially responsible for consumer rejection. Figure 3.7 shows an example relevant to the evolution of the quality index, set by *consensus* by a trained panel, of ready-to-eat salad.

Although providing extensive information about the changes in quality attributes during storage, data about the intensity of the sensory attribute could have no relation to the consumer's decision to eat or buy the product. In other words, trained panels do not allow any sound suggestion about the potential risk of consumer sensory

BOX 3.1

The risk of food unacceptability is the probability that an acceptable food becomes unacceptable at a certain time t. Because food unacceptability is a detrimental event, its risk is generally referred to as a hazard.

The *hazard rate* is defined as the rate at which acceptable food items at any given instant t are becoming unacceptable. The hazard rate (or failure rate) is denoted by $h(t)$ and calculated as

$$h(t) = \frac{f(t)}{1 - F(t)}$$

where $f(t)$ is the probability of a food becoming unacceptable at time t, and $F(t)$ is the probability that a food becomes unacceptable before the time t.

Table 3.1.1 shows how to calculate the hazard rate of unacceptability of a set of 100 food samples becoming unacceptable during storage.

Figure 3.1.1 shows the hazard rate as a function of storage time for the food items considered in Table 3.1.1.

Table 3.1.1 Computation of Hazard Rate of a Set of 100 Food Items

Storage Time (months)	Unacceptable Food Items	$f(t)$	$F(t)$	$h(t)$
	A	$B = \frac{A}{100}$	$C = \sum_{i=1}^{n} B_i$	$D = \frac{B}{1-C}$
0	0	0	0	0
1	2	0.02	0.02	0.0204
2	5	0.05	0.07	0.0538
3	7	0.07	0.14	0.0814
4	9	0.09	0.23	0.1169
5	10	0.1	0.33	0.1493
6	11	0.11	0.44	0.1964
7	10	0.1	0.54	0.2174
8	10	0.1	0.64	0.2778
9	9	0.09	0.73	0.3333
10	7	0.07	0.8	0.35
11	6	0.06	0.86	0.4286
12	5	0.05	0.91	0.5556
13	3	0.03	0.94	0.5
14	2	0.02	0.96	0.5
15	2	0.02	0.98	1
16	1	0.01	0.99	1
17	1	0.01	1	
18	0	0	1	

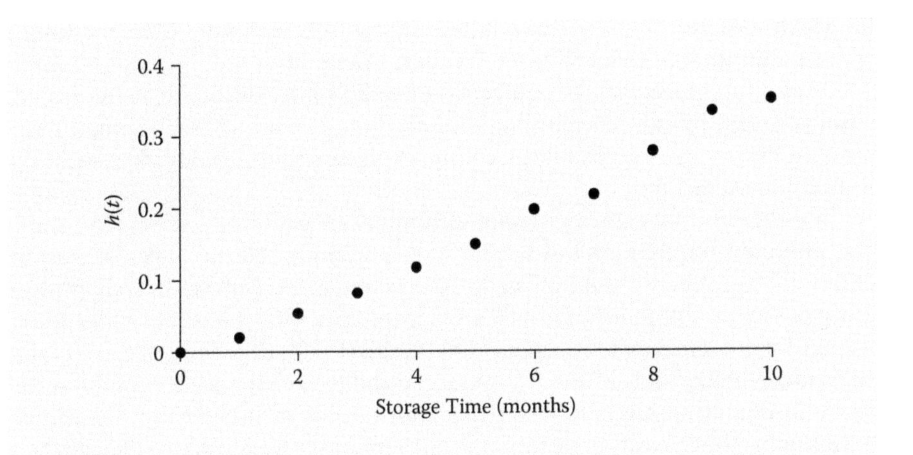

Figure 3.1.1 Hazard rate as a function of storage time for the food items considered in Table 3.1.1.

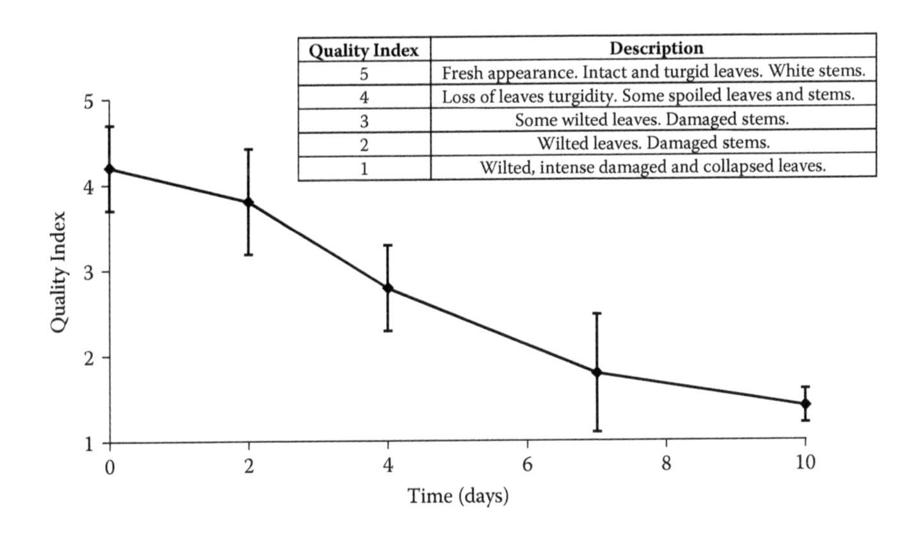

Quality Index	Description
5	Fresh appearance. Intact and turgid leaves. White stems.
4	Loss of leaves turgidity. Some spoiled leaves and stems.
3	Some wilted leaves. Damaged stems.
2	Wilted leaves. Damaged stems.
1	Wilted, intense damaged and collapsed leaves.

Figure 3.7 Quality index of ready-to-eat salad stored at 4°C. Inset: Description of salad appearance in relation to the 1–5 quality index scores

dissatisfaction to be achieved. The questions, Where should the cutoff be set? and, What is the acceptability limit? remain unsolved.

When available sensory data are relevant to the evolution of sensory attributes evaluated by trained assessors on intensity scales, the cutoff value adopted to discriminate acceptable from unacceptable products is generally chosen based on largely subjective and arbitrary considerations. A number of different examples for which acceptability values of selected critical indicators were used to estimate product shelf life can be found in the literature (Table 3.6).

It can be noted that these acceptability limits are widely subjective. In this regard, it is noteworthy that any choice of an acceptability limit of a sensory attribute without clear knowledge of its relation to consumer dissatisfaction could induce mistakes in shelf life estimation.

In other cases, as shown in Figure 3.8, affective analysis is performed by asking consumers for their level of satisfaction on different hedonic scales. It must be observed that affective data obtained when consumers evaluate overall product acceptability on a hedonic scale only provide intuitive information about the potential risk for consumer dissatisfaction. Muñoz et al. (1992) suggested using a 6.0 value on an acceptability scale of 1 to 9 as the acceptability limit for quality control specification. Although potentially sound, this limit value is subjected to a wide individual discretion since no apparent justification of its choice is provided. Although there must be a relation between consumer acceptability expressed on a scale of 1 to 9 and consumer dissatisfaction risk, the latter cannot be simply computed based only on these data unless the relation between the two is known. Details about the methodology to find such a relation are provided in Chapter 4.

Consumer sensory dissatisfaction can be identified using survival analysis methodology (Gacula and Kubala, 1975; Gacula and Singh, 1984; Hough et al., 2003; Hough et al., 2010). The product is analyzed during storage by asking the consumers about acceptability or unacceptability, and data are elaborated by survival analysis. This methodology can be used to estimate the shelf life of the product directly, and extensive details on its application are presented in Chapter 7. However, such methodology can also be applied to support the choice of the acceptability limit to be subsequently used for shelf life computation. In practical terms once data for food sensory acceptability or unacceptability are gathered, a risk function of consumer rejection of the food product over the storage time is obtained. While the cutoff is represented by the food acceptability or unacceptability expressed by consumers, the acceptability limit is selected as the percentage of food failure considered tolerable by the producer.

Figure 3.9 shows an example of consumer rejection function obtained by survival analysis of acceptability data for fresh-cut lamb's lettuce.

A good support for the acceptability limit choice can derive from the availability of such functions. In this case, the food company can choose to be exposed to more or less risk by selecting, as the acceptability limit, a certain percentage of consumers dissatisfied by the product stored up to the end of shelf life. In other words, the acceptability limit becomes the maximum percentage of consumers the company can tolerate to dissatisfy. It can be observed that 50% consumer dissatisfaction is generally chosen as an acceptability limit to estimate product shelf life. Based on product turnover on the shelves, the probability that a consumer actually eats the product at the end of its shelf life is low. For this reason, the assumption of an acceptability limit corresponding to 50% consumer rejection implies that only half of the consumers, incidentally eating the product at that time, will reject it. However, even such a low rejection probability may represent for food companies an excessive risk.

Table 3.6 Acceptability Value and Reference Scale for Critical Indicators Used in the Literature to Calculate Shelf Life of Different Foods

Food	Critical Indicators	Scale	Cutoff Value	References
Lettuce	Color, dryness, texture	1–5	3	McKellar et al., 2004
Lettuce	Color, dryness, texture	1–5	3	Zhou et al., 2004
Cucumber slices and mixed lettuce	Taste, smell/flavor, crispness, color of the cut surfaces, dryness/transparency, general appearance	1–10	5	Jacxsens et al., 2002
Papaya fruit	Firmness, shriveling, chilling injury	1–5	3	Nunes et al., 2006
Lemon verbena leaves infusion	Acceptability	0–15	7.5	Infante et al., 2010
Chicken breast	Odor, taste	1–9	6	Patsias et al., 2006
Rainbow trout	Odor	0–10	5	Pyrgotou et al., 2010
Sea bass fillets (roe samples)	Color	1–4	3	Provincial et al., 2010
	Fresh color	1–7	4	
	Fermented odor	1–6	3	
Sea bass fillets (steamed samples)	Aspect of the fillet, characteristic fresh sea bass odor, firmness, juiciness, intensity of fresh sea bass flavor, intensity of sea bass off-flavor, global appreciation	0–10	5	
Salmon (roe)	Color	0–4	2	Gimenez et al., 2005
	Odor	0–7	3	
Salmon (cooked)	Appearance, intensity of fresh salmon odor, firmness, intensity of fresh salmon flavor, off-flavor intensity, overall acceptability	0–10	5	
Sea bream fillets	Odor, flesh color, texture	0–10	6	Goulas and Kontominas, 2007a
Norway lobster	Odor, melanosis, color of the head (cephalothorax), cephalothorax-tail junction and the tail-parapodos junction, general acceptability	0–5	2	Gomez-Guillen et al., 2007
Chub mackerel	Odor, fresh color, texture	0–10	6	Goulas and Kontominas, 2007b
Chub mackerel	Odor, taste, fresh color, texture	0–10	6	Goulas and Kontominas, 2005
Sea bream	Color	0–3	2	Gimenez et al., 2004
	Odor	0–6	3	
Sea bass	Raw fish odor, cooked fish odor, cooked fish taste	0–3	2	Koutsoumanis et al., 2002
Chub mackerel, yellow gurnard, hake, and cuttlefish	Color, odor, texture	0–5	2	Speranza et al., 2009

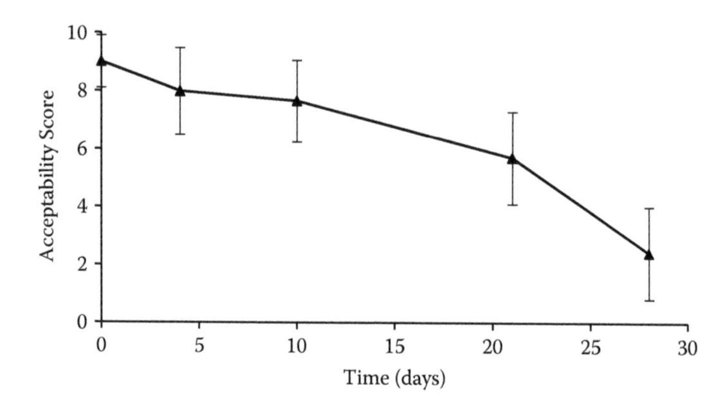

Figure 3.8 Acceptability score for fresh-cut lamb's lettuce stored at 4°C. (Modified from Manzocco, L., Foschia, M., Tomasi, N., Maifreni M., Dalla Costa, L., Marino, M., Cortella, G., and S. Cesco. 2011. Influence of hydroponic and soil cultivation on quality and shelf life of ready-to-eat lamb's lettuce (*Valerianella locusta* L. Laterr). *Journal of the Science of Food and Agriculture*, 91, 1373–1380.)

For this reason, it has been suggested that lower percentages of consumer dissatisfaction could be much more reliable. To this regard, Table 3.7 shows the percentages of consumer rejection used by different authors to calculate the shelf life of different food products.

Actually, there are also companies preferring to apply a risk tolerance approaching zero. In this case, the company decides that consumers eating the product at

Table 3.7 Percentage of Consumer Rejection Used in the Literature to Calculate Shelf Life of Different Foods

Food	Acceptability Limit (% Consumer Rejection)	Reference
Minced meat	50	Hough et al., 2006
Fuji apples	50	Varela et al., 2005
Recently harvested pears	50	Salvador et al., 2007
Milk	50	Duyvesteyn et al., 2001
Roasted and ground coffee	50	Cardelli and Labuza, 2001
Biscuits	50	Calligaris et al., 2007
Breadsticks	50	Calligaris et al., 2008
Ready-to-eat lettuce	25, 50	Araneda et al., 2008
Probiotic yogurt	25, 50	Cruz et al., 2010
White pan bread	25, 50	Gambaro et al., 2004
Coffee brew	25, 50	Manzocco and Lagazio, 2009
Muffins	20, 50	Baixauli et al., 2008
Stored pears	25	Salvador et al., 2007
Lamb's lettuce	25	Manzocco et al., 2011
Apple-baby food	25	Gambaro et al., 2006
Brown pan bread	25	Gimenez et al., 2007

Table 3.8 Percentage of Consumer Dissatisfaction Used as Acceptability Limit and Relevant Proposed Risk Level

Acceptability Limit (% Consumer Dissatisfaction)	Risk Level	
0	Negligible	Reasonable
10	Very low	
25	Low	
50	Medium	Excessive
75	High	

the end of shelf life must still find it acceptable. Even more restrictive acceptability limits can be applied when the company decides that the product must be not only acceptable to all consumers but also not significantly different from the fresh one. In general terms, it could be inferred that a percentage of consumer dissatisfaction lower than 25 could correspond to a reasonable risk level. In this regard, Table 3.8 shows the relation between the percentage of consumer dissatisfaction used as the acceptability limit and the proposed risk level.

It should be noted that the development of plots describing the evolution of consumer rejection, such as that shown in Figure 3.9, requires a large sample size, assembling consumer panels, and the application of appropriate statistical techniques to perform survival analysis of acceptability data. These conditions make survival analysis, despite being powerful and accurate, still difficult to apply by company operators. To our knowledge, there is no literature indication about procedures able to accelerate or simplify such a process.

To meet industrial needs for routine estimation, instrumental or sensory attributes, whose evolution is correlated to sensory dissatisfaction expressed by

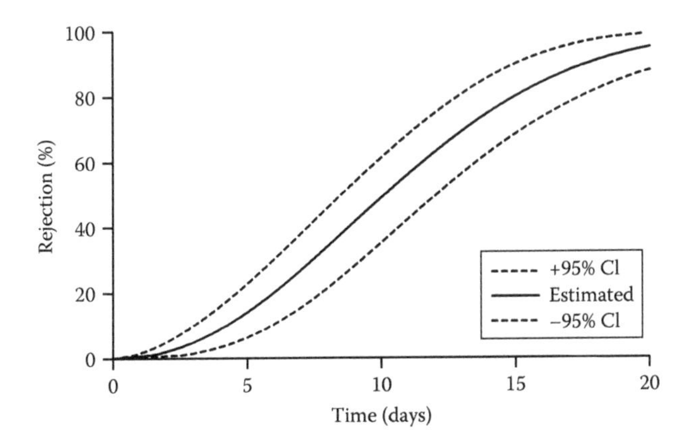

Figure 3.9 Percentage of consumer rejection of fresh-cut lamb's lettuce stored at 4°C. CI, confidence interval.

consumers, could be identified. The methodology to identify analytical quality indices that are easily assessable and potentially correlated to consumer dissatisfaction during food storage is presented in Chapter 4.

3.6 RISK FOR LABEL CLAIM DEFAULT

Similar to the risk for consumer sensory dissatisfaction, the risk for label claim default can be identified using the survival analysis methodology. In particular, the number of products not respecting the claim is monitored during storage, and data are elaborated to estimate the relevant risk function. In this case, the cutoff is actually the claim default, and the acceptability limit is the tolerable risk of food exceeding such a cutoff.

Figure 3.10 shows an example of risk functions for the default of a vitamin C claim in a fruit nectar stored under different atmospheric conditions in low- and high-barrier packaging. In this case, the acceptability limit was chosen to correspond to a 10% risk of product failure since the company decided to run a minor risk of claim default at the end of shelf life.

3.7 ACCEPTABILITY LIMIT BETWEEN HISTORICAL EXPERIENCE AND SENSORY DATA

Given the probabilistic nature of the risk of sensory dissatisfaction or label claim default and the difficulty of its accurate assessment, some tolerance in food quality at the end of shelf life is generally admitted. However, it is evident that producers must be well aware of potential adverse effects deriving from being too restrictive or

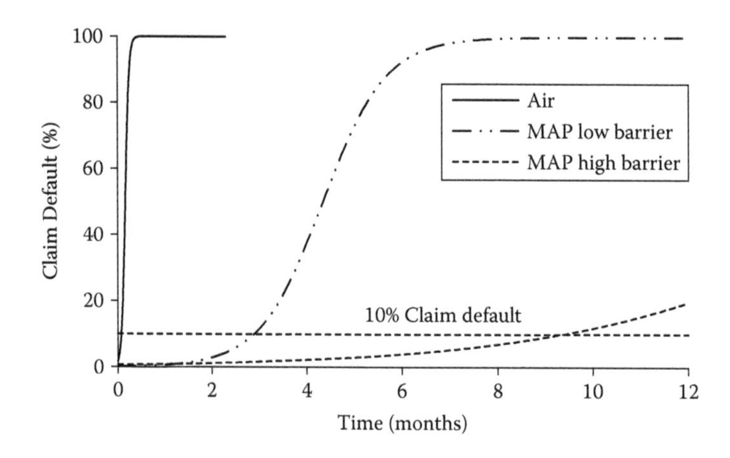

Figure 3.10 Percentage of default of a vitamin C claim in a fruit nectar stored under different atmosphere conditions in low and high barrier package.

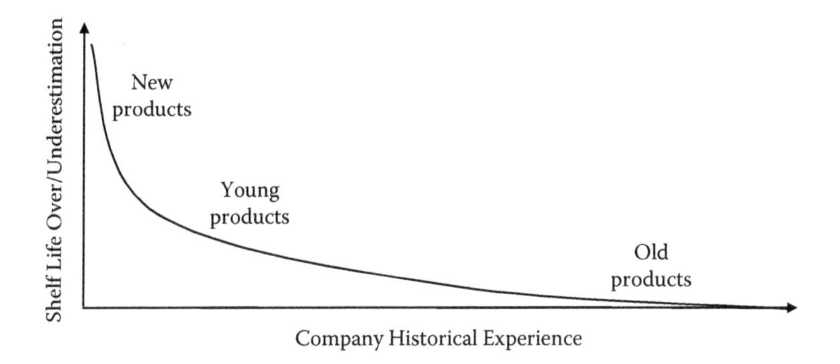

Figure 3.11 Risk for over-/underestimation of shelf life as a function of the level of historical experience of the producer.

too permissive regarding the risk level of consumer dissatisfaction the company is willing to tolerate at the end of shelf life.

The choice of the risk for food failure to be adopted as the acceptability limit not only can be chosen based on analytical data but also can be rationally derived from the historical experience developed by the company for the product. It is evident that the risk for over-/underestimation of shelf life decreases with the increase in the historical experience of the producer (Figure 3.11).

When historical experience for the product is high, the acceptability limit is chosen based on product performance data collected over years of commercialization. In this case, a shelf life value is already attributed to the product. Processing historical data can thus allow previously assigned shelf life to be confirmed or corrected. In this case, performing sensory or instrumental analysis is generally not necessary since consistent information on consumer satisfaction or dissatisfaction or label claim default are directly obtained from sales, complaint records, and the quality control database of the company. An emblematic case is that related to foods susceptible to microbial spoilage due to the development of alterative microorganisms. Even if not dangerous for health, these microorganisms are well known to be associated with potential undesired sensory defects that could be perceived by the consumers. Although the presence of alterative nonpathogenic microorganisms is not regulated, producer associations, retail consortia, and standard agencies have provided cutoff/acceptability limit values to discriminate acceptable from unacceptable food. For instance, levels of spoilage bacteria (total viable count, *Pseudomonas* spp., lactic acid bacteria, yeasts, or molds depending on the product) in fully cooked products or in products intended for eating cold are often suggested to be below 10^6 cfu (colony-forming units) per gram during the entire shelf life of the product. This acceptability limit derives from the consideration that higher levels of alterative nonpathogenic microorganisms are known to exert a significant impact on the sensory quality of the product and could thus be related to consumer dissatisfaction. In other words, it is historically recognized that consumer dissatisfaction due to the development of spoilage microorganisms tends not to be negligible when counts exceed the limit of 10^6 cfu/g.

There are cases when the company has low historical experience. This generally happens when a company is starting to produce a food that is already available on the market. Emulation of historical experience of competitors can be a possible strategy. Such a procedure, however, is fraught with the risk of critical overestimations or disadvantageous underestimation of final shelf life. To decrease this hazard, it is convenient to perform sensory and instrumental analysis to make sure that the acceptability limit historically chosen by competitors is also applicable to the new production.

Unfortunately, in the case of completely new foods, historical indications are not available due to the absence of any previous experience regarding product performance. In this case, consumer acceptability or label claim default data relevant to products stored for increasing time are necessarily required to support the acceptability limit decision and make the risk for over-/underestimation of shelf life tolerable to the company.

3.8 THE DECISION TREE OF THE ACCEPTABILITY LIMIT

Figure 3.12 schematically shows a possible decision tree to segment the different consumer and marketing aspects and support the process leading to the identification of the acceptability limit. When dealing with consumer aspects, it is necessary to analyze the level of company experience with the product. If the product performance on the market has been recorded during the previous commercialization period, a rational estimation of acceptability limit, and hence of shelf life, could be made with minor error risk. If these data are not available, two possibilities may arise depending on the existence of a specific label claim. If a label claim exists, the risk of unconformity of the product to what is declared on the label should be evaluated by a proper experimental plan. When no claim is declared by the company, the estimation of the acceptability limit could be performed based on market performance of analogous foods or emulation of competitors. It is clear that such a procedure may be risky in terms of over-/underestimation of shelf life. To make this risk tolerable to the company, it is convenient to produce sensory data estimating the risk of consumer dissatisfaction with the product as a function of storage time. This step is obligatory in the case of completely new foods due to the total absence of any previous experience of the company regarding product performance.

Finally, information about risk of product unacceptability should be merged with marketing considerations relevant to production, delivery system, and business aspects. The matching of these two visions should determine where it is more convenient for the company to set the acceptability limit.

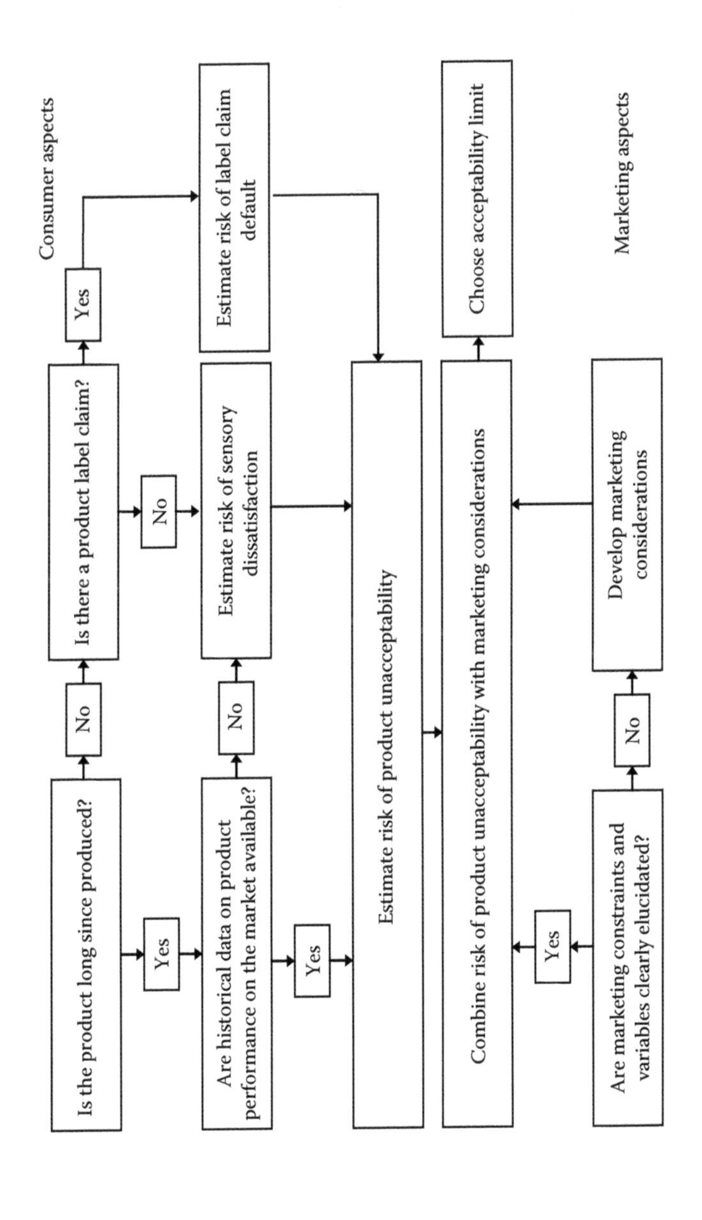

Figure 3.12 Decision tree for the identification of the acceptability limit.

3.9 REFERENCES

Araneda, M., Hough, G., and E.W. De Penna. 2008. Current-status survival analysis methodology applied to estimating sensory shelf life of ready-to-eat lettuce (*Lactuca sativa*). *Journal of Sensory Studies* 23: 162–170.

Baixauli, R., Salvador, A., and S.M. Fiszman. 2008. Textural and colour changes during storage and sensory shelf life of muffins containing resistant starch. *European Food Research and Technology* 226: 523–530.

Calligaris S., Da Pieve S., Kravina G., Manzocco L., and M.C. Nicoli. 2008. Shelf-life prediction of bread sticks by using oxidative indices: a validation study. *Journal of Food Science* 73: E51–E56.

Calligaris, S., Manzocco L., Kravina G., and M.C. Nicoli. 2007. Shelf-life modeling of bakery products by using oxidation indices. *Journal of Food Chemistry and Agriculture* 55: 2004–2009.

Cardelli, C., and T.P. Labuza. 2001. Application of Weibull hazard analysis to the determination of the shelf life of roasted and ground coffee. *LWT—Food Science and Technology* 34: 273–278.

Costell, E., Tárrega, A., and S. Bayarri. 2010. Food acceptance: the role of consumer perception and attitudes. *Chemosensory Perception* 3: 42–50.

Cruz, A.G., Walter, E.H.M., Cadena, R.S., Faria, J.A.F., Bolini, H.M.A., Pinheiro, H.P., and A.S. Sant'Ana. 2010. Survival analysis methodology to predict the shelf-life of probiotic flavored yogurt. *Food Research International* 43: 1444–1448.

Duyvesteyn, W.S., Shimoni, E., and T.P. Labuza. 2001. Determination of the end of shelf-life for milk using Weibull hazard method. *LWT—Food Science and Technology* 34: 143–148.

Gacula, M.C., and J.J. Kubala. 1975. Statistical models for shelf-life failures. *Journal of Food Science* 40: 404–409.

Gacula, M.C., and J. Singh. 1984. *Statistical methods in food and consumer research.* New York: Academic Press.

Gambaro, A., Ares, G., and A. Gimenez. 2006. Shelf-life estimation of apple-baby food. *Journal of Sensory Studies* 21: 101–111.

Gambaro, A., Fiszman, S.M., Gimenez, A., Varela, P., and A. Salvador. 2004. Consumer acceptability compared with sensory and instrumental measures of white pan bread: sensory shelf-life estimation by survival analysis. *Journal of Food Science* 69: S401–S405.

Gimenez, A., Varela, P., Salvador, A., Ares, G., Fiszman, S., and L. Garitta. 2007. Shelf life estimation of brown pan bread: A consumer approach. *Food Quality and Preference* 18: 196–204.

Gimenez, B., Roncales, P., and J. Beltran. 2004. The effects of natural antioxidants and lighting conditions on the quality characteristics of gilt-head sea bream fillets (*Sparus aurata*) packaged in a modified atmosphere. *Journal of the Food Science and Agriculture* 84: 1053–1060.

Gimenez, B., Roncales, P., and J. Beltran. 2005. The effects of natural antioxidants and lighting conditions on the quality of salmon (*Salmo salar*) fillets packaged in modified atmosphere. *Journal of the Science of Food and Agriculture* 85: 1033–1040.

Gomez-Guillen, M.C., Lopez-Caballero, M.E., Martinez-Alvarez, O., and P. Montero. 2007. Sensory analyses of Norway lobster treated with different antimelanosis agents. *Journal of Sensory Studies* 22: 609–622.

Goulas, A.E., and M.G. Kontominas. 2005. Effect of salting and smoking-method on the keeping quality of chub mackerel (*Scomber japonicus*): biochemical and sensory attributes. *Food Chemistry* 93: 511–520.

Goulas, A.E., and M.G. Kontominas. 2007a. Combined effect of light salting, modified atmosphere packaging and oregano essential oil on the shelf-life of sea bream (*Sparus aurata*): biochemical and sensory attributes. *Food Chemistry* 100: 287–296.

Goulas, A.E., and M.G. Kontominas. 2007b. Effect of modified atmosphere packaging and vacuum packaging on the shelf-life of refrigerated chub mackerel (*Scomber japonicus*): biochemical and sensory attributes. *European Food Research and Technology* 224: 545–553.

Guerra, S., Lagazio, C., Manzocco, L., Barnabà, M., and R. Cappuccio. 2008. Risks and pitfalls of sensory data analysis for shelf life prediction: data simulation applied to the case of coffee. *LWT—Food Science and Technology* 41: 2070–2078.

Hough, G. 2010. *Sensory shelf life estimation of food products.* Boca Raton, FL: Taylor & Francis.

Hough, G., Garitta, L., and G. Gomez. 2006. Sensory shelf-life predictions by survival analysis accelerated storage. *Food Quality and Preference* 17: 468–473.

Hough, G., Langohr, K., Gomez, G., and A. Curia. 2003. Survival analysis applied to sensory shelf life of foods. *Journal of Food Science* 68: 359–362.

Infante, R., Rubio, P., Contador, L., and V. Moreno. 2010. Effect of drying process on lemon verbena (*Lippia citrodora* Kunth) aroma and infusion sensory quality. *International Journal of Food Science and Technology* 45: 75–80.

Jacxsens, L., Devlieghere, F., and J. Debevere. 2002. Temperature dependence of shelf-life as affected by microbial proliferation and sensory quality of equilibrium modified atmosphere packaged fresh produce. *Postharvest Biology and Technology* 26: 59–73.

King, S.C., and H.L. Meiselman. 2009. Development of a method to measure consumer emotions associated with foods. *Food Quality and Preference* 21: 168–177.

Koutsoumanis, K., Giannakourou, M.C., Taoukis, P.S., and G.J.E. Nychas. 2002. Application of shelf life decision system (SLDS) to marine cultured fish quality. *International Journal of Food Microbiology* 73: 375–382.

Manzocco, L., Foschia, M., Tomasi, N., Maifreni M., Dalla Costa, L., Marino, M., Cortella, G., and S. Cesco. 2011. Influence of hydroponic and soil cultivation on quality and shelf life of ready-to-eat lamb's lettuce (*Valerianella locusta* L. Laterr). *Journal of the Science of Food and Agriculture* 91(8): 1373–1380.

Manzocco, L., and C. Lagazio. 2009. Coffee brew shelf life modelling by integration of acceptability and quality data. *Food Quality and Preference* 20: 24–29.

McKellar, R.C., Odumeru, J., Zhou, T., Harrison, A., Mercer, D.G., Young, J.C., Lu, X., Boulter, J., Piyasena, P., and S. Karr. 2004. Influence of a commercial warm chlorinated water treatment and packaging on the shelf life of ready-to-use lettuce. *Food Research International* 37: 343–354.

Muñoz, M., Civille, V.G., and B.T. Carr. 1992. *Sensory evaluation in quality control.* New York: Van Nostrand Reinhold.

Nunes, M.C., Emond, J.P., and J.K. Brecht. 2006. Brief deviations from set point temperatures during normal airport handling operations negatively affect the quality of papaya (*Carica papaya*) fruit. *Postharvest Biology and Technology* 41: 328–340.

Official Journal of the European Communities. 1991. n.L. 2568 of 11 July. EEC Regulation no. 2568/1991.

Official Journal of the European Communities. 2002. n.L. 72 of 16 January. EEC Regulation no. 72/2002.

Official Journal of the European Communities. 2005. n.L. 2073 of 15 November. EEC Regulation no. 2073/2005.

Patsias, A., Chouliara, I., Badeka, A., Savvaidis, I.N., and M.G. Kontominas. 2006. Shelf-life of a chilled precooked chicken product stored in air and under modified atmospheres: microbiological, chemical, sensory attributes. *Food Microbiology* 23: 423–429.

Provincial, L., Gil, M., Guillen, E., Alonso, V., Roncales, P., and J.A. Beltran. 2010 Effect of modified atmosphere packaging using different CO_2 and N_2 combinations on physical, chemical, microbiological and sensory changes of fresh sea bass (*Dicentrarchus labrax*) fillets. *International Journal of Food Science and Technology* 45: 1828–1836.

Pyrgotou, N., Giatrakou, V., Ntzimani, A., and I.N. Savvaidis. 2010. Quality assessment of salted, modified atmosphere packaged rainbow trout under treatment with oregano essential oil. *Journal of Food Science* 75: 406–411.

Salvador, A., Varela, P., and S.M. Fiszman. 2007. Consumer acceptability and shelf life of "Flor de Invierno" pears (*Pyrus communis* L.) under different storage conditions. *Journal of Sensory Studies* 22: 243–255.

Speranza, B., Corbo, M.R., Conte, A., Senigaglia, M., and M.A. Del Nobile. 2009. Microbiological and sensorial quality assessment of ready-to-cook seafood products packaged under modified atmosphere. *Journal of Food Science* 74: 473–478.

Varela, P., Salvador, A., and S.M. Fiszman. 2005. Shelf-life estimation of "Fuji" apples: sensory characteristics and consumer acceptability. *Postharvest Biology and Technology* 38: 18–24.

Zhou, T., Harrison, A.D., McKellar, R., Young, J.C., Odumeru, J., Piyasena, P., Lu, X., Mercer, D.G., and S. Karr. 2004. Determination of acceptability and shelf life of ready-to-use lettuce by digital image analysis. *Food Research International* 37: 875–881.

CHAPTER 4

Critical Indicators in Shelf Life Assessment

Sonia Calligaris and Lara Manzocco

CONTENTS

4.1 DETERIORATIVE EVENTS

As highlighted in Chapter 2, the first step of the shelf life assessment process implies the identification of the event judged to have the most critical impact on food quality at the storage conditions foreseen for the packed product. A good understanding of this event is imperative to face a shelf life study correctly.

When a fresh-made product is put on the shelf, a number of biological, chemical, and physical deterioration events take place, and as a consequence, different food quality attributes are expected to change simultaneously or consecutively during storage time. The prevalence of one deteriorative phenomenon over the others depends on integrated effects of product characteristics, process operations, packaging, and storage conditions.

This book is not specifically designed to describe the main biological, chemical, or physical reactions at the basis of food degradation. For more extended coverage of these topics, refer to books by Robertson (2010), Kilcast and Subramaniam (2011), Man and Jones (2000), Eskin and Robinson (2000), Damodaran et al. (2007), and Taub and Singh (1997). A summary of the main deteriorative events responsible for food failure during storage is reported in Table 4.1. In this table, food products are divided into three groups based on the applied storage temperature: frozen, chilled, and ambient foods. This partition indirectly gives indications of the preservation

Table 4.1 Food Products and Main Expected Deteriorative Events

Products	Deteriorative Event	Effect	Example
Frozen foods	Enzymatic reactions	Color, flavor, texture changes, nutrient and bioactive compound loss	Fruits and vegetables, ready meals containing fruit and vegetable derivatives
	Oxidative reactions	Off-flavor formation, color changes, nutrient and bioactive compound loss Toxic compound formation	Ready meals, meat and fish derivatives, colored fruit and vegetable derivatives (e.g., tomato)
	Freezer burn	Visual aspect	Pasta, fish, meat, fruit, and vegetable derivatives
Chilled foods	Bacterial, yeast, and mold growth	Off-flavor formation, visible colony formation, texture and color changes	Ready meals, milk, cheeses, meat and fish derivatives
	Enzymatic reactions	Color, flavor, and texture changes, nutrient and bioactive loss	Fruit and fresh-cut fruit and vegetables, fresh meat
	Senescence	Color and texture changes	Fruits and vegetables
	Physical changes	Phase separation, water migration	Fruit juices and smoothies, emulsions, bakery products
Ambient foods	Oxidative reactions	Off-flavor formation, nutrient and bioactive compound loss, loss of flavors, color changes, toxic compound formation	Oils and fats, bakery products, dried foods and ingredients, canned products, fried snacks, soft drinks, dried foods, and ingredients
	Nonenzymatic browning	Changes in color and aroma, toxic compound formation, nutrient and bioactive compound loss	Dried foods and ingredients, canned products
	Physical changes	Phase separations, gelling, loss of crispness, caking	Emulsions, fruit juices, milk drinks, breakfast cereals, powders

processes applied to products and relevant stability on the shelves. It must be pointed out that the indications reported in Table 4.1 make sense if the following conditions are satisfied: (a) no processing failures have occurred during food manufacture; (b) the applied storage conditions are kept constant; (c) the original packaging of the product is perfectly sealed and undamaged.

Since at frozen temperatures microbial growth is inhibited, the shelf life of frozen foods is mainly limited by the development of enzymatic or chemical reactions. The former take place in fruit and vegetable derivatives containing active enzymes. Enzyme systems actually remain active even at subzero temperatures and can cause rapid quality depletion if inappropriate pretreatments (e.g., blanching) are performed. Chemical reactions, above all oxidative reactions, are the main deteriorative events leading to the quality depletion of frozen foods. This is particularly true for foods containing compounds highly prone to react with oxygen, such as unsaturated lipids, nutrients, and bioactive compounds. It is a matter of fact that oxidative reactions develop at subzero temperatures at a considerable rate, sometimes higher than expected, for the following reasons: (a) the reactant concentration in the liquid phase increases as a consequence of ice formation; and (b) the oxygen solubility into the food matrix increases as the temperature decreases. Finally, frozen foods can undergo surface dehydration, causing the formation of superficial pale spots known as "freezer burn."

In the case of chilled foods, the growth of spoilage microorganisms, which affect the sensory proprieties and the overall acceptance of the product, is generally the main cause of quality decay. This is the case for fresh fish and meat products as well as for the wide variety of ready-to-eat foods. Moreover, enzymatic reactions or senescence events could cause quality depletion of fruit and vegetable derivatives. Physical events, such as sedimentation or phase separation, could also become critical in multiphase foods, such as fruit juices, smoothies, or dairy products.

Chemical and physical deterioration events are the main causes of the quality depletion of shelf-stable products stored at ambient temperatures. Most of these events are triggered by the applied thermal preservation or sanitization processes. Oxidations and nonenzymatic browning are the most important chemical deteriorative reactions of this product category. The advanced stages of oxidative reactions generally cause the formation of off-flavors, which are easily recognized by consumers. Additional "silent" changes, not affecting the sensory properties of the food product (i.e., the formation of potentially toxic compounds, the degradation of nutrients and bioactive molecules), can take place. As previously mentioned, lipid-containing products are those expected to mainly undergo oxidative reactions. This is obviously the case for bulk fats and oils, but even the shelf life of canned and dried foods, fried snacks, dry bakery products, and ingredients is frequently affected by lipid oxidation. Moreover, oxidations of polar compounds such as ascorbic acid and other bioactive compounds are of key importance in shelf life assessment of functional foods. On the other hand, nonenzymatic browning development leads to the formation of brown compounds responsible for color changes. Even in this case, some silent changes involving nutrient loss and toxicant formation can take place. Dried and canned foods are particularly prone to develop nonenzymatic browning reactions. These are not necessarily fast reactions but become critical in consideration of the long life of this product category. Besides these events, physical changes can be frequently responsible for quality decay of ambient foods during storage. There is a wide variety of physical events; some of them imply phase separation and gelling phenomena, such as those responsible for viscosity changes in sterilized milk and milk-based drinks. Volatile partition from the food matrix into the headspace should also be taken into account in flavor-rich products, such as coffee and tea derivatives. In other cases, moisture adsorption causes caking of powders or changes in peculiar food sensory characteristics, such as crispness and texture. These changes are often ascribed to transitions from the glass to the rubber state, which can cause a cascade of other chemical and physical events due to the modification of the molecular mobility.

4.2 QUALITY INDICATORS

The evolution of a given event could be monitored during storage following the changes of different quality indicators using chemical, physical, biological, and sensory methodologies.

Chemical, physical, and microbiological indicators can be measured by using instrumental analysis. The instrumental analysis presents the undoubted advantage of good reproducibility and objective results. Since the same quality changes could be followed by different analytical techniques, making the correct choice is of key importance for saving time and money. This is particularly true in *routine* shelf life studies. When selecting the proper methodology, the first aspects to be considered are the required accuracy; methodology quickness, easiness, and cost; amount of sample needed for the analysis; and availability of internal resources (instruments and operators). For instance, food color changes could be assessed using a spectrophotometer or a colorimeter as well as by image analysis, which implies the acquisition of the food image and its subsequent elaboration. While spectrophotometric analysis requires preliminary sample preparation and dilution, the use of a colorimeter gives quick information about the chromatic characteristics of the product with minor sample preparation issues. However, the colorimetric analysis might not be the appropriate technique when a multicolor product is under study. Food image analysis, by contrast, is an effective technique to assess color, even for chromatically inhomogeneous food products. However, it requires the use of dedicated software and the initial development of methodologies by skilled operators.

Oxidative reactions could be monitored using simple and low-cost analytical methodologies (i.e., peroxide value, conjugated dienes, anisidine value) as well as more sophisticated ones, such as those addressed to quantify specific oxidation markers via gas chromatographic (GC) analysis, eventually coupled with mass spectrometry (GC-MS). Obviously, the two approaches allow different information to be acquired. The advice is to select the most appropriate methodology by merging analytical requirements with factory needs and resource availability.

Table 4.2 Nature and Type of Deteriorative Events with Some Relevant Indicators

Nature of Deteriorative Events	Type of Deteriorative Event	Indicator
Chemical	Fat and oil oxidation	Peroxide value Conjugated dienes (CDs) Volatile carbonyl compounds Anisidine value Thiobarbituric acid (TBA) index Hydrocarbons and fluorescent products
	Pigment oxidation	Color Image properties Selected compound concentration
	Nonenzymatic browning	Color Absorbance Image properties Selected compound concentration
	Vitamin/bioactive degradation	Selected compound concentration
Biological	Enzymatic activity	Color Mechanical properties Rheological properties Selected compound concentration
	Microbiological	Total microbial count Defined spoilage microorganism count
Physical	Crystallization	Image properties Calorimetric properties
	Moisture loss/gain	Moisture content Water activity Texture analysis Mechanical properties Rheological properties
	Starch retrogradation	Mechanical properties Calorimetric properties
	Texture changes	Mechanical properties Rheological properties
	Phase separation	Particle size Visual properties Image properties

Table 4.2 gives some examples of the main chemical, physical, and biological indicators that can be used to monitor the quality decay associated with a specific deterioration event during storage. The list is obviously not exhaustive but is intended to give practical indications in the task of looking for a suitable critical indicator.

It is well known that instrumental analysis is powerful. However, in many cases, the application of sensory analysis cannot be disregarded. In fact, the spoilage of the majority of foods is appreciable by consumer senses; consequently, in many cases, food tasting results in the most appropriate tool to monitor quality depletion during storage. Different sensory methodologies can be applied. For a more

Table 4.3 Sensory Methodologies Potentially Applicable in Shelf Life Testing

Panel	Methods	Example
Trained judges	Discrimination testing	Paired comparison tests
		Triangle tests
		Duo-trio tests
		A-not-A tests
Trained judges	Descriptive analysis	Quantitative descriptive analysis®
		Texture profile®
		Sensory spectrum
		Generic descriptive analysis
Consumers	Preference tests	Paired preference tests
		Nonforced preference tests
	Acceptance tests	Hedonic scaling
		Survival analysis

in-depth description, refer to dedicated textbooks (Meilgaard et al., 2006; Kilcast, 2010). Table 4.3 summarizes the main sensory methodologies potentially applicable in shelf life testing.

The choice could fall among discrimination testing, descriptive analysis involving trained judges, and preference or acceptance tests involving consumers.

Trained judges (internal or external to the factory) are required when discrimination or descriptive sensory methodologies are applied. When discrimination testing is used, the choice among the different methodologies depends on the specific question to be answered. As is well known, discrimination tests are used to determine whether there is a difference between two samples (overall difference or descriptor difference). In shelf life studies, valuable information can be obtained by these methodologies only when the end of shelf life is defined as the time when the stored product is recognized as different from the fresh one. In other terms, in this case there is a binary response, and the acceptability limit discriminates products perceived equal to the fresh one from those perceived as different.

On the contrary, descriptive analysis is used for recording the perceived intensity of well-defined food sensory attributes. If descriptive methodologies are properly applied, food sensory attributes are actually analytical indicators, similar to those obtained by any other instrumental analysis.

When sensory analysis is carried out by consumers, it requires the recruitment of a large number of people. Two different kinds of tests can be applied: preference or acceptance tests. In these cases, data acquired give no information about the evolution of specific sensory attributes of food during storage time. Indeed, these methodologies allow the evaluation of the interactions between food and consumers to be monitored during storage. Such interactions are actually the result of the overall food sensory attributes affecting consumer acceptability or preference. In this context, survival analysis of food acceptability expressed by consumers is a powerful tool in shelf life studies. As mentioned in previous chapters, survival analysis of food acceptability expressed by consumers can be used to assess product shelf life or may represent an effective technique for the identification of the acceptability limit. Specific indications about the application of consumer sensory analysis to assess the evolution of food acceptability during storage are described in Chapter 7.

4.3 FROM QUALITY INDICATORS TO THE CRITICAL INDICATOR

As previously observed, different quality indicators may contemporaneously change in food during storage. At this stage of the shelf life assessment process, a question arises: Among all quality indicators potentially exploitable, which is the best for the shelf life study?

The first and easy criterion of choice is the earliness. The critical indicator could be the earlier indicator describing food quality decay over storage time. However, the question is quite complex and requires additional consideration by analyzing the role of the critical indicator in a shelf life assessment process.

As mentioned in Chapter 2, once the critical indicator I is selected, a given I value, discriminating acceptable products from ones no longer acceptable, needs to be identified. This I value has been defined as the cutoff. However, as discussed in Chapter 3, sometimes the cutoff identification comes before the selection of the critical indicator. Apparently, this could sound strange, but it is what frequently happens when the cutoff is defined by regulations or when it depends on consumer acceptability. In these cases, the selection of the critical indicator is consequent to the cutoff identification. As mentioned in Chapter 2, the cutoff can be quantitative or qualitative. Quantitative cutoffs are those obtained by the measurement of a well-defined food quality attribute (i.e., sensory score, peroxide value, viscosity, color, nutrient concentration, etc.), whereas qualitative ones are those obtained by assessing consumer acceptability. As previously mentioned, cutoff is not necessarily synonymous with the acceptability limit. It depends on the modeling approach used to estimate or predict the shelf life. The cutoff value corresponds to the acceptability limit only when the kinetic approach is used. In the case of the survival analysis approach, the cutoff is the basic requirement for discriminating acceptable products from the failed ones. Further data elaboration allows the risk function of food unacceptability as a function of storage time to be obtained. In this case, the acceptability limit is a given percentage of food failure.

Based on these considerations, the critical indicator and cutoff are closely related, and the choice of the critical indicator mainly depends on the nature of the cutoff. Figure 4.1 schematizes the links between the nature of the cutoff and that of the critical indicator.

When the cutoff has a quantitative value assigned on the basis of label claims or compulsory indications, the quality index to be used as a critical indicator is directly defined.

Unfortunately, there are only rare cases in which the identification of the critical indicator is carried out thanks to compulsory indications (Tables 3.2 and 3.3). In all other situations, the cutoff is voluntarily defined whether it has a quantitative

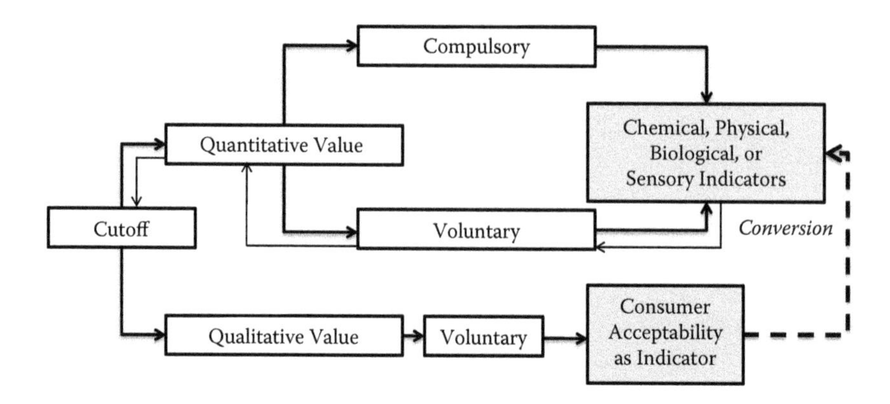

Figure 4.1 Relationship between the nature of the cutoff value and the nature of critical indicator.

or qualitative feature. In the first case, the critical indicator may be chosen among the different chemical, physical, biological, and sensory quality indicators changing during storage time. When previous information on food product stability is available (previous company experience or literature data), the identification of the critical indicator can be rationally made using these data, and a cutoff value can be voluntarily defined.

On the contrary, when it is difficult to find a definite critical indicator or when no previous information is available and the only evidence is the amount of food rejection by consumers, the best shelf life indicator becomes the evolution of consumer acceptability. In this case, the consumers are asked to judge whether the food product is acceptable. Here, the cutoff has a subjective and qualitative feature, being in the mind of the consumers. Acceptability data are then elaborated by applying the survival analysis that allows the risk function of consumer dissatisfaction to be obtained (Chapters 3 and 7).

4.3.1 Conversion of Consumer Rejection Critical Indicator into Instrumental or Sensory Indicators

The procedure based on the assessment of the evolution of consumer rejection as a function of storage time is a time-consuming and expensive process because it requires a wide testing plan and the application of appropriate statistical techniques (Chapter 7). These conditions make survival analysis of consumer dissatisfaction, despite being powerful, hardly applicable by company operators to conduct shelf life assessment routinely. To overcome these pitfalls and find methods that can be applied in the daily management of the shelf life issue, food operators could make an effort to develop internal quality levels, described by proper instrumental or sensory quality indicators, accounting for the risk of consumer rejection (Garitta et al., 2004; Calligaris et al., 2007). This means that instrumental or sensory attributes, whose evolution is correlated to sensory dissatisfaction expressed by consumers, should be identified and used to monitor the evolution of product quality changes during storage. In other terms, food quality decay has to be followed using two different critical indicators that are related to each other. The first one is consumer acceptability; the second one should be an analytical indicator. Once the risk function is obtained and the percentage of consumers rejecting the product is selected as an acceptability limit, this value is used to identify the corresponding limit expressed by the analytical indicator. In this way, further *routine* shelf life studies can be done using only the instrumental index as a critical indicator.

The possible procedural steps to solve this task are

1. Evaluation of the evolution of consumer rejection as a function of storage time
2. Definition of the acceptability limits based on consumer dissatisfaction risk and the relevant storage time
3. Evaluation of deteriorative events leading to product quality depletion
4. Selection of the instrumental or sensory indicators best describing the product quality loss
5. Identification of the analytical indicator best correlating with consumer rejection by correlation analysis
6. Computation of the value of the analytical indicator corresponding to the maximum tolerable risk of consumer rejection

To describe how to proceed practically, consider the case study of a frozen salami pizza. The following example refers to a product packed in a cardboard box and stored at -18°C. Life testing and further survival analysis were applied to assess risk function of consumer dissatisfaction over storage time. Figure 4.2 shows the evolution of the percentage of consumer rejection as a function of storage time. As previously reported in Chapter 3, the acceptability limit can be identified in correspondence to different percentage values of consumer rejection based on the quality policy of the food company. In the given example, the

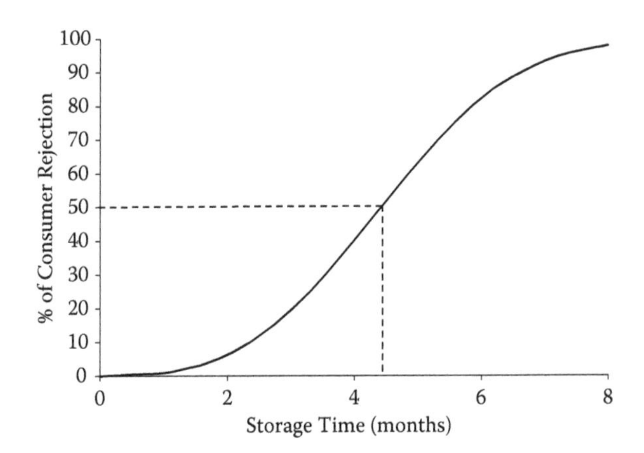

Figure 4.2 Percentage of consumers rejecting frozen salami pizza as a function of storage time at -18°C. Before analysis, the product was thawed and heated at 80°C.

Table 4.4 Most Probable Deteriorative Events, Relevant Indicators, and Methodologies to Assess Quality Depletion during Storage of Frozen Salami Pizza

	Most Probable Deteriorative Event	Indicator	Methodology
Whole pizza	Dehydration	Weight loss	Gravimetric analysis
Salami	Lipid oxidation	Off-flavor	Sensory descriptive analysis
Tomato sauce	Pigment bleaching	Yellowness	Tristimulus colorimetry

acceptability limit was defined as 50% of rejection, which corresponded to a shelf life of about 4.5 months.

To find a quality indicator related to consumer acceptability, the evolution of some instrumental or sensory indicators may be evaluated. Obviously, it does not make sense to deal with all possible deteriorative phenomena and relevant quality indicators. On the contrary, efforts should be made to identify the main degradation reactions involved in food failure to identify the early index to be used as a critical indicator. Literature data or previous experiences with similar products can be used to make a first selection of possible degradation reactions to plan stability tests efficiently aiming to select the best critical indicator.

Based on literature data, a number of possible alterative reactions expected to affect the acceptability of a frozen salami pizza could be identified. Results are reported in Table 4.4. Potential quality indicators and relevant assessment methodologies are also indicated.

Figure 4.3 shows the evolution of selected alterative events expressed as percentage of their normalized intensity. Based on these results, it is evident that weight loss was very slow under the storage conditions used in the study, and thus it could be excluded as a critical indicator. However, the choice between off-flavor formation

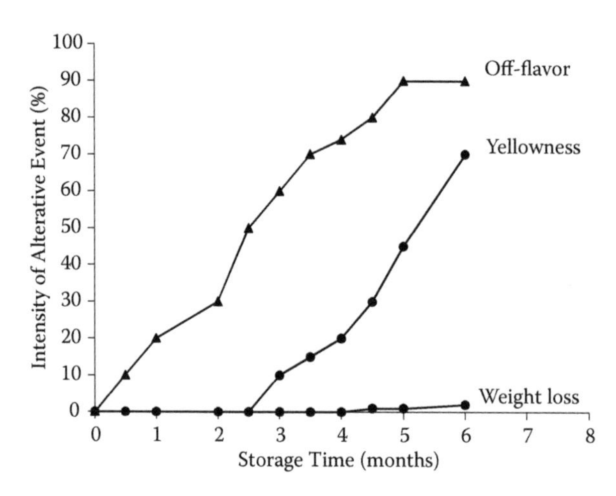

Figure 4.3 Changes of the percentage of intensity of the alterative events relevant to frozen salami pizza during storage at -18°C.

Table 4.5 Results of Correlation Analysis between Quality Indicators and Consumer Acceptability

Quality Index	*R*
Off-flavor	0.904
Yellowness	0.967

and tomato sauce yellowness could not be taken for granted. Instinctively, one would choose the development of off-flavor as a critical indicator since it shows the most rapid behavior. However, although off-flavor formation was the earliest event, it may not be linked to consumer acceptability. If this is the case, the selection of this indicator would make no sense.

In this context, only the study of consumer-product interactions could help to make a rational decision of the most suitable critical indicator. In fact, the quality indicator best correlated to consumer acceptability could be considered the critical indicator. Thus, it is necessary to answer the following question: Which is the analytical indicator best correlated to consumer rejection? The answer can be found by performing correlation analysis between consumer rejection and quality indicators. Results are reported in Table 4.5. Data highlight that tomato yellowness was better correlated to consumer rejection as compared to off-flavor development. This could be attributed to the fact that off-flavors, despite early detection by trained judges, are not easily perceived by the consumers. Thus, tomato sauce yellowness could be reasonably selected as a critical indicator accounting for consumer acceptability.

Once the analytical indicator best related with consumer acceptability is defined, its value in correspondence to the acceptability limit defined based on consumer rejection can be computed. As regards the example of the frozen salami pizza, Table 4.6 shows the estimated values of yellowness corresponding to given percentages of consumers rejecting the product. The intensities of yellowness reported in Table 4.6 may thus be used as acceptability limits and routinely applied to evaluate the shelf life of the product in the industry quality control programs.

The advantages of the exploitation of instrumental or sensory indicators, which are easier, faster, and more cheaply assessed compared to those necessary to obtain a consumer dissatisfaction risk, are undoubted. Once the critical indicator correlating

Table 4.6 Percentage of Yellowness Corresponding to Increasing Risk Levels of Consumers Rejecting Frozen Salami Pizza

Risk Level	Consumers Rejecting the Sample (%)	Yellowness (%)
Very low	10	1.6
Low	25	13.5
Medium	50	33.3
High	75	53.2
Very high	90	65.1

Table 4.7 Examples of Critical Indicators Demonstrated to Correlate Well with Consumer Dissatisfaction during Storage of Different Foods

Food	Critical Indicator	Methodology	Reference
Sunflower oil	Oxidized flavor	Sensory	Ramirez et al., 2001
Powdered milk	Oxidized flavor	Sensory	Hough et al., 2002
Dulce de leche	Off-flavor	Sensory	Garitta et al., 2004
Human milk replacement formula	Dark color	Sensory	Curia and Hough, 2009
Biscuits	Peroxide value	Chemical	Calligaris et al., 2007
Breadsticks	Peroxide value	Chemical	Calligaris et al., 2008
Coffee brews	pH	Physicochemical	Manzocco and Lagazio, 2009
Fresh-cut fruit salad	Percentage of brown	Sensory	Manzocco et al., 2011

with consumer dissatisfaction has been identified, further time-consuming consumer tests can be skipped. Unfortunately, limited indications are reported in the literature about quality indicators potentially exploitable to estimate consumer rejection in shelf life studies. Table 4.7 reports some examples of critical indicators that were well correlated to consumer dissatisfaction during storage.

The possibility of simplifying considerably the complex process leading to shelf life estimation will also depend on the availability of basic information about the relation between the evolution of quality indicators and consumer rejection during food storage. Not all quality indicators are actually critical indicators. Only quality indicators that are important for the consumers to form their attitude toward food consumption may be regarded as critical indicators. When the cutoff is voluntarily defined, these critical indicators can represent efficient tools to perform shelf life studies easily on a routine basis.

4.4 REFERENCES

Calligaris, S., Da Pieve, S., Kravina, G., Manzocco, L., and M.C. Nicoli. 2008. Shelf-life prediction of bread sticks by using oxidation indices: a validation study. *Journal of Food Science* 73(2): E51–E56.

Calligaris, S., Manzocco, L., Kravina, G., and M.C. Nicoli. 2007. Shelf-life modeling of bakery products by using oxidation indexes, *Journal of Agricultural and Food Chemistry* 55: 2004–2009.

Curia, A.V., and G. Hough. 2009. Selection of a sensory marker to predict the sensory shelf life of a fluid human milk replacement formula. *Journal of Food Quality* 32: 793–809.

Damodaran, S., Parkin, K.L., and O.R. Fennema. 2007. *Fennema's food chemistry*. 4th ed. Boca Raton, FL: Taylor and Francis.

Eskin, N.A., and D.S. Robinson. 2000. *Food shelf life stability.* Boca Raton, FL: CRC Press.

Garitta, L., Hough, G., and R. Sánchez. 2004. Sensory shelf-life of dulce de leche. *Journal of Dairy Science* 87: 1601–1607.

Hough, G., Sanchez, R.H., de Pablo, G.G., Sanchez, R.G., Villaplana, S.C., Gimenez, A.M., and A. Gambaro. 2002. Consumer acceptability versus trained sensory panel scores of powdered milk shelf-life defects. *Journal of Dairy Science* 9: 2075–2080.

Kilcast, D. 2010. *Sensory analysis for food and beverage quality control: a practical guide.* Cambridge, UK: Woodhead.

Kilcast, D., and P. Subramaniam. 2011. *Food and beverage stability and shelf life.* Cambridge, UK: Woodhead.

Man, D., and A. Jones. 2000. *Shelf-life evaluation of foods.* 2nd ed. Gaithersburg, MD: Aspen.

Manzocco, L., Foschia, M., Tomasi, N., Maifreni, M., Dalla Costa, L., Marino, M., Cortella, G., and S. Cesco. 2011. Influence of hydroponic and soil cultivation on quality and shelf life of ready-to-eat lamb's lettuce (*Valerianella locusta* L. Laterr). *Journal of the Science of Food and Agriculture* 91(8): 1373–1380.

Manzocco, L., and C. Lagazio. 2009. Coffee brew shelf life modelling by integration of acceptability and quality data. *Food Quality and Preference* 20: 24–29.

Meilgaard, M.C., Carr, B.T., and G.V. Civille. 2006. *Sensory evaluation techniques.* Boca Raton, FL: Taylor and Francis.

Ramirez, G., Hough, G., and A. Contarini. 2001. Influence of temperature and light exposure of sensory shelf life of commercial sunflower oil. *Journal of Food Quality* 24: 195–204.

Robertson, G.L. 2010. *Food packaging and shelf life. A practical guide.* Boca Raton, FL: Taylor and Francis.

Taub, I.A., and R.P. Singh. 1997. *Food storage stability.* Boca Raton, FL: Taylor and Francis.

CHAPTER 5

Modeling Shelf Life Using Chemical, Physical, and Sensory Indicators

Sonia Calligaris, Lara Manzocco, and Corrado Lagazio

CONTENTS

5.1 INTRODUCTION

As highlighted in previous chapters, once an appropriate critical indicator for shelf life has been identified, it is necessary to estimate the evolution of this indicator over time. This step, generally defined as shelf life testing, implies the continuous monitoring of the changes of the critical indicator during storage of food under controlled environmental conditions.

Data relevant to the evolution of the critical indicator during storage are then modeled to obtain proper parameters describing/predicting its kinetics. These parameters are necessary to obtain the shelf life value once knowing the acceptability limit.

When the critical indicator is represented by chemical, physical, or sensory attributes, the monitoring of food deterioration can be performed by following two different strategies: (a) real-time shelf life testing performed under storage conditions that reasonably foresee the situation experienced by the product on the shelf; or (b) accelerated shelf life testing (ASLT) performed under environmental conditions able to speed up the quality loss.

5.2 REAL-TIME SHELF LIFE TESTING

Real-time shelf life testing is the methodology that allows the experimental data relevant to the critical indicator *I* over time under reasonably foreseeable storage conditions to be gathered. This approach is theoretically applicable for shelf life estimation of any food category. However, it becomes profitable in the case of perishable foods, for which quality decay occurs in a rather short time. In fact, no speeding up of the shelf life assessment process is required since the time needed to perform the trials is reasonable and usually fits industrial needs.

The basic requirement to perform a reliable real-time shelf life test is that the environmental conditions during storage (e.g., temperature, moisture, light) are kept constant and reproduce those that the product experiences during storage on the shelf. It is appropriate to store the product under its normal storage conditions, controlling not only temperature but also keeping constant other environmental conditions (e.g., humidity and light), if they are likely to have a significant impact on shelf life. For instance, shelf-stable products, such as bakery, canned, dried foods, should be stored at 20–25°C; chilled products at 3–4°C and frozen foods at -18°C (E.U. conditions) or -12°C (U.S. conditions). However, since temperature oscillation is frequent during food storage, it could be profitable to perform the shelf life test under the worst situation that one could expect during storage. For instance, a shelf life test of chilled products could be performed at 8–10°C instead of 3–5°C. Similarly, for ambient stable products, 30–35°C could represent a reasonable room temperature in summer. In all cases, if the package is a see-through container, the light exposure suffered by the product on the shelf could become a critical factor in determining its shelf life. For this reason, it is necessary to take into account this factor during shelf life testing. For instance, the light intensity commonly found on market shelves is about 600–800 lux.

5.2.1 Experimental Design of Real-Time Shelf Life Testing

The time-costing and labor-consuming nature of real-time shelf life testing requires careful planning of the experiments that must be developed imagining, as much as possible, the eventual occurrence of issues potentially mining test success.

After choosing the most appropriate storage conditions, it is necessary to design a detailed experimental plan in terms of sample size, resources, and costs needed to obtain reliable shelf life dating.

To calculate the number of samples needed, sampling frequency, number of products to be analyzed at each sampling time (e.g., single or multipackages and replications) as well as total number of sampling times should be carefully estimated. In fact, it would be disastrous to be out of samples before the product spoiled; the experiments have to be carried out until the acceptability limit has been overcome by at least 20%. In this way, the complete evolution of the critical indicator can be observed. A minimum of six time intervals can be suggested to improve the predictive ability of the shelf life model. At the same time, the sampling frequency should be properly defined to collect data that effectively describe the behavior of quality depletion. The distribution of the sampling times should be as much as possible uniform within the testing interval to avoid dangerous interpretation mistakes.

The design of experiments could be an easy task when enough information on the product is available. For instance, an experimental plan should be based on experience from prior stability studies, or indications could come from literature data as well as from distribution time of a similar or competitive product on the marketplace. On the contrary, if this information is not exhaustive, the setup of experiments cannot be clearly defined at the beginning, and online rearrangement could be necessary based on the results progressively acquired during testing. It is thus convenient to store extra samples to be able to correct the experimental plan on time, avoiding the possibility of being out of samples before the acceptability limit has been reached.

After these considerations, the storage strategies to be adopted during a shelf life test should be defined. The most common and easy approach consists of storing a single batch of product under the desired storage conditions and periodically removing samples from storage to analyze them. The advantage of this approach is that it can be easily and quickly designed and allows the continuous monitoring of the changes of the critical indicator. In this way, it is possible to have an "online" vision of the shelf life evaluation. This approach becomes the most appropriate when little information on the product is available.

The inclusion of additional "control samples" is absolutely necessary when the methodology to assess the quality depletion of the product requires the comparison of samples stored for increasing time with the fresh one. This often happens when sensory methodologies are applied (Chapter 7). Table 5.1 reports, from indications of Taoukis et al. (1997), the most appropriate temperatures to be used for storing a control sample during shelf life testing of different product categories. Unfortunately, keeping a "fresh" control sample could not be so easy, and data reported in Table 5.1 can only be considered a suggestion that should be carefully tested.

Depending on the specific criticism of the product and the critical indicator involved, keeping a representative control sample could be relatively simple or almost impossible. Considering shelf-stable foods undergoing oxidation (i.e., bakery products), even storage at -18°C does not allow the development of oxidation to be hindered. For instance, after some months of storage at -18°C, biscuits

Table 5.1 Suggested Storage Temperature for Keeping a Control Sample during Shelf Life Testing

Product Type	Control Sample Storage Temperature (°C)
Canned	4
Dehydrated	-18
Chilled	0
Frozen	<-40

Source: Modified from Taoukis P.S. et al., 1997. Kinetics of food deterioration and shelf-life prediction. In *The handbook of food engineering practice*, K.J. Valentas, E. Rotstein, and R.P. Singh, Eds., 363–405. New York: CRC Press.

showed a significant level of oxidation, indicating that reference samples cannot be obtained by storage at frozen temperatures (Calligaris et al., 2007a). These considerations highlight the necessity of verifying the eventual modification occurring in the control sample during storage to understand the reliability of the selected control temperature. If adequate environmental conditions that allow the control of deteriorative reactions cannot be identified, the experimental plan should be modified so that a just-prepared batch of product can be regarded as a control sample.

In some circumstances, it could be more useful that samples with different storage times are available for evaluation at the same moment. For instance, if the sample has to be evaluated by sensory analysis, the possibility of recruiting the panel on a single occasion could greatly simplify the experiments (Hough, 2010). In this case, a reverse storage design could be applied. Essentially, two different approaches can be used:

1. Use of different batches of product. If the differences among batches of product are expected to be negligible, it is possible to store different product batches for increasing storage times as schematically represented in Figure 5.1. Batch 1 is stored at time zero t_0 and will have the longest storage time; batch 2 is stored at $t_0 + t_i$ and will have $t_f - t_i$ storage time at the end of the test; and so on for batch 3. In such a way, batch 4 becomes the fresh sample.
2. Use of a single batch of product. In this case, a batch of product is stored under environmental conditions at which the quality changes are negligible (control sample). An aliquot of this batch is put at time zero at the shelf life testing conditions, and this will become the sample with the longest storage time t_f. After a defined time t_i, a new aliquot of sample is placed under actual storage conditions, and this will have $t_f - t_i$ storage time. This procedure is followed repeatedly, obtaining samples with decreasing storage time (Figure 5.2).

It should be noted that this approach could be followed only if enough information on the product is available to define the timing correctly.

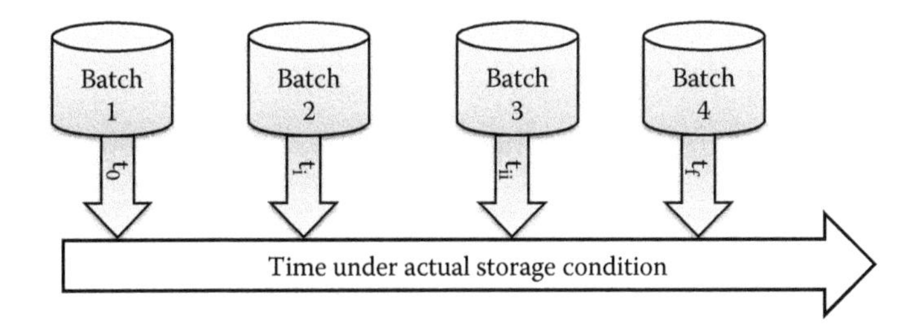

Figure 5.1 Reverse storage experimental design with different product batches.

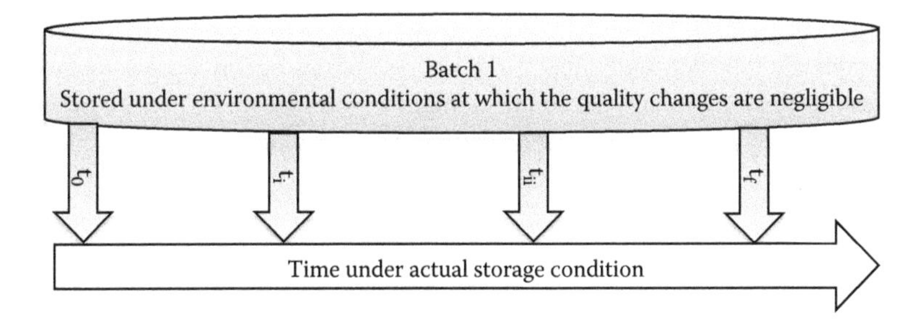

Figure 5.2 Reverse storage experimental design with a single product batch.

One other aspect that should not be underestimated is the confirmation of the availability of the resources required to perform the test during its entire duration. Before scheduling the starting date for the shelf life test, the availability of all the resources needed, including equipment, personnel, or other facilities, should be carefully guaranteed. For instance, the possibility to prepare the samples without interfering with the routine activities of the factory should be preliminarily checked. Similarly, the presence of employees to perform analyses or to participate in a sensory panel is crucial for the success of the tests; thus, holidays should be preliminarily excluded from the scheduling calendar.

At the end of this process, the test could start, and the changes of the critical indicator as a function of storage time under controlled environmental conditions can be collected and submitted to modeling to obtain a shelf life estimate.

Practical examples of experimental planning of shelf life studies are presented in Chapter 9.

5.2.2 Shelf Life Modeling

5.2.2.1 Basic Principles

As mentioned in Chapters 1 and 2, food quality moves dynamically to lower levels during storage until reaching a cutoff value chosen as the acceptability limit.

According to Chapter 2, the kinetics of the critical indicator I during storage time t can be described by the function g:

$$I = g(t,\ \ ) \tag{5.1}$$

where ϑ is the kinetic parameter of the model function. The inverse function of Equation 5.1 is the following:

$$t = f(I,\lambda) \tag{5.2}$$

where λ are the parameters of the model function f.

According to its definition, shelf life is the value assumed by the variable t in correspondence to a selected value of I, which is chosen in correspondence with the acceptability limit I_{lim}. Thus, SL can be defined as follows:

$$SL = f(I_{lim,}\ \lambda) \tag{5.3}$$

I_{lim} is an extrinsic variable linked to the product but is not a property of the food itself.

As highlighted in Chapter 3, the acceptability limit is not a constant value but a complex variable depending on regulatory and market constraints. For instance, the acceptability of food may depend on emotional reasons that are strictly conditioned by geographical and social factors. For this reason, I_{lim} is a complex variable that results from different inputs and, if necessary, can be represented by opportune models.

On the other hand, the kinetics of the critical indicator, described by the kinetic parameters of the mathematical function g, and thus of its inverse function f, is an intrinsic factor that is typically linked to the specific quality instability of the food. The values of the kinetic parameters ϑ strictly depend on the food instability properties and may significantly change with environmental inputs. Several factors, including temperature, moisture, light intensity, and gas pressure have been widely demonstrated to affect their value. A number of different models describing the effect of these variables on kinetic parameters of quality decay have been developed. They are indeed necessary to set up accurate shelf life models, but they are not shelf life models themselves. In fact, it is evident that any model aiming to be a shelf life model must clearly include a function operator f, the acceptability limit I_{lim}, and appropriate kinetic parameters λ.

Based on these preliminary considerations, and given a previously decided acceptability limit for the critical indicator, a data modeling strategy should be identified to obtain a proper function operator and relevant mathematical parameters describing the kinetics of quality depletion.

5.2.2.2 *Shelf Life Modeling Based on the Classical Kinetic Approach*

In experimental practice, the first unavoidable approach to have an overall idea of the evolution of the critical indicator I is to plot the I value as a function of storage time, keeping constant packaging and environmental variables. Two possible examples regarding the evolution of the apparent viscosity of a UHT milk drink and of the peroxide values of frozen fish fingers are given in Figures 5.3a and 5.3b, respectively.

As previously reported, the kinetics of the critical indicator I during storage time t can be described by Equation 5.1. Different behaviors could be obtained moving from simple straight lines to a number of complex evolutions depending on the phenomena involved.

As mentioned, the term ϑ in Equation 5.1 assumes a kinetic meaning since it accounts for the parameters that mathematically express the rate of the degradation event described by the indicator I. The most common method to analyze experimental data, and hence to measure the rate of food quality decay, is to apply the principles of classical kinetic theory. According to the well-known approach (Fu and

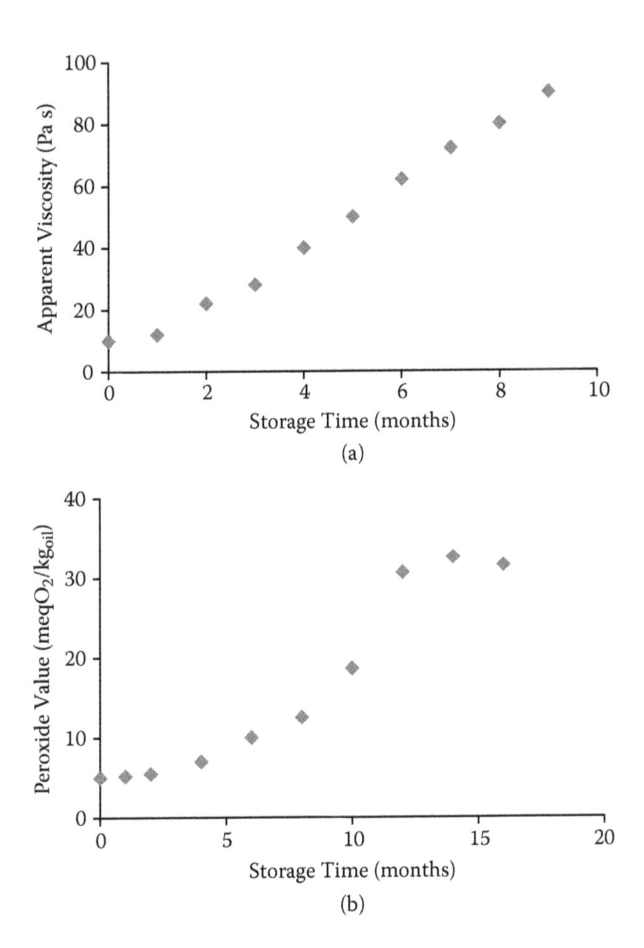

Figure 5.3 Evolution of apparent viscosity of a UHT milk drink (a) and of peroxide value of frozen fish fingers (b) during their storage.

Labuza, 1993; Taoukis et al., 1997), the rate r of change of the critical indicator I is defined as:

$$r = \frac{dI}{dt} = kI^n \quad (5.4)$$

where k is the rate constant, t is the storage time and n is the reaction order.

Equation 5.4 can also be written as follows:

$$\frac{dI}{I^n} = kdt \quad (5.5)$$

It can be noted that Equation 5.5 is the differential form of Equation 5.1. Therefore, it can be inferred that the general term ϑ of the equation includes the rate constant k. Thus, as reported in Chapter 2, since ϑ is related to food product characteristics C_i, packaging properties P_i, and environmental factors E_i, similar considerations can be drawn for the rate constant k to satisfy the following expression:

$$k = h'(C_i, P_i, E_i, \alpha) \quad (5.6)$$

The term h' indicates the mathematical function, and α is the relevant parameter. In other terms, the rate of food quality decay k depends on product characteristics and composition, packaging properties, and the environmental conditions suffered by the product during storage.

Although the general rate law has been developed theoretically for chemical reactions, it has been shown to hold empirically for a wide range of complex chemical, biochemical, and physical phenomena occurring in foods. Since food quality changes are highly complex and there are many factors potentially affecting the quality depletion rate, it should be stressed that the evolution of any critical indicator versus time is frequently the result of different reactions taking place simultaneously or consecutively. Thus, the reaction order n does not give any indications of the true reaction mechanisms involved, and k is therefore considered an "apparent" rate constant. In food science literature, n usually varies from zero to two (Van Boekel, 2008). However, higher orders can also be found. Figure 5.4 shows the quality indicator changes as a function of storage time for an event having the same initial value but varying the order n.

When C_i, P_i, and E_i factors are kept constant, the general rate law (Equation 5.5) can be integrated considering the time range from 0 to t that corresponds to a variation of I from I_0 to I_i to obtain the equations of the pseudozero ($n = 0$), first ($n = 1$), second ($n = 2$), or n order:

$$\int_{I_o}^{I_i} \frac{dI}{I^n} = \int_{o}^{t} k\,dt \quad (5.7)$$

Table 5.2 shows the integrated equations for $n = 0$, $n = 1$, and $n = 2$ and the general equation that can be used for $n \neq 1$. In classical kinetics, the reaction rate constant k is included in the ϑ parameter expressed in Equation 5.1. Rate constants may

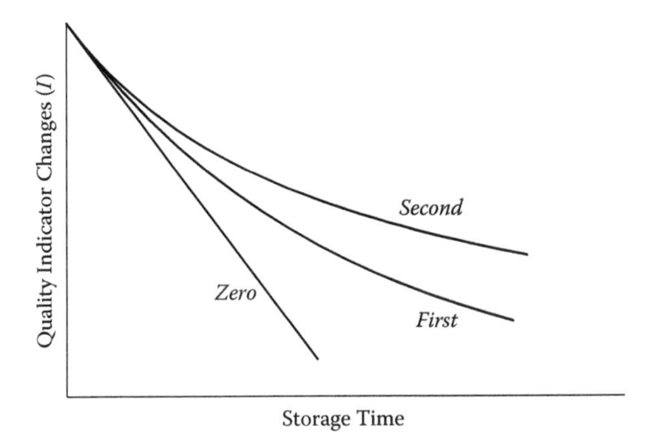

Figure 5.4 Quality indicator changes as a function of storage time for an event having the same initial value but varying the order *n*.

have positive or negative values depending on whether the quality indicator increases or decreases versus time, respectively.

Such data elaboration presents an extraordinary computational advantage. When kinetic theory was developed (Waage and Guldberg, 1864), it was not possible to fit functions other than those described by a straight line. Indeed, kinetic equations reported in Table 5.2 represent straight lines on an opportune axis. Even now, when regression software packages are easily available to fit complex functions, classical kinetic analysis is chosen for its friendly fitting.

The first problem to be faced is the identification of the proper reaction order. To do this, usually the experimental values expressed as I, ln I, $1/I$, $1/I^{n-1}$ are plotted versus storage time. The appropriate order corresponds to a seemingly linear pattern of the points. After that, the two parameters of the kinetic function, the rate constant and I_0, must be estimated from experimental data. The problem is simply solved with linear regression for the kinetic function of order zero. In Box 5.1, a numerical example of the linear regression procedure to be followed to determine the kinetic parameters and the relevant statistical parameters is reported.

Table 5.2 Zero-, First-, Second-, and *n*-Order Integrated Kinetic Equations

Reaction Order	Integrated Rate Law
$n = 0$	$I = kt + I_o$
$n = 1$	$\ln I = kt + \ln I_o$
$n = 2$	$\frac{1}{I} = kt + \frac{1}{I_o}$
$n \neq 1$	$I^{1-n} = (n-1)kt + I_0^{1-n}$

BOX 5.1 LINEAR REGRESSION

Linear regression analysis is a cornerstone of statistics. Its application will be exemplified with reference to a zero-order kinetic equation.

The "classical" simple linear regression model is

$$I_i = I_0 + k\,t_i + \varepsilon_i$$

where I_i indicates the observed values of the critical indicator (the dependent variable); t_i are the experimental storage times (the independent variable); I_0 and k are the regression parameters that represent, respectively, the intercept and the slope of the line; and ε_i is the experimental or random error.

The classical regression model is completed by the following assumptions on the behavior of the experimental errors:

- The experimental errors are independent
- The experimental error acts additively on the model function that relates the dependent and the independent variables
- The experimental errors are normally distributed
- The experimental errors have constant (homogeneous) variance

The first steps of the analysis are based on the hypothesis that the model holds; however, each assumption needs to be checked with proper techniques.

We illustrate how to conduct a regression analysis using the data taken from Varela et al. (2005), describing a descriptive sensory analysis of Fuji apples stored for increasing time (Table 5.1.1). Data may be graphically represented with a so-called scatterplot, a Cartesian diagram in which each observation is represented with a point whose coordinates are given by the values of the independent variable (*X* axis) and of the dependent variable (*Y* axis) (Figure 5.1.1).

Although the model assumes that the relation between t and I is linear, the points are not on a straight line because of the experimental error. However, the seemingly linear pattern of points indicates that the order zero kinetic

Table 5.1.1 Alcoholic Odor Sensory Score of Fuji Apples at Different Storage Times

Time (days)	Alcoholic Odor Score
0	1.2
7	1.4
13	2.5
18	3.2
24	4.3
28	4.6

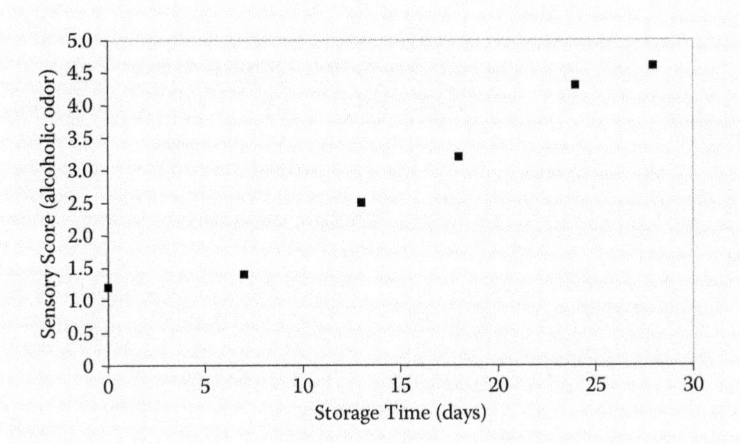

Figure 5.1.1 Alcoholic odor sensory score of Fuji apples as a function of storage time.

function is adequate. If the pattern exhibits a curvature, kinetic functions of higher order must be considered.

The regression line cannot be observed and must be estimated from data with the method of (ordinary) least squares. The method roughly consists of finding the line that minimizes the sum of squared distances of observed and predicted I values given by

$$\hat{I}_i = \hat{I}_0 + \hat{k}\, t_i$$

The quantities $\hat{I}_0$ and $\hat{k}$ indicate, respectively, the least squares estimators of I_0 and k.

Regression analysis may be performed with Excel (a simple regression tool is included in the Analysis Toolpack that may be added to Excel through Add-ins in the Tools menu) or specific software for statistical analysis (e.g., R, Stata, SPSS, SAS, etc.). The output is standard and includes a table (Table 5.1.2) with parameter estimates.

The two lines of the table refer respectively to the intercept and the slope of the model. The first column ("Coefficients") contains the estimates of the parameters, that is, the values of $\hat{I}_0$ and $\hat{k}$. The rate constant is the parameter associated with storage time and is then equal to 0.134. The value is positive, indicating an increase of alcoholic odor with increasing storage times.

The second column ("SE") reports the standard error of estimates, that is, a measure of the uncertainty on the parameter values due to the experimental error. The use of the standard error is twofold:

Table 5.1.2 Estimates of the Parameters of the Kinetic Function Obtained by Linear Regression Carried Out Using Excel

	Coefficients	SE	Stat t	*p*-value
Intercept I_0	0.858	0.221	3.887	0.0177
Rate constant (*k*, alcohol odor/day)	0.134	0.012	10.79	0.0004

1. It may be used to construct a (100 – α)% confidence interval for the two parameters of interest, according to the following formulas:

$$\left[\hat{I}_o - t_{\alpha/2;n-2} SE(\hat{I}_o); \hat{I}_o + t_{\alpha/2;n-2} SE(\hat{I}_o)\right]$$

and

$$\left[\hat{k} - t_{\alpha/2} SE(\hat{k}); \hat{k} + t_{\alpha/2} SE(\hat{k})\right]$$

where $t_{\alpha/2;n-2}$ is the critical value of the t distribution with n - 2 degrees of freedom (4 degrees of freedom in our example). For the data considered here, the 95% confidence interval for the rate constant is

$$[0.134 - 2.78 * 0.012; 0.134 + 2.78 * 0.012] = [0.099; 0.168]$$

Confidence intervals for the parameters are frequently included in the output table (e.g., Excel includes them).

2. It may be used to test if the parameters (mainly the rate constant) are equal to zero. The test is based on the ratio between the parameter value and its standard error (reported in the third column from the left of the table). The obtained value is then compared with the critical value of the t distribution or is used to compute the p value (rightmost column of the table). If the p value is lower than the chosen significance level (usually 0.05), the corresponding parameter is declared to be "significantly different" from 0, that is, the slope of the regression line quantifies a true relation between the variables and has not merely been determined by chance. The p value for storage time is largely lower than 0.05, indicating that alcoholic odor has a strong evolution with time.

The output of a regression procedure is usually completed by some numerical measures of goodness of fit (i.e., how well a straight line approximates the observed data) and some plots, which may be used for diagnostic purposes.

Both the evaluation of the goodness of fit and the diagnostic tools are based on the residuals, which are the differences between the observed and the predicted values:

$$e_i = I_i - \hat{I}_i$$

Regarding the goodness of fit, the most important index is the coefficient of determination R^2, the proportion of the total variation of the response variable accounted for by the regression line. It ranges between 0 (no fit; the regression line is horizontal) and 1 (perfect fit; all the points are on a straight line), with higher values indicating better fit. In the example, $R^2 = 0.97$. It must be said, however, that a good fit does not mean that the order of the kinetic function has been correctly chosen. The appropriateness of the reaction order must be evaluated with other diagnostic tools that will be described later.

Another important indicator of the goodness of fit commonly included in the output of regression analysis by software (including Excel) is the residual standard error, given by

$$\hat{\sigma} = \sqrt{\frac{\sum e_i^2}{n-2}}$$

The smaller the value of residual standard error is, the more precise are estimates based on regression analysis. Indeed, all the measures of uncertainty of parameter estimates (e.g., the standard error previously defined) and predicted values are derived from the residual standard error.

We use the residual standard error in further discussion to derive a measure of the uncertainty of shelf life estimates. In this case, the value of the residual standard error is 0.29.

The main diagnostic graphical tools are used to evaluate if the assumptions of the regression model (i.e., linearity, normality, and variance homogeneity of the errors) are valid.

The plot of the residuals versus the predicted values or the explanatory variable is used for linearity and homogeneity of variances. The cloud of points should not exhibit any clear pattern. If the residuals show a curvilinear trend, this indicates that the linear model is not appropriate, and a kinetic function of higher order should be used. If the assumption of variance homogeneity is inappropriate, the residual plot may show a fan-shaped pattern of increasing or decreasing residuals as $\hat{I}_0$ increases. The plot of residuals for the example is shown in Figure 5.1.2. The residuals do not show any evident pattern, confirming the evaluations made based on the scatterplot of the data.

Normality of the residuals may be evaluated with the normal probability plot, a plot that compares the observed quantiles of the residuals with the corresponding ones of the Gaussian distribution. If the normality assumption is satisfied, the points should approximately be along a 45° line. The normal probability plot for the data on alcoholic odor is not shown because the sample size was too small to have useful indications.

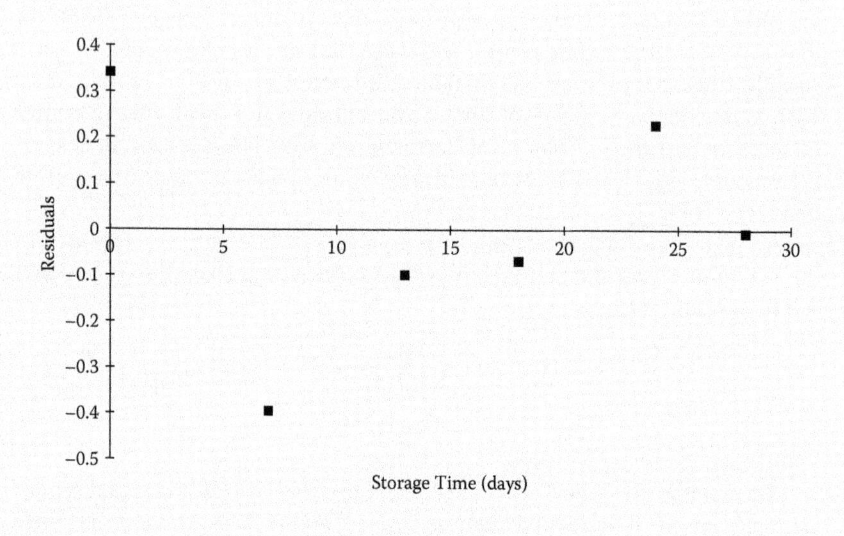

Figure 5.1.2 Plot of the residuals as a function of storage time.

When the reaction order is not zero, there are two possibilities to estimate the parameters of the kinetic function. The first is again to use linear regression, working with the values of the critical indicator on the transformed scale (logarithm, reciprocal). The second one is to use nonlinear regression to fit a curve (e.g., exponential) to the original data.

Although the first method is much simpler, the choice between the two cannot be based only on convenience. As we have seen previously, linear regression (and nonlinear regression) is based on several assumptions of the experimental error that must be satisfied to have proper inferences (Van Boekel, 2009; Draper & Smith, 1998).

Data transformation acts also on the error structure, so that if the assumptions are satisfied on the original scale, they could be not respected on the transformed one. On the other hand, data transformation is a commonly used tool to make the experimental error characteristics more similar to the assumptions.

As an example, we consider a situation with the critical indicator expressed in terms of concentration and the reaction order is one. Two models may be adopted on the original scale, depending on the nature of the experimental error. The first one is the additive model:

$$I_i = I_0 \exp\left(k\, t_i\right) + \varepsilon_i \tag{5.8}$$

while the second is the multiplicative one:

$$I_i = I_0 \exp\left(k\, t_i\right) \varepsilon_i \tag{5.9}$$

In the first case, data transformation may give biased results, but in the second one the application of logarithms gives

$$\ln I_i = \ln I_0 + k\, t_i + \varepsilon_i^* \tag{5.10}$$

Provided that $\varepsilon_i^* = \ln \varepsilon$ satisfies the assumptions of normality and variance homogeneity, linear regression on the transformed scale is, in this case, more appropriate.

The choice of the appropriate model should be based on a priori knowledge of the characteristics of the critical indicator (e.g., the multiplicative model is frequently appropriate for concentrations, for which the error term may be thought of as a dilution coefficient) and on linear regression diagnostics.

Regarding the computational aspects, nonlinear regression requires the use of techniques of iterative maximization, which are usually implemented in statistical software but are not available in Excel. Furthermore, these algorithms critically depend on the specification of some parameters (e.g., the initial values), whose choice strongly influences the convergence of the iterative procedure. In Box 5.2, an example of linear and nonlinear regression procedure is reported.

A number of articles dealing with the application of the classical kinetic approach to describe the deteriorative kinetics of several foods in different environmental conditions have been produced. Zero and first reaction orders are the most frequently applied kinetics to describe the changes of quality indicators in foods; few examples of second-order kinetics could be found in the literature. Table 5.3 shows some literature examples reporting apparent reaction orders describing the evolution of chemical, physical, or sensory critical indicator changes during storage of different foods. Due to the huge volume of literature data available, Table 5.3 should not be considered exhaustive.

Once the most appropriate apparent order of quality depletion has been defined, the reaction rate constant k must be computed. It is obvious that the k units depend on the reaction order as shown in Table 5.4.

As the reaction rate constant has been calculated, the shelf life can be computed by solving the integrated forms of Equation 5.7 as a function of time:

$$SL = \frac{1}{k} \int_{Io}^{I_{\lim}} \frac{dI}{I^n} \tag{5.11}$$

where I_0 is the value of the critical indicator just after food production, and I_{lim} is the critical indicator value corresponding to the previously defined acceptability limit (Chapter 3). To get an estimate of shelf life, the parameters I_0 and k must

BOX 5.2 FIRST-ORDER REACTION LINEAR AND NONLINEAR REGRESSION

Data reported in Table 5.2.1 were obtained from the work of Tiwari et al. (2009) and refer to the study of ascorbic acid degradation kinetics in orange juice during storage at 10°C.

The plot on the original scale with the regression line (Figure 5.2.1) clearly shows that zero-order kinetics is not adequate. Indeed, the regression line tends to underestimate the concentration of ascorbic acid for short and long storage times, while it gives overestimation for intermediate storage times. On the contrary, the plot on the logarithmic scale (Figure 5.2.2) shows that the pattern is linear, and that a first-order kinetic function must be used. Linear regression carried out with the Excel package on the transformed scale gives the results shown in Table 5.2.2. Similarly, with nonlinear regression, the data shown in Table 5.2.3 were obtained.

Table 5.2.1 Ascorbic Acid Concentration in Orange Juice at Different Storage Times at 10°C

Storage Time (days)	Ascorbic Acid Concentration (mg/100 ml)
0	43.29
7	32.08
15	24.36
22	18.67
33	11.22

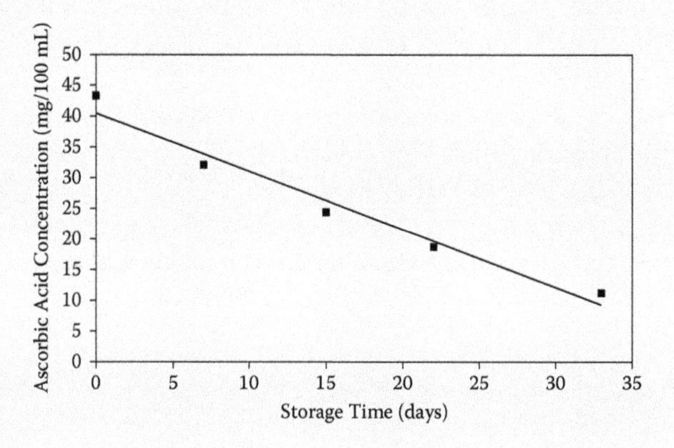

Figure 5.2.1 Ascorbic acid concentration of orange juice as a function of storage time at 10°C and corresponding regression line.

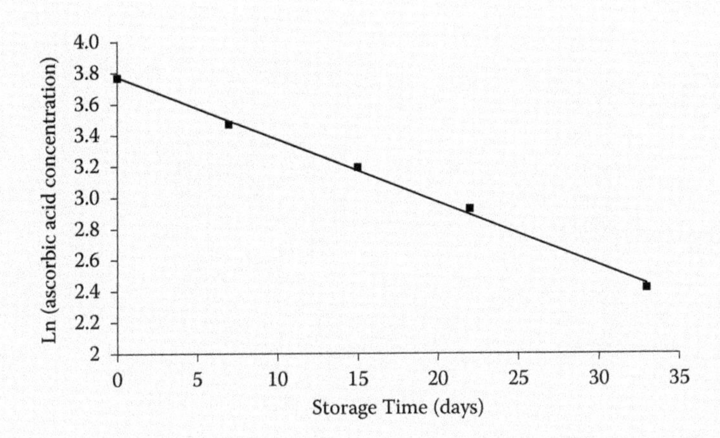

Figure 5.2.2 Ascorbic acid concentration, expressed in logarithmic values, of orange juice as a function of storage time at 10°C and corresponding regression line.

Table 5.2.2 Parameters Estimates with Linear Regression on the Transformed Scale

	Coefficients	SE	Stat *t*	*p*-Value
Intercept (Ln I_0)	3.773	0.025	150.26	6.50e-07
Rate constant (k, 1/day)	-0.040	0.001	-30.75	7.56e-05
$R^2 = 0.99$				

Table 5.2.3 Parameters Estimates with Nonlinear Regression on the Observed Scale

	Coefficients	SE	Stat *t*	*p*-Value
Intercept (I_0)	3.763	0.013	284.29	9.60e-08
Rate constant (k, 1/day)	-0.039	0.001	-33.05	6.09e-05
$R^2 = 0.99$				

Goodness-of-fit measures cannot be directly compared because they are evaluated on different scales. A comparison of the parameter values, however, shows that the two procedures in this case gave similar results. A plot of the two curves on the original scale, together with the observed data (Figure 5.2.3), shows that they are overlapping and that the goodness of fit is high.

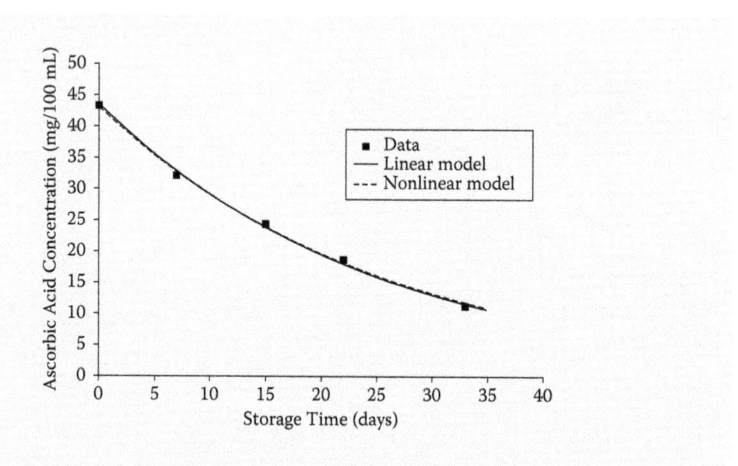

Figure 5.2.3 Observed and predicted values of ascorbic acid concentration in orange juice at 10°C under the linear and nonlinear regression models.

be substituted by the corresponding estimates derived with linear regression (see Box 5.3). In other words, although I_0 is the experimental value obtained at the beginning of the shelf life experiment, for a better estimation of the kinetic equation, it would be convenient to substitute it with that obtained by linear regression. Table 5.5 shows the shelf life equations relevant to different reaction orders.

5.2.2.3 When Classical Kinetic Modeling Does Not Work

Even though the classical kinetic approach is frequently applied for shelf life determination due to its computational simplicity, reactions involved in food quality depletion are often complicated, showing evolutions of the critical indicators that may be difficultly fitted by exploiting the classical kinetic approach.

A case in point is the evolution of phenomena characterized by an induction period (lag time), with no significant changes of the critical indicator, followed by the progressive increase or decrease of the indicator until a maximum or minimum value is reached. This behavior is typically observed when monitoring microbial spoilage during food storage (Chapter 6), but it may often be detected even as a consequence of other quality depletion phenomena. Actually, the changes of a number of chemical, physical, and sensory indicators as a function of time have been shown to evolve according to this kind of behavior (Table 5.6). Among these phenomena, typical events characterized by the presence of a lag phase are lipid oxidation, enzymatic reactions, crystallization, nonenzymatic browning (NEB), and sensory consumer acceptability.

Two different approaches could be followed to model complex behaviors in critical indicators. The first one can be called the "segmentation approach" and consists of modeling the changes of the critical indicator by separately considering each

Table 5.3 Literature Examples of Apparent Zero and First Reaction Order Used to Describe the Evolution of Chemical, Physical, Sensory Indicators during Storage of Different Food Products, Including Relevant Type of Event and Analytical Indexes Used in the Kinetic Studies

Phenomena	Apparent Reaction Order	Critical Indicator	Type Of Event	Product
Chemical	Zero	Peroxide value	Oxidation	Potato chips (Houhoula and Oreopolou, 2004); biscuits (Calligaris et al., 2007a); extra virgin olive oil (Calligaris et al., 2006; Mancebo-Campos et al., 2008); salmon oil (Huang and Sathivel, 2008)
		Absorbance at 420 nm	Nonenzymatic browning	Apple juice concentrates (Burdurlu and Karadeniz, 2003); milk powder (Patel et al., 1996); dehydrated carrots (Koca et al., 2007)
	First	Concentration changes	Vitamin C degradation	Frozen vegetables (Giannakourou and Taoukis, 2003); orange juice (Polyedra et al., 2003; Tiwari et al., 2009); fresh-cut strawberries (Odriozola-Serrano et al., 2009)
		Oxygen consumption	Oxidation	Extra virgin olive oil (Gutierrez and Fernandez, 2002; Mancebo-Campos et al., 2008); soybean oil (Colakoglu, 2007)
		Concentration changes	Carotenoids oxidation	Pumpkin puree (Dutta et al., 2006); beverages (Manocco et al., 2008); dehydrated carrots (Koca et al., 2007)
		Concentration changes	Chlorophyll loss	Dehydrated broccoli florets (Sanjuan et al., 2004)
Physical	Zero	Firmness	Staling	Cakes (Seyhun et al., 2005)
	First	Viscosity	Coalescence	UHT-sterilized peanut beverage (Rustom et al., 1996)
		Texture	Staling	Frozen dough bread (Giannou and Tzia, 2007)
		Color	Carotenoids degradation	Beverages (Manzocco et al., 2008); tomato derivatives (Manzocco et al., 2006)
		Color	Pigment oxidation	Frozen shrimps (Tsironi et al., 2009)
		Number of droplets	Coalescence	Oil-in-water emulsions (Das and Kinsella, 1993)
Sensory	Zero	Consumer acceptability	Staling	Whole pan bread (Gimenez et al., 2008); chocolate cakes (Villanueva and Trinidade, 2010)
		Consumer acceptability	Nonenzymatic browning	Apple baby food (Gambaro et al., 2006)
	First	Sensory attributes	Senescence	Butterhead lettuce (Lareo et al., 2009); Fuji apples (Varela et al., 2005)

Table 5.4 Units of Rate (k) According to Different Reaction Orders Given Specific Units of Critical Indicator (I Unit) and Time (t)

Reaction Order	Units of k
$n = 0$	I Unit^{-1}time^{-1}
$n = 1$	Time^{-1}
$n = 2$	I Unit^{-1} time^{-1}
$n \neq 1$	I Unit^{1-n} time^{-1}

evolution step. Figure 5.5 shows, as an example, the elaboration by the segmentation approach of data from the work of Calligaris et al. (2008) describing the changes of peroxide value of breadsticks as a function of storage time at 30°C.

The assumption is that the overall evolution is the result of consequential reactions, each following a fixed apparent kinetic order, leading to a characteristic apparent rate constant. For instance, most data showing a lag phase followed by an increase or decrease in the critical indicator can be elaborated to separately estimate the lag time (*lag*) value and the rate of increase or decrease of the critical indicator. The lag duration can be computed by finding the time corresponding to the intercept between the tangent line to the peroxide curve at $t = 0$ and that describing the increasing or decreasing part of the curve.

When the end of product life has been previously chosen in correspondence with the storage time at which the first detectable change in the critical indicator appears, the computation of the sole lag phase allows by itself the definition of shelf life. On the contrary, if the acceptability limit corresponds to a certain critical indicator value, other than that of the freshly prepared food, and thus is set within the increase or decrease phase, it is necessary to model this part of the overall curve using an appropriate kinetic model. If, for instance, in the case of breadsticks (Figure 5.5), the acceptability limit is set at 11–14 meq O_2/kg_{oil} and thus within the growth phase of peroxide value, both the lag phase and growing step have to be taken into account to obtain an adequate shelf life model. For a peroxide increase described by a zero-order kinetic, the shelf life equation becomes as follows:

$$SL = \frac{I_0 - I_{\lim}}{k} + lag \tag{5.12}$$

The second approach, which can be called "the fitting approach," is to consider the entire evolution of the curve during the overall product life. For instance, Figure 5.6 shows the fitting elaboration of peroxide changes as a function of time of breadsticks stored at 30°C according to a sigmoidal equation (Calligaris et al., 2008):

$$PV = PV_o + \frac{PV_{\max} - PV_o}{1 + e^{\left(\frac{4k_{\max}(\lambda - t)}{PV_{\max} - PV_o} + 2\right)}} \tag{5.13}$$

BOX 5.3 SHELF LIFE COMPUTATION AND UNCERTAINTY EVALUATION

The estimation of shelf life is exemplified using the data on changes of peroxide value of extra virgin olive oil as a function of storage time at 25°C as reported by Calligaris et al. (2006) (Table 5.3.1). The shelf life acceptability limit was chosen to be 20 meq O_2/kg_{oil}. Figure 5.3.1 shows that a zero-order kinetic is appropriate. Parameter estimates are reported in Table 5.3.2, together with the values of R^2 and residual standard error.

Shelf life is then given by

$$\overline{SL} = \frac{I - \hat{I}_0}{\hat{k}} = \frac{20.00 - 10.17}{0.07} = 140.43 \text{ days}$$

Based on this computation, shelf life was defined as 140 days.

This shelf life estimate was affected by the uncertainty on the parameters of the reaction kinetic equation, whose estimates were affected by random variation. In most literature articles and, to our knowledge, in common practice, the shelf life uncertainty is generally not calculated due to the difficulty of estimation of the propagation error. This is a critical point since the shelf life uncertainty can provide the operators with an important indication of the goodness of the overall shelf life testing process.

Since the relation between the shelf life estimator and the parameters of the model is not linear, approximation techniques are required. The most used, although not the most accurate, is based on error propagation (delta method in statistics). Error propagation consists of a first-order Taylor approximation

Table 5.3.1 Peroxide Value of Extra Virgin Olive Oil at Different Storage Times at 25°C

Storage Time (days)	Peroxide Value (meq O_2/kg_{oil})
0	8.6
11	10.7
21	12.5
40	13.9
53	13.5
62	14.1
102	18.2
123	20.0
144	20.5
200	23.8
250	27.3

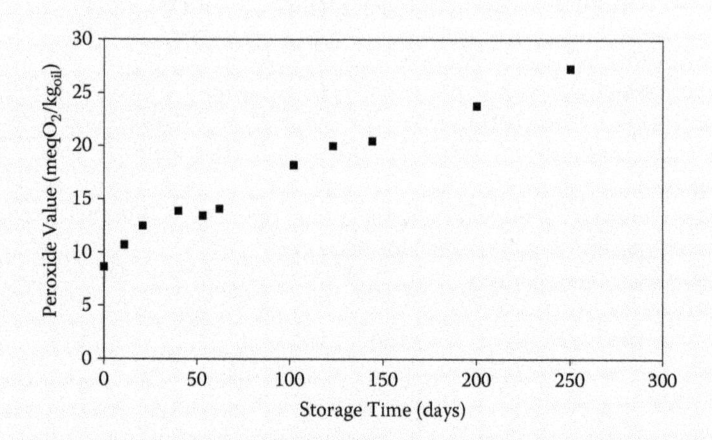

Figure 5.3.1 Peroxide value of extra virgin olive oil as a function of storage time at 25°C.

Table 5.3.2 Parameters Estimates with Linear Regression

	Coefficients	SE	Stat *t*	*p*-Value
Intercept (I_0)	10.174	0.410	24.77	1.36 10^{-9}
Reaction rate (*k*, PV/days)	0.070	0.003	20.59	7 10^{-9}
$R^2 = 0.98$				
$\hat{\sigma}^2 = 0.878$				

of the following expression (the method is shown for the order-zero kinetics; similar considerations apply for other orders) (Tanner, 1996):

$$\overline{SL} = \frac{I_{\text{lim}} - \hat{I}_0}{\hat{k}} = \bar{t} + \frac{I_{\text{lim}} - \bar{I}}{\hat{k}}$$

where $\overline{SL}$ is the estimator of SL, $\bar{t}$ is the mean of sampling times, and $\bar{I}$ is the mean of the observed values of the critical indicator.

The application of error propagation gives

$$VAR(SL) = \left(\frac{\partial\,\overline{SL}}{\partial \bar{I}}\right)^2 VAR(\bar{I}) + \left(\frac{\partial \overline{SL}}{\partial \hat{k}}\right)^2 VAR(\hat{k}) + \left(\frac{\partial \overline{SL}}{\partial \bar{I}\,\partial \hat{k}}\right) COV(\bar{I}, \hat{k})$$

Usually, the term involving the covariance is omitted, and the preceding expression develops into

$$VAR(\overline{SL}) = \frac{\hat{\sigma}^2}{\hat{k}^2}\left(\frac{1}{n} + \frac{1}{\hat{k}^2}\frac{\left(I_{\text{lim}} - \bar{I}\right)^2}{\sum (t_i - \bar{t})^2}\right)$$

which is the usual formula to compute the variance of calibrated measures. To derive the variance of shelf life when the order of the kinetic function is not zero, $\bar{I}$ must be substituted with the appropriate mean value (e.g., the mean of ln I for first-order kinetics).

An approximate (100 - α)% confidence interval for shelf life may be derived, assuming a Gaussian distribution for the shelf life estimator, in the following way:

$$\overline{SL} \pm t_{\alpha/2;n-2}\sqrt{VAR(\overline{SL})}$$

The accuracy of the approximation depends on two sources of error; the first is the application of error propagation rules, which may give over- or underestimates of the true variance; the second is the use of the Gaussian distribution, which gives symmetric confidence intervals even if the true distribution is not symmetric. However, the error implied by these approximations is usually negligible.

To determine the variance of the shelf life estimate, the mean of the critical indicator $\bar{I}$ and the deviance of storage times $\sum (t_i - \bar{t})^2$ must be computed from the data. In the example, $\bar{I} = 16.63818$ and $\sum (t_i - \bar{t})^2 = 655580.73$. The approximate variance of shelf life is then

$$VAR(\overline{SL}) = \frac{\hat{\sigma}^2}{\hat{k}^2}\left(\frac{1}{n} + \frac{1}{\hat{k}^2}\frac{\left(I_{\text{lim}} - \bar{I}\right)^2}{\sum (t_i - \bar{t})^2}\right) =$$

$$\frac{0.88^2}{0.07^2}\left(\frac{1}{11} + \frac{1}{0.07^2}\frac{\left(20 - 16.64\right)^2}{65580.73}\right) = 19.38$$

An approximate 95% confidence interval for shelf life may be computed in the following way:

$$\left[\overline{SL} - t_{\alpha/2;n-2}\sqrt{VAR(\overline{SL})}; \overline{SL} + t_{\alpha/2;n-2}\sqrt{VAR(\overline{SL})}\right]$$

The value of $t_{\alpha/2;n-2}$ is 2.26. The 95% confidence interval is then

$$\left[139.02-2.26\sqrt{19.92};139.02+2.26\sqrt{19.92}\right]=[129;149]$$

This means that the shelf life estimate is equal to 140 days with a 95% confidence interval going from 129 to 149 days.

Table 5.5 Shelf Life Equations Relevant to Different Reaction Orders

Reaction Order	Shelf Life Equation
$n = 0$	$SL=\frac{I_{\lim}-I_o}{k}$
$n = 1$	$SL=\frac{\ln I_{\lim}-\ln I_o}{k}$
$n = 2$	$SL=\frac{\frac{1}{I_{\lim}}-\frac{1}{I_o}}{k}$
$n \neq 1$	$SL=\frac{\frac{1}{I_{\lim}^{n-1}}-\frac{1}{I_o^{n-1}}}{k}$

where PV_0 is the initial peroxide value, PV_{max} is the peroxide value at the maximum of the curve, λ is the lag phase defined as the crossing point between the tangent through the inflection, and $k_{\max}$ is the maximum reaction rate. This is an example in which a more complex model than the classical kinetic equations has been used to describe peroxide evolution over time. Equation parameters are obviously all included into the ϑ term of Equation 5.1. In this case, the fundamental kinetic concepts are skipped, and other empirical descriptive models can be applied to describe the changes of the critical indicator as a function of storage time. Multiple linear regression, polynomial regression, and nonlinear regression can be used to fit these models to shelf life data.

The choice between the segmentation and fitting approach depends not only on the specific criticism of the reaction involved but also on the availability of specific data analysis software and the statistical skills of the operator. In most cases, the segmentation approach may provide satisfactory results in terms of shelf life and is easier to pursue as compared to the fitting one. In fact, it does not require the application of specific statistical procedures and the use of advanced statistical software to perform the best-fitting analysis.

Table 5.6 Literature Examples of Sigmoidal Evolution of Chemical, Physical, Sensory Indicators during Storage of Different Food Products, Including Relevant Critical Phenomena, Food Involved, and Critical Indicator Used in the Studies

Critical Phenomena	Food	Critical Indicator	Reference
Lipid oxidation	Fat and oils	Peroxide value	Özilgen and Özilgen, 1990; Cunha et al., 1998; Corradini and Peleg, 2007; Calligaris et al., 2008; Imai et al., 2008; Odriozola-Serrano et al., 2009
Lycopene chemical and enzymatic oxidation	Frozen tomato derivatives	Color	Calligaris et al., 2002
Crystallization	Fat	Solid fat content	Herrera et al., 1999
Nonenzymatic browning	Honey	Absorbance at 420 nm	Vaikousi et al., 2009
Myoglobin oxidation	Minced beef	Consumer acceptability	Hough et al., 2006
Staling	Coffee powder	Consumer acceptability	Guerra et al., 2008
Enzymatic activity	Frozen model systems	Enzymatic activity	Terefe et al., 2004

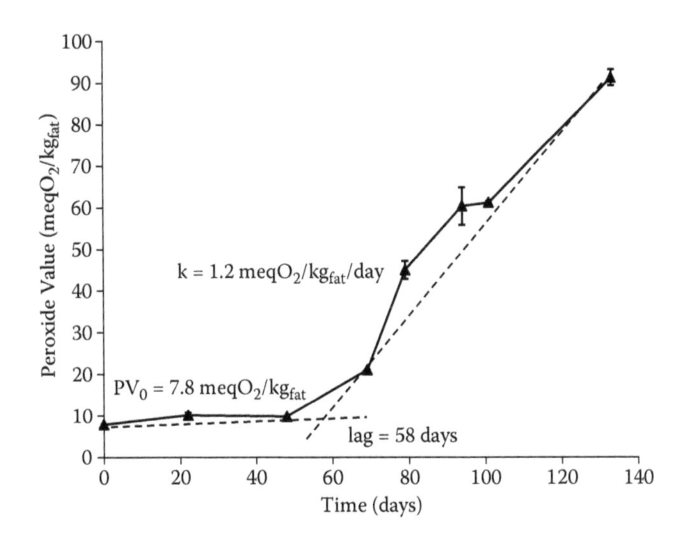

Figure 5.5 Elaboration of peroxide changes of breadsticks as a function of storage time at 30°C using the segmentation approach. (Elaborated from Calligaris, S., Da Pieve, S., Kravina, G., Manzocco, L., and M.C. Nicoli. 2008. Shelf-life prediction of bread sticks by using oxidation indices: a validation study. *Journal of Food Science* 73(2): E51–E56.)

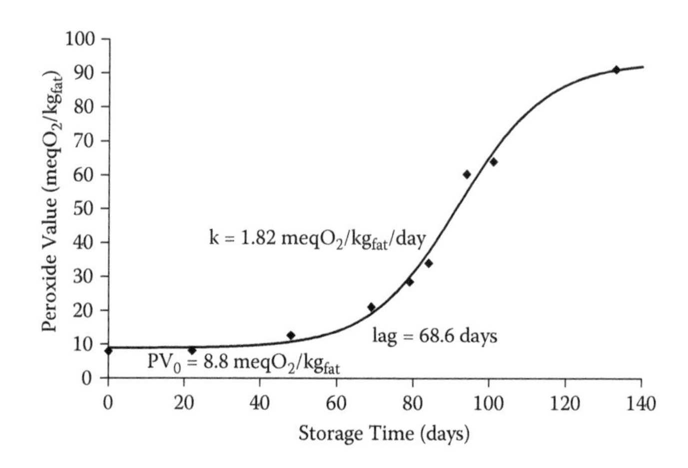

Figure 5.6 Elaboration of peroxide changes of breadsticks as a function of storage time at 30°C using the fitting approach. (Elaborated from Calligaris, S., Da Pieve, S., Kravina, G., Manzocco, L., and M.C. Nicoli. 2008. Shelf-life prediction of bread sticks by using oxidation indices: a validation study. *Journal of Food Science* 73(2): E51–E56.)

5.2.2.4 The Decision Tree for Real-Time Shelf Life Testing

Figure 5.7 summarizes the consequential steps to be followed to compute a reliable shelf life through modeling of data relevant to real-time testing. The starting point is the collection of data describing the changes of the quality indicator as a function of storage time under the actual storage conditions. These data are then plotted to evaluate the shape of the function g. Based on the observed results, data may show an evolution described using the classical kinetic approach (see Section 5.2.2.2) or a complex behavior. In the latter case, data can be elaborated following the segmentation approach or the fitting one (see Section 5.2.2.3). Once defined the most appropriate approach to be pursued, the model best describing the changes of I versus t, should be selected based on the statistical response. Using this model, the shelf life under actual storage conditions can be computed.

5.3 ACCELERATED SHELF LIFE TESTING

When the quality depletion proceeds fairy slowly under actual storage conditions, it is convenient to accelerate shelf life experiments by testing food under environmental conditions that speed up the quality deterioration and then extrapolating the results to milder conditions usually experienced by the product (Mizrahi, 2000). This kind of test is generally called accelerated shelf life testing (ASLT).

It is useful to remember that the reaction rate is a function of compositional C_i, packaging P_i, and environmental E_i factors:

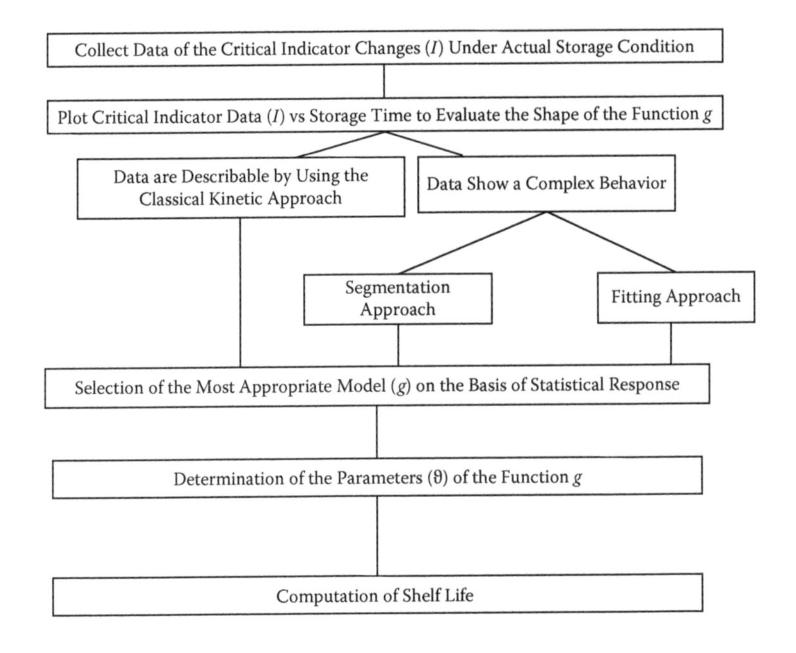

Figure 5.7 Decision tree for shelf life modeling of real-time data.

$$k = h'(C_i, P_i, E_i, \alpha) \tag{5.6}$$

When performing ASLT, compositional variables must be kept constant, and a proper factor to be exploited in the test must be chosen. To this regard, it is noteworthy that a number of different factors potentially applicable in ASLT are known, including temperature (T), relative humidity (ERH%), light intensity (L), or gas partial pressure (PO_2). Independent of the accelerating factor selected, the absolute requirements for the ASLT application are

- The quality depletion rate varies only as a function of the selected accelerating factor, while other environmental, packaging-related, and compositional variables are kept constant; packaging permeability does not change when the accelerating factor changes
- A model describing quality depletion during food storage is available
- A model describing the effect of the accelerating factor on quality depletion kinetics is known

5.3.1 Temperature as Accelerating Factor

Among all environmental factors that may be used in principle as an accelerating factor in ASLT, temperature is certainly the most widely used. This is due not only to the fact that temperature is one of the most critical factors affecting reaction kinetics in food but also to the availability of a theoretical basis for the development of a mathematical description of the temperature sensitivity of quality loss rates. In fact, the Arrhenius equation (5.14) (Arrhenius, 1901), developed theoretically on the molecular basis for reversible chemical reactions, has been shown to hold empirically for a wide range of complex chemical, physical, and sensory changes occurring in foods (Labuza and Riboh, 1982):

$$k = k_o \cdot e^{-\frac{E_a}{RT}} \tag{5.14}$$

Its linearized form is

$$\ln k = \ln k_o - \frac{E_a}{RT} \tag{5.15}$$

in which k is the reaction rate constant; R is the molar gas constant (8.31 J/K/mol), T is the absolute temperature (K), E_a is the apparent activation energy (J/mol), and k_o is the so-called preexponential factor. Also in this case, the term *apparent* is used since the reaction mechanisms of the alterative phenomena are not elucidated.

To estimate the effect of temperature on the reaction rate of a specific quality deterioration event, values of k are estimated at different temperatures in the range of interest, and ln k is plotted against the reciprocal of absolute temperature. If a linear relation between these variables can be observed, the Arrhenius behavior is fulfilled. Hence, by measuring the rate of quality depletion at least at three different temperatures, the reaction rate at a desired temperature can be extrapolated.

It should be noted that the Arrhenius equation in its original form implies that k_o is the frequency factor representing the number of successful collisions between reactants at 0 K, which is a temperature of no practical interest. From a statistical point of view, it is recommended to reparameterize the Arrhenius equation by inserting a reference temperature T_{ref}, corresponding to the average of the temperature range used during ASLT (an example of the application of this procedure is reported in Box 5.4). The reparameterized Arrhenius equation would then be written as

$$k = k_{ref} \cdot e^{-\frac{E_a}{R}\left(\frac{1}{T}-\frac{1}{T_{ref}}\right)} \tag{5.16}$$

where k_{ref} is the rate constant at the reference temperature.

Its linearized form becomes

$$\ln k = \ln k_{ref} - \frac{E_a}{R}\left(\frac{1}{T} - \frac{1}{T_{ref}}\right) \tag{5.17}$$

The reparameterization process is performed to enhance the characteristics of parameter estimates; indeed, it reduces the correlation between the reaction rate constant and the activation energy parameter estimation. In addition, its

BOX 5.4 ARRHENIUS EQUATION AND SHELF LIFE ESTIMATION

The application of ASLT to predict shelf life is illustrated using data taken from the work of Manzocco and Nicoli (2007) (Table 5.4.1). Data refer to the chemical instability of a coffee brew, monitored by the pH changes and expressed as H_3O^+ molar concentration, as a function of storage temperature.

Figure 5.4.1 shows the evolution of H_3O^+ concentration in coffee brew during storage at different temperatures. Shelf life value at 25°C was predicted using two-step procedure. For most temperatures, there is no evident deviation from linearity, allowing the use of zero-order reaction kinetics.

First, values of the rate constant specific for each temperature must be obtained by applying regression analysis separately to each temperature-specific dataset. The estimates of the parameters of the reaction kinetic equation obtained with linear regression at the different temperatures are reported in Table 5.4.2 together with a measure of the goodness of fit.

The estimated values of the rate constant may be used to estimate the parameters of the Arrhenius equation. First, a plot of $\ln(\hat{k})$ versus 1/temperature (measured in K) (Figure 5.4.2) is needed to evaluate if the Arrhenius equation is appropriate to describe the relation between the rate constant and temperature. The pattern should be approximately linear.

Table 5.4.1 H_3O^+ Concentration (M) of a Coffee Brew during Storage at Different Temperatures

Temperature (°C)							
60°C		45°C		30°C		20°C	
Time (days)	H_3O^+ ($M \cdot 10^{-6}$)	Time (days)	H_3O^+ ($M \cdot 10^{-6}$)	Time (days)	H_3O^+ ($M \cdot 10^{-6}$)	Time (days)	H_3O^+ ($M \cdot 10^{-6}$)
0.00	5.50	0.00	5.50	0.00	5.50	0.00	5.26
0.02	5.62	0.09	5.37	0.17	5.13	1.00	5.28
0.08	6.46	0.31	6.03	1.00	6.03	2.00	5.20
0.17	7.08	1.00	6.92	2.00	6.03	3.29	5.20
0.31	8.91	1.28	7.94	3.17	7.08	7.00	5.10
0.94	11.48	2.00	9.12	6.00	9.77	9.00	4.98
1.00	13.49	2.28	10.72	7.00	10.47	11.00	4.96
1.28	13.80	5.00	14.45	8.00	10.47	15.00	4.95
				10.00	12.59	18.00	4.92
				14.00	14.13	21.00	4.90
				17.00	14.45	24.00	4.80
				20.00	15.14		

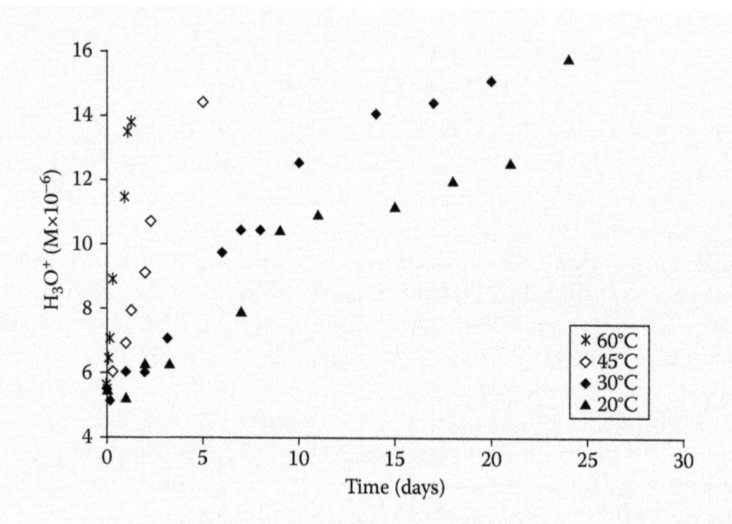

Figure 5.4.1 H_3O^+ concentration of coffee brews as a function of storage time at different temperatures.

Table 5.4.2 Rate Constant Estimates at Different Temperatures

Temperature (°C)	$\hat{k}$ ($M \cdot 10^{-6}$/days)	R^2
60	6.421	0.98
45	1.870	0.98
30	0.540	0.95
20	0.399	0.94

In the second stage of the procedure, parameters of the Arrhenius equation may be estimated with linear regression applied to the transformed data: rates must be taken as logarithms, while temperature must be transformed in the following way:

$$-\frac{1}{R}\left(\frac{1}{T}-\frac{1}{T_{ref}}\right)$$

where T_{ref} is set equal to the mean value of the temperatures tested (in our case, 45°C).

The regression analysis gives the estimates shown in Table 5.4.3.

Given that a zero-order kinetic equation has been used, shelf life may be estimated accordingly:

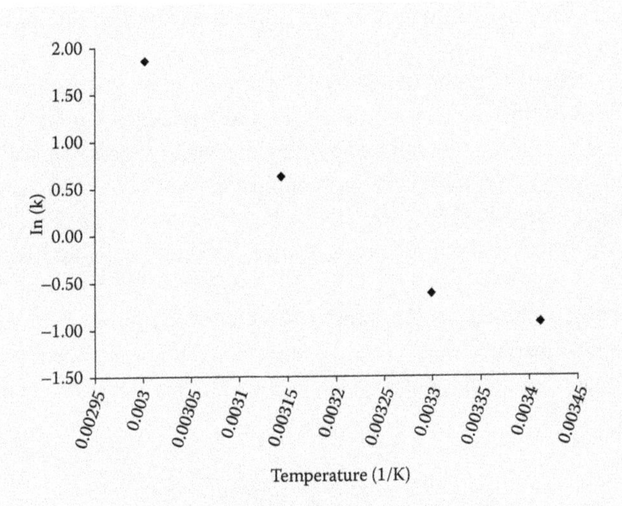

Figure 5.4.2 Rate constant $\hat{k}$, expressed in logarithmic values, as a function of the reciprocal of absolute temperature.

Table 5.4.3 Estimates of the Arrhenius Equation Parameters

	Coefficients	SE	Stat *t*	*p* Value
$\ln(k_{ref})$	0.73	0.14	4.8854	0.0394
E_a	58,032.63	7,287.99	7.9627	0.0154
$R^2 = 0.97$				

$$SL = \frac{I_{\lim} - I_o}{k_{ref} \exp\left[-\frac{E_a}{R}\left(\frac{1}{T^*} - \frac{1}{T_{ref}}\right)\right]}$$

where I_0 is the experimental value of I at time zero, k_{ref} and E_a must be substituted by the corresponding estimates, and T^* is the temperature (298 K) at which to predict shelf life. Assuming an acceptability limit for $H_3O^+ \cdot 10^{-6}$ equal to 8.0, the estimate of the shelf life is

$$SL = \frac{8.0 - 5.5}{2.076 \exp\left[-\frac{58032.63}{8.31}\left(\frac{1}{298} - \frac{1}{328}\right)\right]} = 11.5 \text{days}$$

Unfortunately, different sources of uncertainty affect the final shelf life value, begetting a wide confidence interval. Its computation may be demanding from the mathematical point of view and is generally not performed in common practice.

actual result is that the frequency factor is replaced by the rate constant at the reference temperature.

From the statistical point of view, the estimation of the Arrhenius equation parameters raises some relevant points. First, the Arrhenius equation is nonlinear. Parameters may then be estimated by applying nonlinear regression to Equation 5.16 or linear regression to its linearized version in Equation 5.17. Again (see the case of first-order reaction and Box 5.2), the choice between the two approaches should be based on considerations about the error structure and on available linear regression diagnostic tools.

Alternatively, a one-step regression method, considering all data versus storage time for all tested temperatures, could be applied. In this case, the reparameterized Arrhenius equation is integrated into the rate equations. For a zero-order reaction, this would give

$$I = I_o + t \cdot k_{ref} \exp\left[-\frac{E_a}{R}\left(\frac{1}{T} - \frac{1}{T_{ref}}\right)\right] \tag{5.18}$$

and for a first-order reaction

$$I = I_o \cdot \exp\left[t \cdot k_{ref} \cdot \exp\left[-\frac{E_a}{R}\left(\frac{1}{T} - \frac{1}{T_{ref}}\right)\right]\right] \tag{5.19}$$

Among the advantages of the one-step procedure, there is the fact that all the different sources of uncertainty are accounted for by a single equation and this increases the number of residual degrees of freedom. This last point is relevant because the size of confidence intervals depends on the precision of estimates and on the residual degrees of freedom: With only three temperatures (as is frequently done in practice), a wide confidence interval is expected with a two-step procedure because only one residual degree of freedom is available. In the same case, the residual degrees of freedom are equal to $3*n_t$ - 3 (n_t being the number of times used in the experiment) with the one-step procedure.

On the other side, the one-step procedure necessarily implies the use of nonlinear estimation techniques, at least when the reaction order is not zero. For this reason, it is difficult to find applications in the daily management of shelf life estimation.

Table 5.7 reports some examples of literature data relevant to the application of the Arrhenius equation to describe the temperature dependence of chemical, physical, and sensory indicator changes during storage. The relevant E_a values as well as the temperature range considered in the study are also reported.

The temperature dependence of the shelf life index changes can be integrated in the shelf life equation (5.11), and a general shelf life predictive model can be obtained as follows:

Table 5.7 Examples of the Application of Arrhenius Equation to Describe the Temperature Dependence of the Changes of Chemical, Physical, and Sensory Critical Indicators during Food Storage, Including Estimated Values of Apparent E_a and the Relevant Temperature Range Considered

Index	Critical Indicator	Type of Event	Temperature Range (°C)	E_a (kJ/mol)	Product
Chemical	Peroxide value	Oxidation	-18/60 5/60 25/60	51 59 44	Sunflower oil (Calligaris et al., 2004) Encapsulated rapeseed oil (Orlien et al., 2006) Extra virgin olive oil (Calligaris et al., 2006)
	Lycopene loss	Oxidation	-5/-20	25/48	Osmo-dehydrofrozen tomatoes (Dermesonlouoglou et al., 2007)
	Vitamin C	Oxidation	-5/-15 28/45 -5/20	97 53/79 87/94	Frozen vegetables (Giannakourou and Taoukis, 2003) Citrus juices (Burdurlu et al., 2006) Osmo-dehydrofrozen tomatoes (Dermesonlouoglou et al., 2007)
	HMF formation	Nonenzymatic browning	28 to 45	181/335	Citrus juices (Burdurlu et al., 2006)
	Hydrogen ion	Hydrolysis	-18/60	50	Coffee liquids (Manzocco and Nicoli, 2007)
Physical	Color	Carotenoid oxidation	60/99 20/40 -30/0	100/113 25/30 5/10	Paprika (Topuz, 2008) Beverages (Manzocco et al., 2008) Tomato puree (Manzocco et al., 2006)
	Color	Nonenzymatic browning	5/65	21/33	Apple juice concentrates (Burdurlu and Karadeniz, 2003)
	Color	Enzymatic reactions	10/25	29	Fresh-cut asparagus (Sothornvit and Kiatchanapaibul, 2009)
Sensory	Sensory attributes	Quality attributes changes	5/15 -5/-15	52/61 143	Butterhead lettuce leaves (Lareo et al., 2009) Frozen shrimp (Tsironi et al., 2009)

$$SL = \frac{\int_{I_0}^{I_{\lim}} \frac{dI}{I^n}}{k_{ref} \exp\left[-\frac{E_a}{R}\left(\frac{1}{T} - \frac{1}{T_{ref}}\right)\right]} \quad (5.20)$$

It must be underlined that with the two-step procedure, multiple values of I_0 are estimated (one for each temperature), and no criterion is given to choose among them. To solve this problem, it is customary to use the experimental I value measured at time zero as I_0 to be included in the equation. This represents a drastic simplification, however, allowing the application of the two-step procedure methodology, which remains the most frequently used in common practice. Box 5.4 reports a numerical example of the application of the two-step procedure for shelf life computation.

5.3.1.1 Deviations from the Arrhenius Behavior

The successful application of the Arrhenius model is that food is able to withstand the increase in temperature without leading to dramatic development of phenomena other than the event responsible for product unacceptability at usual storage temperatures. The choice of the temperature interval for ASLT is crucial to avoid the prevalence of degradation reactions other than that of interest at actual storage temperature. For instance, the quality decay of foods containing tomato stored below 0°C is generally due to carotenoid oxidation, which causes the product color to turn from red to yellow. When trying to accelerate the quality depletion kinetics by increasing storage temperature, pigment bleaching can be masked by the concomitant development of other color-effective reactions, such as NEB (Manzocco et al., 2010). An example relevant to tomato powder is reported in Figure 5.8, which shows the evolution of the oxidation rate, described by the increase in yellowness (b*), as a function of storage temperature. Above -7°C, a clear deviation from the Arrhenius equation is observed due to the prevalence of NEB. The color index (b*), chosen as the critical indicator, describes the evolution of both oxidation and NEB reactions. The latter prevails at a temperature higher than -7°C. For this reason, in the case of shelf life accelerated tests of frozen foods containing tomato powder, storage temperatures above -7°C should be avoided in ASLT.

Additional and even more intricate complications could arise in multicomponent foods when reactions occurring at high temperatures lead to the formation of new compounds that could join the reaction pathway. For instance, the formation of Maillard reaction products with pro- or antioxidant activity could interfere with the kinetics of the oxidative reaction in lipid-containing foods. For this reason, it is advisable always to maintain a sample stored at the actual temperature experienced on the shelf to ensure complete understanding of the temperature-dependent profile (Waterman and Adami, 2005).

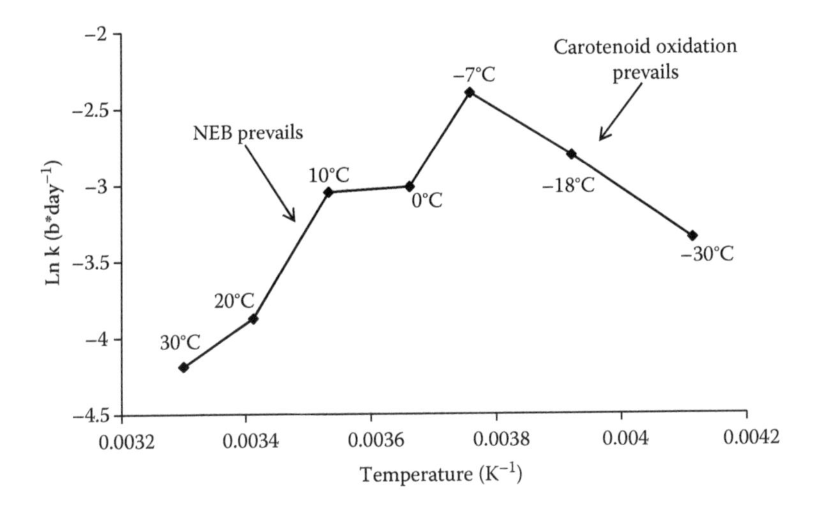

Figure 5.8 Oxidation rate, described by the increase in yellowness and expressed in logarithmic values, as a function of storage temperature of tomato powder.

Unfortunately, even though the temperature dependence of quality depletion is adequate to perform a proper ASLT, many other pitfalls could arise in practice, leading to deviations from the Arrhenius equation and potentially causing significant shelf life prediction errors. Figure 5.9 schematically shows possible deviations from Arrhenius behavior. There is a temperature range in which the Arrhenius equation can be used to describe the temperature dependence of the reaction rate. Below the critical temperature, an abrupt change in the temperature dependence of reaction rate is appreciated. The situation depicted in Figure 5.9a describes a positive deviation from the Arrhenius behavior. This kind of deviation is particularly dangerous since it can cause an overestimation of the reaction rate at the actual storage temperature of the product. The reaction rate predicted based on the Arrhenius equation is lower than that experimentally observed. On the contrary, Figure 5.9b shows an example of a negative deviation from the Arrhenius equation. In this case, the use of the Arrhenius equation as a predictive tool can cause an underestimation of the reaction rate at the actual storage conditions.

Based on these observations, it is evident that extrapolation of reaction rates at the usual storage temperatures from accelerated data must be performed only within the temperature range experimentally proven to conform to the Arrhenius model. In other words, the Arrhenius methodology requires adaptation to the specific circumstances of the product considered. There are different possible causes for Arrhenius deviations:

- Changes in reaction pathway. For a complex reaction in which multistep pathways are involved, the overall reaction rate is directly dependent on the slowest step, which is obviously the rate-determining one. As temperature changes, different activation energies and preexponential terms for these steps can lead to

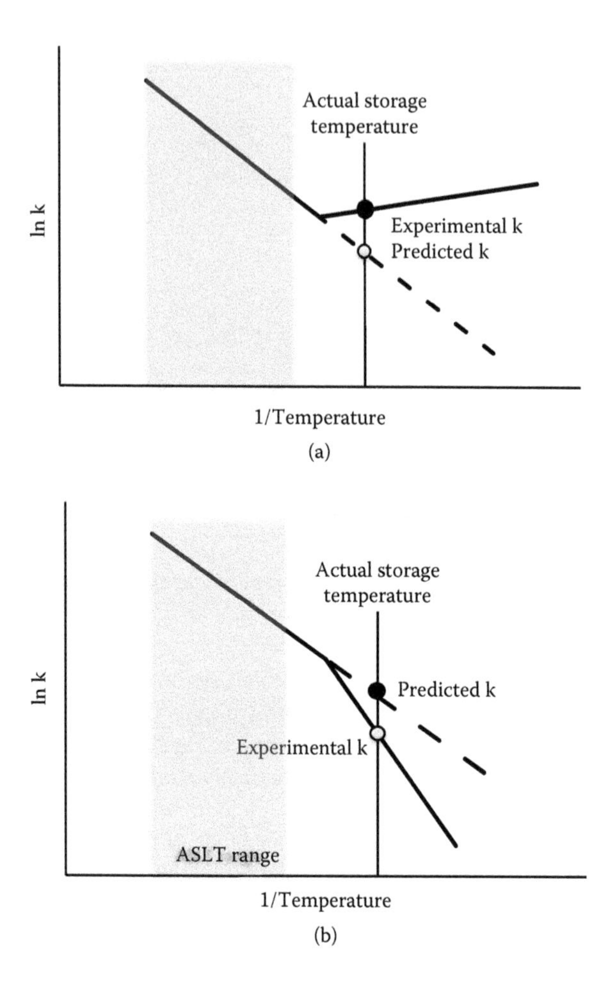

Figure 5.9 Schematization of possible (a) positive or (b) negative deviations from the Arrhenius behavior.

non-Arrhenius behavior. In addition, a switch in the rate-determining steps or a shift in the reaction pathway can occur. When a pathway dominates at lower temperatures while another dominates at higher temperatures, the prediction based on high-temperature behavior could underestimate or overestimate the instability.

This situation could be observed for foods susceptible to oxidative reactions. As stated by Frankel (2005) for food lipids, the use of temperatures higher than 100°C in ASLT is questionable because samples develop excessive levels of rancidity, which are not relevant to what happens under normal storage conditions. For instance, Kaya et al. (1993) evidenced an overestimation of the induction periods at room temperature when performing the tests at temperatures above 100°C. In addition, a kinetic study of olive oil triacylglycerol oxidation indicated that the Arrhenius equation could be employed to describe the temperature dependence of primary and secondary oxidation product formation only between 25°C and 75°C

(Gomes-Alonso et al., 2004). The more polyunsaturated the oils, the lower the temperatures that should be used to test their oxidative stability; for instance, vegetable oils should be tested at temperatures lower than 60°C, while for fish oils the value is only below 40°C (Frankel, 2005). Besides the rancidity level reached in ASLT, the eventual thermal degradation of minor compounds with pro- or antioxidant activity could become critical since they can modify the temperature dependence of the overall oxidation rate of lipids.

- Changes in gas solubility. As is well known from thermodynamic laws, the solubility of gases in solvents tends to decrease with increasing temperature. In this context, particularly critical could be the oxygen concentration in the matrix. In fact, when the main driving force involved in the alterative reaction is the oxygen availability, deviations from the Arrhenius equation can be observed. For example, the rate of peroxide formation in oil contained in emulsions stored at -30°C was higher than that expected based on the Arrhenius equation (Calligaris et al., 2007b). The Arrhenius equation was fulfilled at temperatures from 60°C to -18°C. It was hypothesized that at -30°C the role of oxygen concentration becomes critical in affecting primary oxidation product formation. Similarly, oxygen concentration was critical in determining the rate of carotenoid bleaching in tomato derivatives stored below zero (Manzocco et al., 2006). For these reasons, other models accounting for such effect have been developed.
- Changes in physical structure. If a phase transition occurs in the temperature range considered for an ASLT, there can be an unexpected modification in the temperature dependence of the reaction rate. In these cases, deviations from Arrhenius equation have been frequently observed and were attributed to the occurrence of first-order or second-order phase transitions of food constituents as a consequence of temperature changes (Parker and Ring, 1995; Kristott, 2000; Calligaris et al., 2004, 2006; Manzocco et al., 2007; Calligaris et al., 2007b, 2008; Slade and Levine, 1991). As examples, Figure 5.10 shows the Arrhenius deviations observed for hexanal formation in sunflower oil (Calligaris et al., 2004).

In these cases, the deviations from the Arrhenius behavior may be the result of the occurrence of a cascade of temperature-dependent events, such as solute concentration and changes in physicochemical properties (i.e., reactant solubility, pH, ionic strength, water activity, viscosity) in the liquid phases surrounding crystals (Parker and Ring 1995; Fennema 1996; Champion et al., 1997). These compositional modifications could counterbalance or even oppose the direct effect of temperature on the reaction rate, giving reason to the observed deviations. Extensive work has been published regarding the effect of physical state changes on lipid oxidation rate. When differences in the physical state of lipids are expected as a consequence of temperature modification, the changes in the relative reactant concentration have been indicated as the main driving force affecting the oxidation rate (Calligaris et al., 2004). The progressive separation of fat crystals gradually leads to an increase in the reactant concentration (i.e., unsaturated triacylglycerols, O_2, antioxidants, and prooxidants) in the liquid phase surrounding fat crystals, causing an unexpected acceleration of the oxidation rate. Similarly, deviations from Arrhenius equations can be observed in foods containing water undergoing crystallization (Fu and Labuza, 1993). This is obviously the situation that could often happen in frozen foods, in which the freeze–concentration effect can cause rates to increase dramatically just below the freezing point. For instance, curvature in the Arrhenius model has been

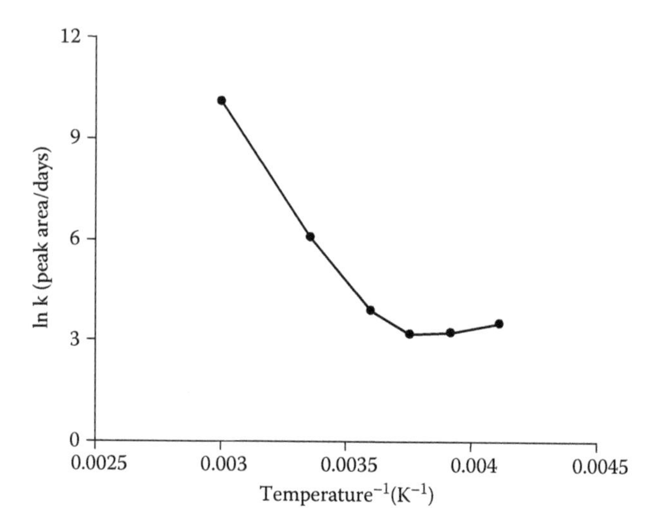

Figure 5.10 Apparent zero-order rate constants of hexanal formation as a function of temperature^{-1}. (Modified from Calligaris, S., Manzocco, L., Conte, L.S., and M.C. Nicoli. 2004. Application of a modified Arrhenius equation for the evaluation of oxidation rate of sunflower oil at subzero temperatures. *Journal of Food Science* 69: 361–366.)

reported for enzyme-catalyzed reactions in the subfreezing temperature range and oxidative reactions (e.g., ascorbic acid degradation, carotenoid bleaching) (Terefe et al., 2004; Manzocco et al., 2006). It has been proposed that reactions in frozen systems are diffusion-limited processes governed by the physicochemical properties of the freeze-concentrated matrix surrounding the ice crystals, including enzyme-catalyzed reactions in frozen systems (Slade and Levine, 1991). According to these authors, in the freeze-concentrated "rubbery" matrix above the glass transition temperature T_g, the Arrhenius equation is not applicable.

5.3.1.2 *Other Temperature Dependence Models*

When the Arrhenius equation is not applicable, other models should be identified. Such models may be simply descriptive ones or built up starting from the understanding of the physicochemical phenomena leading to the Arrhenius deviation.

In such a context, the Williams-Landel-Ferry (WLF) model (1955) has been stated as appropriate to describe the rate–temperature relation for diffusion limited reactions:

$$\ln\frac{k_{ref}}{k}=\frac{C_1\left(T-T_{ref}\right)}{C_2+\left(T-T_{ref}\right)} \tag{5.21}$$

where C_1 and C_2 are system-dependent constants, T is the temperature and T_{ref} is a reference temperature ($T_{ref}>T_g$), k is the rate constant, and k_{ref} is the rate constant at T_{ref}.

The WLF approach should theoretically apply to reactions within rubbery systems. The equation was used to model the kinetics of enzymatic browning in dried foods and model systems, flowability of fructose and melted cheese, and kinetics of microbial and enzymatic inactivation with satisfactory results (Roos, 1995).

However, it is surprising that the WLF equation is only rarely applied in scientific literature as well as at the industry level to predict food shelf life. This may be because the Arrhenius relationship has also been shown to describe the temperature dependence of chemical reactions adequately over narrow temperature ranges even within the rubbery state. Data over a broad temperature range are generally required to differentiate between the WLF and the Arrhenius models. Both may be used successfully in describing the temperature dependence of chemical reactions over limited temperature ranges, but the WLF approach also provides a physical explanation for the effect of temperature on reaction rates within rubbery systems (Nelson and Labuza, 1992). For instance, Giannakourou and Taoukis (2003) reported that both Arrhenius and WLF models are adequate to represent temperature dependence of ascorbic acid degradation within the rubbery state of a frozen vegetable matrix.

Other models have also been proposed to describe the temperature dependence of the rate constant. Waterman and Adami (2005) reported that to take into account nonlinear Arrhenius behavior, in many cases it is possible to use the modified Arrhenius relationship:

$$k = AT^{n}e^{-\frac{E_a}{RT}} \tag{5.22}$$

where A, n, and E_a are parameters determined using nonlinear fitting programs. The use of this equation allows for a better mathematical description of some level of curvature in the Arrhenius plot, but parameters are pure corrective factors with no specific relation to the phenomena causing deviation.

To find out a feasible predictive model taking into account the modifications occurring in the food as a consequence of temperature changes, a modified Arrhenius equation (Calligaris et al., 2004) was proposed:

$$k = k_0 \cdot \quad k \cdot e^{-\frac{E_a}{RT}} \tag{5.23}$$

where Δk is a corrective factor included in the Arrhenius equation to take into account the influence of variables, other than temperature, that significantly affect the reaction rate. Since at a given temperature the rate at which any reaction develops can be considered the result of the ratio between driving forces and resistances, Δk can be defined by the identification of the proper forces and resistances responsible for the Arrhenius deviation:

$$k = \frac{\Sigma(\text{driving forces})}{\Sigma(\text{resistances})} \tag{5.24}$$

In principle, Equation 5.23 is general since it can always be applied when deviations from the Arrhenius equation occur. It obviously implies the understanding of the driving forces and resistances involved as well as the quantification of their ratio. In some cases, this approach is not simple since it requires a deep understanding of the complex phenomena involved and the use of sophisticated analytical techniques for quantifying the Δk value. It is generally a matter for food scientists rather than for food industry operators. However, when this approach is successful and Δk is calculated and included in the Arrhenius equation, the deviation is corrected, and a straight line is obtained. Therefore, the advantage of this approach is the possibility to perform accelerated shelf life tests even in the temperature range precluded due to Arrhenius deviations.

This approach has been efficaciously applied in different food matrixes to describe the temperature dependence of oxidation rate in sunflower oil, extra virgin olive oil, emulsions, biscuits, breadsticks, and tomato derivatives (Calligaris et al., 2004, 2006, 2007a, 2007b, 2008; Manzocco et al., 2006). Examples of Δk are reported in Box 5.5.

5.3.1.3 *The Decision Tree for Temperature-Accelerated Shelf Life Testing*

Figure 5.11 highlights the consequential steps to be followed once dealing with the performance of an ASLT. The initial stages mimic those previously described in Figure 5.7 and aim to find the ϑ parameters of the function g at least at three different temperature conditions. Once the ϑ parameters are defined, their temperature dependence should be evaluated by their Arrhenius plot (reporting ln ϑ versus $1/T$). The best situation is when the fulfillment of the Arrhenius behavior is observed. This directly leads to estimation of the equation parameters, computation of the value of ϑ at the actual storage temperature, and estimation of the shelf life of the product. If deviations from the Arrhenius behavior are detected, one could try to find/develop alternative models able to describe the temperature dependence of ϑ parameters. If the goal is obtained, this alternative model can be used to predict the product shelf life. The opposite case is the worst-possible situation and does not allow the exploitation of ASLT for the prediction of the product shelf life using temperature as an accelerating factor. The only possibility to accelerate the shelf life assessment process could be to find alternative environmental acceleration factors.

5.3.2 Other Accelerating Factors

The application of temperature as an accelerating factor in ASLT may be convenient because it allows the alterative reactions to be efficiently sped up, dramatically decreasing the time required to obtain shelf life data. However, caution should be

BOX 5.5 MANAGING DEVIATIONS BY INTEGRATION OF Δk INTO THE ARRHENIUS EQUATION

As previously stated, Δk has been defined as the ratio between driving forces and resistance affecting the reaction rate (Equation 5.24). The value of Δk has to be defined case by case, taking into account the specific criticism of the product.

For instance, peroxide rate constants of sunflower oil showed a sharp deviation from the Arrhenius equation at subzero temperatures (Calligaris et al., 2004). Since sunflower oil is frequently used as an ingredient in a wide variety of frozen foods, this event can greatly hinder correct estimation of the product shelf life by ASLT at subzero temperatures. The cause at the basis of the deviation at subzero temperatures was identified in the crystallization of the more saturated oil fraction. Thus, the remaining liquid fraction is expected to be formed by unsaturated lipids, which are known to be more prone to oxidation. Based on these considerations, the increase in the relative reactant concentration of unsaturated lipids in the liquid phase surrounding fat crystals has been indicated as the main driving force affecting the oxidation rate. Thus Δk was defined as the relative reactant concentration C, which is the ratio between the liquid fraction originally present in the sample and the liquid fraction at selected temperatures. The C value indicates how many times compounds involved in the oxidative process concentrate in the liquid phase as a consequence of crystallization. A C value equal to 1 indicates that, at the given temperature, crystallization did not occur, and the sample is completely liquid so its concentration is unchanged. By contrast, C values higher than 1 indicate that sunflower oil is partially crystallized at the given temperature, leading to a C-times increase in the concentration of the remaining liquid phase.

The liquid fraction of the fat at selected temperatures has been determined by differential scanning calorimetry. Table 5.5.1 shows, as an example, the liquid fraction and the relevant C factor of the fat extracted from biscuits at different temperatures.

Table 5.5.1 Liquid Fraction (LF, % w/w) and Corresponding Concentration Factor (C) of Oil Extracted from Biscuits as a Function of Temperature (Calligaris et al., 2007a with permission).

Temperature (°C)	LF (% w/w)	Concentration factor (C)
-18	3.21	31.1
20	73.4	1.36
30	90.6	1.10
37	98.5	1.01
45	100	1.00

After the computation of the C factor at different temperatures, such a value can be integrated in the modified Arrhenius equation (5.23), and in its logarithmic form

$$\ln\left(kC^{-1}\right) = \ln k_o - \frac{E_a}{RT}$$

The new dependent variable kC^{-1} was linearly correlated ($R^2 = 0.99$, $p < 0.05$) with the reciprocal of the absolute temperature, indicating that the modified Arrhenius equation efficiently predicts the oxidation rate from -18°C to 45°C (Figure 5.5.1). This means that the obtained modified Arrhenius equation allows ASLT and extrapolation of rate constant values in the whole temperature range of interest.

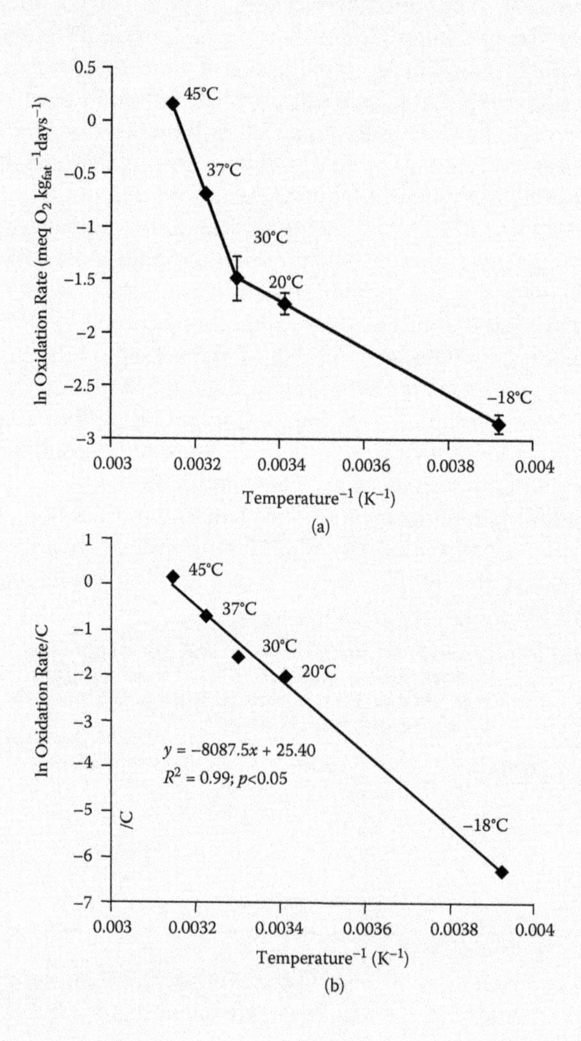

Figure 5.5.1 (a) Arrhenius and (b) modified Arrhenius plots of apparent oxidation rate of biscuits. (From Calligaris et al., 2007a, with permission)

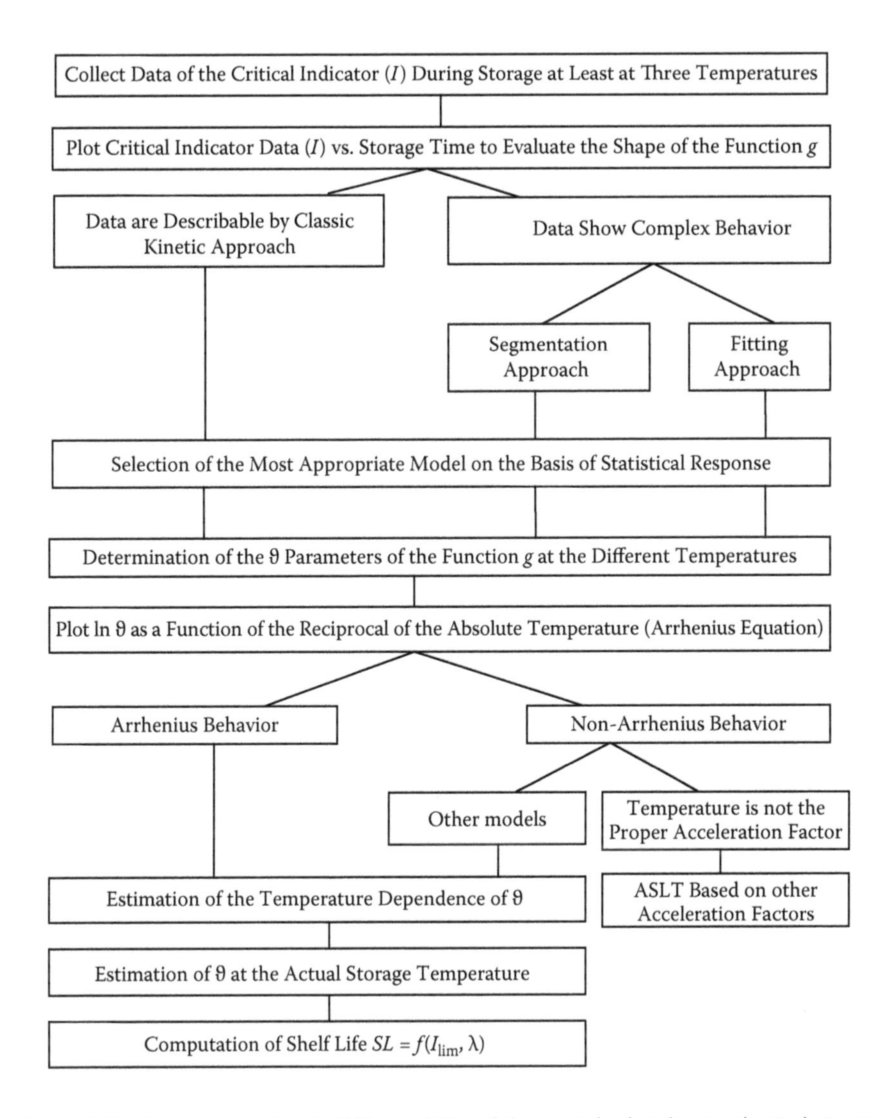

Figure 5.11 Decision tree for shelf life modeling of data acquired under accelerated storage conditions.

recommended when choosing to accelerate a shelf life test by increasing storage temperature. In fact, it may happen that the increase in temperature does not promote any interesting acceleration of the rate of the alterative phenomena (Figure 5.12).

The use of temperature does not allow the time necessary for the shelf life test to be sufficiently decreased. For this reason, such a test would be useless. It would be better to perform the test at the actual storage temperature because testing under accelerated and actual conditions would take similar time.

The existence of different magnitudes of apparent activation energy (Table 5.7) indicates that not all quality depletion phenomena are similarly accelerated by the increase

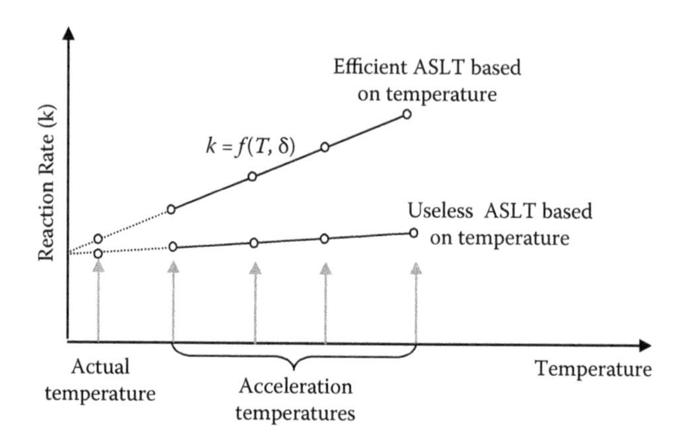

Figure 5.12 Efficient and useless ASLT based on temperature.

in temperature. This means that the time saved when performing the shelf life test under temperature-accelerated conditions may be considerably different. Figure 5.13 shows the percentage of time saved when the temperature of the shelf life test is increased by 10°C as a function of the activation energy of the critical phenomena. It is evident that about 50 kJ/mol activation energy is required to save 50% of the time needed to perform the shelf life test. From a practical point of view, this means that, for critical indicators characterized by E_a values lower than 30–40 kJ/mol, the acceleration obtained by a 10°C increase could be too low to be of interest in the attempt to save time during shelf life assessment. For instance, for a deteriorative phenomenon showing an E_a value around 20 kJ/mol and an expected shelf life of 12 months, a 10°C increase would allow reducing the testing time by 20% to only 10 months. The worst-possible case is when no temperature dependence of the alterative phenomena reaction rate (E_a approaching 0) is observed.

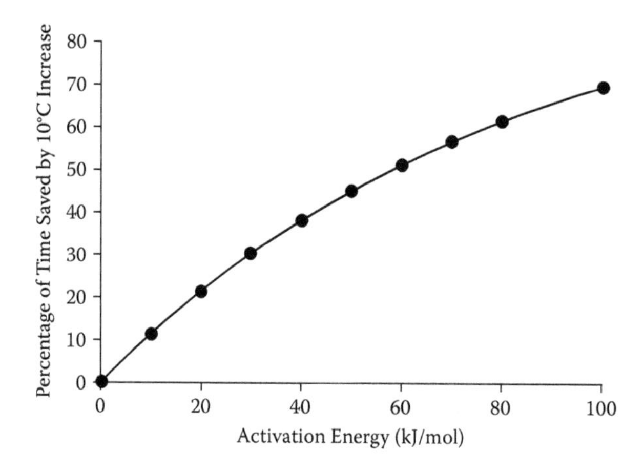

Figure 5.13 Percentage of expected time saved by 10°C temperature increase as a function of activation energy.

The absence of a clear temperature dependence to be exploited in ASLT is typical of reactions in which the number of efficient collisions among reactants does not depend on the molecular mobility within the matrix. This is the case of phenomena including, among reactants, a gas species that can easily reach the reaction site independently on molecular mobility and thus on the temperature of the system. For this reason, some oxidative phenomena, even enzyme catalyzed, for which a molecule is expected to react with gaseous oxygen may present scarce temperature dependence. It is evident that there is no reason to perform an accelerated shelf life test if it does not provide a significant advantage in terms of time saved. In other words, for such products, performing an ASLT based on temperature increase would not help to reach the goal.

A second situation of limited temperature dependence occurs when an environmental factor other than temperature strongly affects the reaction rate, making negligible the contribution of temperature in speeding up the reaction. This is the case of light-induced reactions. For instance, oxidation of some lipids, pigments, and flavors under dark may be slow, even at temperatures higher than the ambient one. By contrast, as illustrated in Figure 5.14, when the product is exposed to the light, oxidation quickly proceeds, showing, in any case, a slight temperature dependence (Kristensen et al., 2001; Manzocco et al., 2008).

Finally, there are foods whose quality depletion is clearly sped up by temperature, but the food itself cannot withstand the temperature increase. For instance, quality depletion of most frozen foods could be strongly accelerated by a temperature increase, but temperatures higher than -7°C are not recommended due to their effect on the physical and physicochemical characteristics of the food (Figure 5.8). This is why ASLT of frozen food is still a matter of concern for both scientists and producers.

When the intrinsic properties of food or the magnitude of the temperature dependence of the alterative phenomena make temperature unsuitable for performing ASLT, other accelerating factors must be identified. Besides temperature, a number of other factors are well known to affect the kinetics of quality depletion. Among them are pressure, relative gas pressure, relative humidity, oxygen concentration,

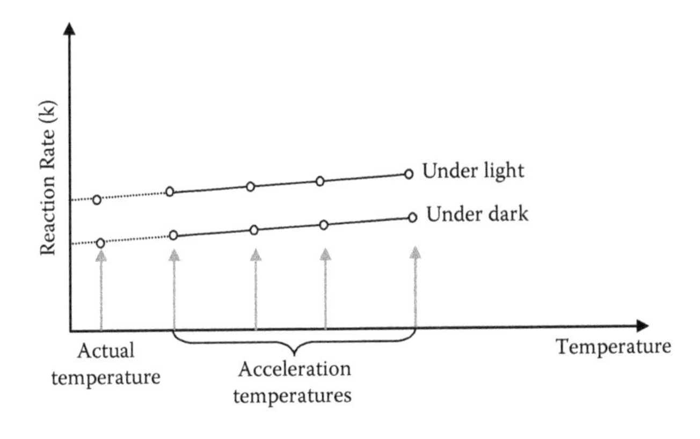

Figure 5.14 Effect of temperature on photooxidation of light-sensitive molecules.

and light intensity. In principle, they all could be of a certain utility. However, to our knowledge, until now little effort has been made in the attempt to rationalize the use of these variables as accelerating factors in shelf life assessment. This is surprising considering that the amount of foods not showing temperature dependence is not negligible. It is our opinion that the lack of use of accelerating factors other than temperature comes from two major difficulties:

- Limited number of exploitable accelerating factors. As stated in Chapter 2, the basic premise to perform an ASLT is that factors accounting for the overall intrinsic product characteristics C_i and packaging properties P_i must be kept constant, while the reaction rate of quality depletion is accelerated by changing the level of a selected environmental variable. Most factors potentially exploitable in ASLT are actually product characteristics and not environmental variables. For instance, relative humidity of the atmosphere in contact with food could be regarded as an environmental factor only for unpacked food or food packed in material without moisture barrier properties. In this case, one could develop an ASLT by conditioning the food under different ERH% and then extrapolating the reaction rate at the actual ERH% condition. Indeed, most foods are packed in moisture barrier materials, and ERH% in the headspace atmosphere of the food is an intrinsic product characteristic. To exploit ERH% for ASLT of a packed food, different sets of samples should be produced, each having increasing ERH% in the headspace atmosphere. This is not impossible and probably may be effective, but is evidently complicated. Similar considerations can be drawn for other "environmental" factors, such as relative gas pressure and oxygen concentration. A peculiar case is that of light, which may have a double role. When food is packed in opaque material, not allowing the product to be reached by environmental light, this variable cannot be used as an accelerating factor in ASLT without modifying the product/packaging combination. As previously observed, its exploitation would require the preparation of peculiar samples packed into a material different from the usual one. However, if food is packed in a see-through material, allowing the product to be exposed to environmental light, this may be an excellent accelerating factor for ASLT. This approach is certainly interesting since it may allow solving the difficult task of predicting shelf life of photosensitive food (Manzocco et al., 2012). In fact, in common practice, most photosensitive foods (e.g., oils, potato chips, beer, soft drinks) are packed in transparent packaging materials and exposed on highly enlightened shelves, dramatically affecting the final shelf life.
- Unavailability of predictive mathematical models. The basic requirement for the successful exploitation of any accelerating factor is the availability of a robust and validated mathematical model that correctly predicts the effect of the selected accelerating factor on the reaction rate leading to quality depletion. From the development of the Arrhenius equation (1901), a wide literature has been produced on the mathematical models predicting the effect of temperature on reaction rate with reference to a number of different situations, including food quality depletion. By contrast, to our knowledge, there are no generally accepted mathematical models predicting the effect of other accelerating factors on quality depletion kinetics. This is an open research field that will certainly require more attention from scientists and researchers in the future.

5.3.2.1 *Decision Tree for Development of ASLT Based on Accelerating Factors Other than Temperature*

Figure 5.15 summarizes the decision tree to be applied in the development of ASLT based on accelerating factors other than temperature. An adequate accelerating factor should be able to decrease the time required for the test significantly. For this reason, as previously reported, temperature may be the right accelerating factor only if activation energy is higher than 50–60 kJ/mol. If this is not the case, a further accelerating factor should be identified among those parameters known to increase the quality depletion rate of the product (e.g., oxygen concentration, light intensity, relative humidity). If none of these factor is able to accelerate the quality depletion of the product, based on present knowledge, there is nothing left to do other than perform the shelf life test at actual storage conditions and wait until its end, even if this may take a long time. However, in the majority of these situations, shelf life dating is not identified based on the rate of food quality decay. Since food quality decay is frequently slow, different criteria, such as product turnover on the shelves, are generally preferred.

When an accelerating factor can be identified, two possible situations arise, depending on whether the factor can be regarded as an environmental factor or a

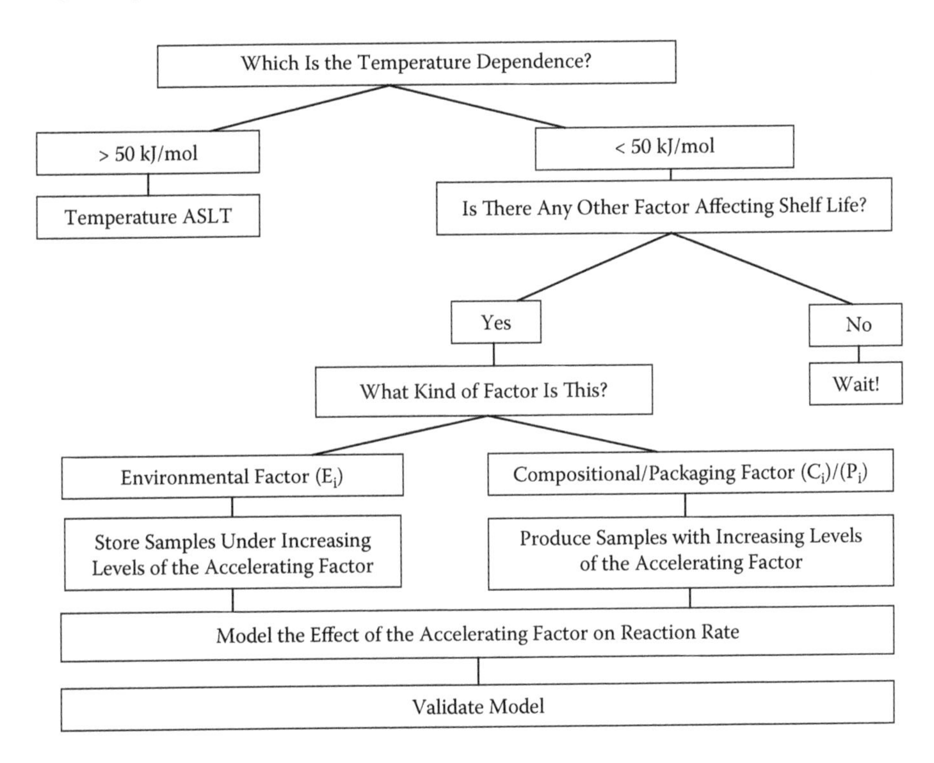

Figure 5.15 Decision tree for the development of ASLT based on accelerating factors other than temperature.

compositional or packaging one. If samples can be stored at different levels of an environmental accelerating factor, the test can start and data relevant to its effect on the quality depletion rate collected. If the accelerating factor cannot be applied as an environmental variable, the product itself must be modified to produce samples conditioned at different levels of the compositional or packaging accelerating factor. Once data relevant to the effect of the accelerating factor on the quality depletion kinetic parameter are available, these data should be elaborated by best fitting to identify an adequate mathematical model. It is evident that this process can be successfully exploited to predict shelf life only after validation of the mathematical model.

5.4 REFERENCES

Arrhenius, S.A. 1901. *Larobok I teoretisk elektrokeni*. Leipzig: Quando and Handel.

Burdurlu, H.S., and F. Karadeniz 2003. Effect of storage on nonenzymatic browning of apple juice concentrates. *Food Chemistry* 80(1): 91–97.

Burdurlu, H.S., Koca, N., and F. Karadeniz. 2006. Degradation of vitamin C in citrus juice concentrates during storage. *Journal of Food Engineering* 74: 211–216.

Calligaris, S., Da Pieve, S., Kravina, G., Manzocco, L., and M.C. Nicoli. 2008. Shelf-life prediction of bread sticks by using oxidation indices: a validation study. *Journal of Food Science* 73(2): E51–E56.

Calligaris, S., Falcone, P., and M. Anese. 2002. Colour changes of tomato purees during storage at freezing temperatures. *Journal of Food Science* 67(6): 2432–2435.

Calligaris, S., Manzocco, L., Conte, L.S., and M.C. Nicoli. 2004. Application of a modified Arrhenius equation for the evaluation of oxidation rate of sunflower oil at subzero temperatures. *Journal of Food Science* 69: 361–366.

Calligaris, S., Manzocco, L., Kravina, G., and M.C. Nicoli. 2007a. Shelf-life modeling of bakery products by using oxidation indexes. *Journal of Agricultural and Food Chemistry* 55: 2004–2009.

Calligaris, S., Manzocco, L., and M.C. Nicoli. 2007b. Modeling the temperature dependence of oxidation rate in water-in-oil emulsions stored at sub-zero temperatures. *Food Chemistry* 101: 1019–1024.

Calligaris, S., Sovrano, S., Manzocco, L., and M.C. Nicoli. 2006. Influence of crystallisation on the oxidative stability of extra virgin olive oil. *Journal of Agricultural and Food Chemistry* 54: 529–535.

Champion, D., Blond, G., and D. Simatos. 1997. Reaction rates at sub-zero temperatures in frozen sucrose solution: a diffusion controlled reaction. *Cryo-Letters* 18: 251–260.

Colakoglu, A. S. 2007. Oxidation kinetics of soybean oil in the presence of monoleion, stearic acid and iron. *Food Chemistry* 101: 724–728.

Corradini, M.G., and M. Peleg. 2007. Shelf life estimation from accelerated data. *Trends in Food Science and Technology* 18: 37–47.

Cunha, L.M., Oliveira, F.A.R., and J.C. Oliveira. 1998. Optimal experimental design for estimating the kinetic parameters of processes described by the Weibull probability distribution function. *Journal of Food Engineering* 37: 175–191.

Das, K.P., and J.E. Kinsella. 1993. Droplet size and coalescence stability of whey protein stabilized milk fat peanut oil emulsion. *Journal of Food Science* 58: 439–444.

Dermesonlouoglou, E.K., Giannakourou, M.C., and P.S. Taoukis. 2007. Kinetic modelling of the degradation of quality of osmo-dehydrofrozen tomatoes during storage. *Food Chemistry* 103: 985–993.

Draper, N.R., and H. Smith. 1998. *Applied regression analysis,* 3rd ed. New York: Wiley.

Dutta, D., Dutta, A., Raychaudhuri, U., and R. Chakraborty. 2006. Rheological characteristics and thermal degradation kinetics of beta-carotene in pumpkin puree. *Journal of Food Engineering* 26: 538–546.

Fennema, O. 1996. Freezing preservation. In *Physical principles of food preservation*, O. Fennema, Ed., 173–215. New York: Dekker.

Frankel, E.N. 2005. Stability methods. In *Lipid oxidation*, E.N. Frankel, Ed., 165–186. Bridgewater, UK: The Oily Press.

Fu, B., and T.P. Labuza. 1993. Shelf-life prediction: theory and application. *Food Control* 4(3): 125–133.

Gambaro, A., Ares, G., and A. Gimenez. 2006. Shelf-life estimation of apple-baby food. *Journal of Sensory Studies* 21(1): 101–111.

Giannakourou, M.C., and P.S. Taoukis. 2003. Kinetic modelling of vitamin C loss in frozen green vegetables under variable storage conditions. *Food Chemistry* 83: 33–41.

Giannou, V., and C. Tzia. 2007. Frozen dough bread: quality behavior during prolonged storage—prediction of final product characteristics. *Journal of Food Engineering* 79: 929–934.

Gimenez, A., Ares, G., and A. Gambaro. 2008. Survival analysis to estimate sensory shelf life using acceptability scores. *Journal of Sensory Studies* 23: 571–582.

Gomes-Alonso, S., Salvador, M.D., and G. Fregapane. 2004. Evolution of the oxidation process in olive oil triacylglycerol under accelerated storage conditions (40–60 degrees C). *Journal of American Oil Chemist's Society* 81: 177–184.

Guerra, S., Lagazio, C., Manzocco, L., Barnabà, M., and R. Cappuccio. 2008. Risks and pitfalls of sensory data analysis for shelf life prediction: data simulation applied to the case of coffee. *LWT—Food Science and Technology* 41:2070–2078.

Gutierrez, F., and J.L. Fernandez. 2002. Determinant parameters and components in the storage of virgin olive oil. Prediction of storage time beyond which the oil is no longer of "extra" quality. *Journal of Agricultural and Food Chemistry* 50: 571–577.

Herrera, M.L., de Leon Gatti, M., and R.W. Hartel. 1999. A kinetic analysis of crystallization of a milk fat model system. *Food Research International* 32(4): 289–298.

Hough, G. 2010. Design of sensory shelf life experiments. In *Sensory shelf-life estimation of food products*, G. Hough, Ed., 63–82. Boca Raton, FL: Taylor & Francis.

Hough, G., Garitta, L., and G. Gómez. 2006 Sensory shelf-life predictions by survival analysis accelerated storage models. *Food Quality and Preference* 17: 468–473.

Houhoula, D.P., and V. Oreopolou. 2004. Predictive study for the extent of deterioration of potato chips during storage. *Journal of Food Engineering* 65: 427–432.

Huang, J., and S. Sathivel. 2008. Thermal and rheological properties and effects of temperature on the viscosity and oxidation rate of unpurified salmon oil. *Journal of Food Engineering* 89: 105–111.

Imai, H., Maeda, T., and H. Shima. 2008. Oxidation of methyl linoleate in oil-in-water micro- and nanoemulsion systems. *Journal of American Oil Chemist's Society* 85: 809–815.

Kaya, A., Tekin, A.R., and D. Oner. 1993. Oxidative stability of sunflower and olive oils: comparison between a modified active oxygen method and long term storage. *LWT—Food Science and Technology* 26: 464–468.

Kristensen, D., Hansen, E., Arndal, A., Tinderup, R.A., and L.H. Skibsted. 2001. Influence of light and temperature on the colour and oxidative stability of processed cheese. *International Dairy Journal* 11: 837–843.

Kristott, J. 2000. Fats and oils. In *The stability and shelf-life of food*, D. Kilcast and P. Subramaniam, Eds., 25–32. Cambridge, UK: Woodhead.

Koca, N., Burdurlu H.S., and F. Karadeniz. 2007. Kinetics of colour changes in dehydrated carrots. *Journal of Food Engineering* 78: 449–455.

Labuza, T.P., and D. Riboh. 1982. Theory and application of Arrhenius kinetics to the prediction of nutrient losses in food. *Food Technology* 10: 66–74.

Lareo, C., Ares, G., Ferrando, L., Lema, P., Gambaro, A., and Soubes, M. 2009. Influence of temperature on shelf life of butterhead lettuce leaves under passive modified atmosphere packaging. *Journal of Food Quality* 32: 240–261.

Mancebo-Campos, V., Fregapane, G., and M. Desamparados-Salvador. 2008. Kinetic study for the development of an accelerated oxidative stability test to estimate virgin olive oil potential shelf life. *European Journal of Lipid Science and Technology* 110: 967–976.

Manzocco, L., Calligaris, S., and M.C. Nicoli. 2006. Modeling bleaching of tomato derivatives at subzero temperatures. *Journal of Agricultural and Food Chemistry* 54: 1302–1308.

Manzocco, L., Calligaris S., and M.C. Nicoli. 2010. Methods for food shelf life determination and prediction. In *Oxidation in foods and beverages and antioxidant application*, E.A. Decker, R.J. Elais, and D.J. McClements, Eds., 196–219. Cambridge, UK: Woodhead.

Manzocco, L., Kravina, G., Calligaris, S., and M.C. Nicoli. 2008. Shelf life modelling of photosensitive food: the case of coloured beverages. *Journal of Agricultural and Food Chemistry* 56: 5158–5164.

Manzocco, L., and M.C. Nicoli. 2007. Modelling the effect of water activity and storage temperature on chemical stability of coffee brews. *Journal of Agricultural and Food Chemistry* 55: 6521–6526.

Manzocco, L., Panozzo, A., and S. Calligaris. 2012. Accelerated shelf life testing by light and temperature exploitation. *Journal of American Oil Chemist's Society.* In press. doi 10.1007/s11746-011-1958-x.

Mizrahi, S. 2000. Accelerated shelf-life tests. In *The stability and shelf-life of foods*, D. Kilcast and P. Subramaniam, Eds., 107–208. Cambridge, UK: Woodhead.

Nelson, K.A., and T.P. Labuza. 1992. Water activity and food polymer science: implications of state of Arrhenius and WLF models in predicting shelf life. *Journal of Food Engineering* 22: 271–289.

Odriozola-Serrano, I., Soliva-Fortuny, R., and O. Martin-Belloso. 2009. Influence of storage temperature on the kinetics of the changes in anthocyanins, vitamin C, and antioxidant capacity in fresh-cut strawberries stored under high-oxygen atmospheres. *Journal of Food Science* 74: C184–C191.

Orlien, V., Risbo, J., Rantanen, H., and L.H. Skibsted. 2006. Temperature-dependence of rate of oxidation of rapeseed oil encapsulated in a glassy matrix. *Food Chemistry* 94: 37–46.

Özilgen, S., and M. Özilgen. 1990. Kinetic model of lipid oxidation in foods. *Journal of Food Science* 55: 498–501.

Parker, R., and S.G. Ring. 1995. A theoretical analysis of diffusion-controlled reactions in frozen solution. *Cryo-Letters* 16: 197–208.

Patel, A.A., Gandhi, H., Sudai, S., and G.R. Patil. 1996. Shelf life modeling of sweetened condensed milk based on kinetics of Maillard browning. *Journal of Food Processing and Preservation* 20(6): 431–451.

Polydera, A.C., Stoforos, N.G., and P.S. Taoukis. 2003. Comparative shelf life study and vitamin C loss kinetics in pasteurised and high pressure processed reconstituted orange juice. *Journal of Food Engineering* 60: 21–29.

Roos, Y.H. 1995. *Phase transition in foods*. San Diego, CA: Academic Press.

Rustom, I.Y.S., Lopez-Leiva, M.M., and B.M. Nair. 1996. UHT-sterilized peanut beverages: kinetics of physicochemical changes during storage and shelf-life prediction modeling. *Journal of Food Science* 61(1): 195–203, 208.

Sanjuan, N., Bon J., Clemente, G., and A. Mulet. 2004. Changes in the quality of dehydrated broccoli florets during storage. *Journal of Food Engineering* 62: 15–21.

Seyhun, N., Sumnu, G., and S. Sahin. 2005. Effects of different starch types on retardation of staling of microwave-baked cakes. *Food and Bioproducts Processing*. 83(C1): 1–5.

Slade, L., and H. Levine. 1991. Beyond water activity: recent advances based on an alternative approach to the assessment of food quality and safety. *Critical Reviews in Food Science and Nutrition* 30: 115–360.

Sothornvit, R., and P. Kiatchanapaibul. 2009. Quality and shelf-life of washed fresh-cut asparagus in modified atmosphere packaging. *LWT—Food Science and Technology* 42: 1484–1490.

Tanner, M.A. 1996. *Tools for statistical inference*. Heidelberg, Germany: Springer-Verlag.

Taoukis, P.S., Labuza, T.P., and I. Saguy. 1997. Kinetics of food deterioration and shelf-life prediction. In *The handbook of food engineering practice*, K.J. Valentas, E. Rotstein, and R.P. Singh, Eds., 363–405. New York: CRC Press.

Terefe, N.S., van Loey, A., and M. Hendrickx. 2004. Modelling the kinetics of enzyme-catalysed reactions in frozen systems: the alkaline phosphatase catalysed hydrolysis of disodium-p-nitrophenyl phosphate. *Innovative Food Science and Emerging Technologies* 5: 335–344.

Tiwari, B.K., O'Donnell, C.O., Muthukumarappan, K., and P.J. Cullen. 2009. Ascorbic acid degradation kinetics of sonicated orange juice during storage and comparison with thermally pasteurised juice. 2009. *LWT—Food Science and Technology* 42: 700–704.

Topuz, A. 2008. A novel approach for color degradation kinetics of paprika as a function of water activity. *LWT—Food Science and Technology* 41: 1672–1667.

Tsironi, T., Dermesonlouoglou, E., Giannakourou, M., and P.S. Taoukis. 2009. Shelf life of frozen shrimps at variable temperature. *LWT—Food Science and Technology* 42: 664–671.

Vaikousi, H., Koutsoumanis, K., and C.G. Biliaderis. 2009. Kinetic modelling of non-enzymatic browning in honey and diluted honey systems subjected to isothermal and dynamic heating protocols. *Journal of Food Engineering* 95: 541–550.

Van Boekel, M.A.J.S. 2008. Kinetic modelling of food quality: a critical review. *Comprehensive Reviews in Food Science and Food Safety* 7: 144–158.

Van Boekel, M.A.J.S. 2009. *Kinetic modelling of reactions in foods*. Boca Raton, FL: Taylor & Francis.

Varela, P., Salvador A., and S.M. Fiszman. 2005. Shelf-life estimation of Fuji apples: sensory characteristics and consumer acceptability. *Postharvest Biology and Technology* 38: 18–24.

Villanueva, N.D.M., and M.A. Trinidade. 2010. Estimating sensory shelf life of chocolate and carrot cupcakes using acceptance tests. *Journal of Sensory Studies*. 25: 260–279.

Waage, P., and C.M. Guldberg. 1864. Studies concerning affinity. *Forhandlinger: Videnskabs-Selskabet i Christinia*. Norwegian Academy of Science and Letters. 35. Oslo, Norway.

Waterman, K.C., and R.C. Adami. 2005. Accelerating testing: prediction of chemical stability of pharmaceuticals. *International Journal of Pharmaceutics* 293: 101–125.

Williams, M.L., Landel, R.F., and J.D. Ferry. 1955. The temperature dependence of relaxation mechanisms in amorphous polymers and other glass-forming liquids. *Journal of American Oil Chemist's Society* 77: 3701–3706.

CHAPTER 6

Modeling Shelf Life Using Microbial Indicators

Judith Kreyenschmidt and Rolf Ibald

CONTENTS

6.1 INTRODUCTION

The most important criteria for food products from a consumer and industry perspective are high safety and quality. From the industry point of view, long shelf life times are also important for economic reasons. These aspects are a considerable and specific challenge for the food industry with regard to perishable products. The prediction of shelf life of the most perishable products is based on microbial indicators since most of these products provide good bacterial growth conditions due to their nutritional composition and physical properties.

Since the 1980s, a large volume of scientific papers and book chapters have been published regarding the microbial growth of certain spoilage bacteria and about predictive microbiology and other related topics. Also, a range of software tools has been developed to predict the growth of certain microorganisms in food. Even if there is much information available about food models, shelf life models are not in common use. In particular, the use of shelf life models for practical applications (e.g., for process control) is expandable. In most companies, shelf life determination is still based on experience and on sensory parameters.

Consequently, this chapter supports the users of the industry or other organizations in determining shelf life and in using predictive food models. This chapter does not focus on precise explanations of different mathematical models; rather, it shows how to transform shelf life determination and modeling issues from a complex and difficult task to an easy, scientifically based process.

The first part contains a short overview about relevant microorganisms in food. Following this, typical growth curves of bacteria are qualitatively explained, as is their dependency on the main influencing factors. In Section 6.3, a brief overview of predictive food modeling and the main parameters are provided as well as the most commonly used mathematical models to calculate bacterial growth. Furthermore, it is explained how shelf life studies should be conducted, which factors should be kept in mind, and how to adapt the models to realistic scenarios. Specifically, this means how to identify and specify all parameters for particular products within specific environments. Topics specified here are existing databases, the realization of experiments, the use of specific mathematical algorithms to fit measured data, and the use of already existing software tools.

The final section demonstrates how models can be used to optimize processes within the food industry, what benefits shelf life models will bring to the industry and to consumers, and how those models will support the application of new devices, like radio-frequency identification (RFID) systems or time–temperature indicators (TTIs).

6.2 MICROBIAL SPOILAGE INDICATORS FOR SHELF LIFE PREDICTION

Generally, microorganisms in food can be categorized in three main groups:

- Technologically useful bacteria. These microorganisms are added to food to improve the technological and sensory characteristics. Selected organisms are added to food and food products for preservation purposes, such as to prolong shelf life due to the release of antimicrobial substances (e.g., bacterocins or organic acids). Typical examples of useful bacteria in this group are *Leuconostoc mesenteroides* spp. *cremoris* and *Leuconostoc lactis*.

- Foodborne pathogens. These microorganisms can cause foodborne diseases due to the production of toxins or the infection of living cells. Pathogens can be part of the flora of the product itself or can be transferred to the food during processing, storage, or transport. Typical pathogenic microorganisms are *Campylobacter jejuni, Salmonella enteritidis, Yersinia enterocolitica,* and *Listeria monocytogenes.*
- Spoilage microorganisms. Spoilage microorganisms lead to changes in the color, odor, flavor, and texture of the product. These sensory changes are caused by the growth and metabolism of the microorganisms (e.g., *Pseudomonas* spp., *Brochothrix thermosphacta, Photobacterium phosphoreum,* lactic acid bacteria).

Spoilage and pathogenic bacteria are relevant to determine shelf life since quality changes and the safety of a product have to be considered. The growth of spoilage and pathogenic bacteria is closely connected due to interactions between the bacteria and their sensitivity to technological methods. Until now, most of the models used to predict microbial growth in food have concentrated on foodborne pathogens, which on its own is a complex subject. A detailed overview regarding pathogenic bacteria in food and predictive modeling in the field of food safety was given by Zwietering and Nauta (2007). The focus of this section is on quality aspects and thus on spoilage organisms.

Microbiological spoilage is a complex issue since a huge variety of spoilage organisms exist. The spoilage flora of perishable products is comprised on the one hand of the natural flora from the product itself, which is mainly influenced by its environmental conditions. For example, the natural microflora of fish is influenced by the habitat, the geography (tropical or arctic water), and the fishing season (Gram and Huss, 1996), whereas the natural flora of fruits and vegetable is influenced by the soil, air, irrigation water, insects, and animals (Lund and Snowdon, 2000). On the other hand, microorganisms will be transferred to food by cross contamination during or after processing via surfaces, machines, humans, or the surrounding atmosphere.

Thus, due to the different sources of contamination, each product has its own characteristic initial microflora, which normally consists of several different microorganisms. Furthermore, the microbiological flora is not static, so the initial flora differs from the flora at the end of shelf life. As an example, the flora of fresh pork meat mainly consists of *Pseudomonas* spp., *Lactobacillus* spp., *Brochothrix thermosphacta*, *Enterobacteriaceae*, *Shewanella putrefaciens*, *Aeromonas* spp., *Carnobacterium* spp., and *Leuconostoc* spp. at the beginning of storage. During storage, some specific bacteria grow much faster than others and dominate the growth, so that the flora is mainly dominated by *Pseudomonas* spp. at the end of the shelf life (Borch et al., 1996; Olsson et al., 2003; Nychas et al., 2008; Bruckner, 2010).

That only a small number of bacteria varieties is responsible for the loss of quality can be explained based on food characteristics, environmental conditions, and interactions between the microorganisms (see Section 6.2.2). Those organisms mainly responsible for spoilage are known as specific spoilage organisms (SSOs) (Gram and Dillard, 2002) or ephemeral spoilage organisms (ESOs) (Nychas et al., 2007). Table 6.1 gives an overview of typical spoilage organisms of different food products from animal origin and fruits and vegetables. Generally, it has to be mentioned that

Table 6.1 Specific Spoilage Organisms in Food of Animal Origin, Fruits, and Vegetables

Product	Atmosphere	Main Spoilage Organisms	Reference
Poultry	Aerobe packaging	*Pseudomonas* spp.	Bruckner, 2010; Gospavic et al., 2008
Pork	Aerobe packaging	*Pseudomonas* spp.	Bruckner, 2010; Gill and Newton, 1977; Coates et al., 1995; Pooni and Mead, 1984
Pork	Modified atmosphere	*Brochothrix thermosphacta, Pseudomonas* spp.	Liu et al., 2006
Beef	Aerobic packaging	*Pseudomonas* spp., *Rahnella* spp., *Carnobacterium divergens*	Ercolini et al., 2006
Beef	Modified atmosphere	*Pseudomonas* spp., *Lactobacillus sakei, Rahnella* spp.	Ercolini et al., 2006
Ground meat	Aerobic packaging	*Pseudomonas* spp.	Koutsoumanis et al., 2006; Jay et al., 2003
Cooked cured ham	Modified atmosphere	*Lactic acid bacteria*	Mataragas et al., 2006; Vasilopoulos et al., 2008; Slongo et al., 2009; Kreyenschmidt et al., 2010b
Cooked sliced ham	Vacuum	*Lactobacillus sakei, Lactobacillus curvatus, Leuconostoc*	Hu et al., 2009
Fresh fish	Aerobic ice packaging	*Pseudomonas* spp.	Gram and Huss, 1996
Cod	Vacuum, modified atmosphere	*Shewanella putrefaciens, Photobacterium phosphoreum*	Dalgaard et al., 1993; Dalgaard, 1995
Salmon	Aerobe	*Pseudomonas* spp., *S. putrefaciens*	Hozbor et al., 2006
Marinated herring	Glass jars	*Lactobacillus alimentarius*	Lyhs et al., 2001
Mung bean sprouts and cut chicory and endive	Aerobe and controlled atmosphere (different O_2 and CO_2 concentration)	*Enterobacteriaceae*, *Pseudomonas* spp.	Bennik et al., 1998; Vankerschaver et al., 1996
Shredded carrots	Modified atmosphere	Lactic acid bacteria	Kakiomenou et al., 1996; Carlin et al., 1990
Ready-to-eat vegetable salad	Aerobe	*Pseudomonas spp.*	Marchetti et al., 1992
Ready-to-eat vegetable salad	Modified atmosphere	*Pseudomonas* spp., *Enterobacteriaceae*, lactic acid bacteria	Rudi et al., 2002
Peeled oranges and grapefruit	Aerobe	*Enterobacter agglomerans*, *Pseudomonas* spp., *Cryptococcus albidus*, *Rhodotorula glutinis*, *Saccharomyces cerevisiae*	Pao and Petracek, 1997

the spoilage process of meat and fish products is mainly caused by bacteria, while the spoilage of fresh unprocessed fruits is mostly induced by yeast and molds and caused by the low pH value. Also, the natural outer protective epidermis of several fruits and vegetables can be attacked more easily by spoilage fungi because of the diversity and greater amount of extracellular depolymerases. In several fresh-cut products like shredded vegetables and sliced fruits, the spoilage is often dominated by bacteria, especially when these products are packed under modified atmosphere (Barth et al., 2009). Typical spoilage bacteria are *Pseudomonas* spp., lactic acid bacteria, *Enterobacteriaceae*. Yeast species like *Cryptococcus laurenti, S. exiguus, Saccharomyces cerevisiae,* and *S. dairensis* also are often present in salads.

6.2.1 Microbial Growth Characteristics

Microorganisms in food can lead to sensory changes in two different ways: If the growth of microorganisms passes a certain value, high numbers of SSOs will lead to turbidity in liquids or sticky slimy surfaces and changes in color, typical for meat and fish products and fresh-cut fruits and vegetables (Sapers et al., 2001; Fehlhaber et al., 2005; Barth et al., 2009). In addition and much more important are sensory changes caused by enzymatic reactions, which will lead to the degradation of proteins, carbohydrates, and fats and thus to the accumulation of metabolites.

Typical enzymes involved in the spoilage process of fruits and vegetables are pectinases, cellulases, cutinases, proteinases, and lipoxygenases. Pectinases, for example, degrade pectic substances and lead to tissue maceration (Chen, 2002). Relevant extracellular enzymes involved in the spoilage process of meat are lipases, proteases, and oxidoreductases (Fehlhaber et al., 2005). Typical end products are sulfur compounds (hydrogen sulfide, methyl mercaptan, methanethiol); esters (methyl esters, ketones, ethyl esters); and alcohols (methanol, ethanol) (Borch et al., 1996; Nychas et al., 1988, 2008). Which end products will be produced during the spoilage process of food mainly depends on the composition of the microflora, nutrient availability, and environmental conditions. Several studies have shown that typical off-odors, caused by some of the spoilage end products mentioned in meat, become apparent if the SSOs attain a value of 7–7.5 $\log_{10}$ cfu (colony-forming units)/g (Jay et al., 2005; McMeekin and Ross, 1996; Bruckner, 2010). The shelf life can then be defined as the time from the beginning of storage until the point at which the SSOs or ESOs reach a certain maximum level (Borch et al., 1996; Huis in't Veld, 1996; Nychas et al., 2007).

For a better understanding of the spoilage process and the determination of shelf life, the basic principles of microbiological growth are illustrated. Microbial growth is characterized by a sigmoid curve. The so-called growth curve can be divided into four main sections: lag phase, exponential phase, stationary phase, and death phase. For the determination of shelf life, only the first three sections are relevant. In the death phase, the number of viable cells decreases exponentially, and the product is already spoiled. Consequently, only the three important phases for shelf life determination are described (Figure 6.1).

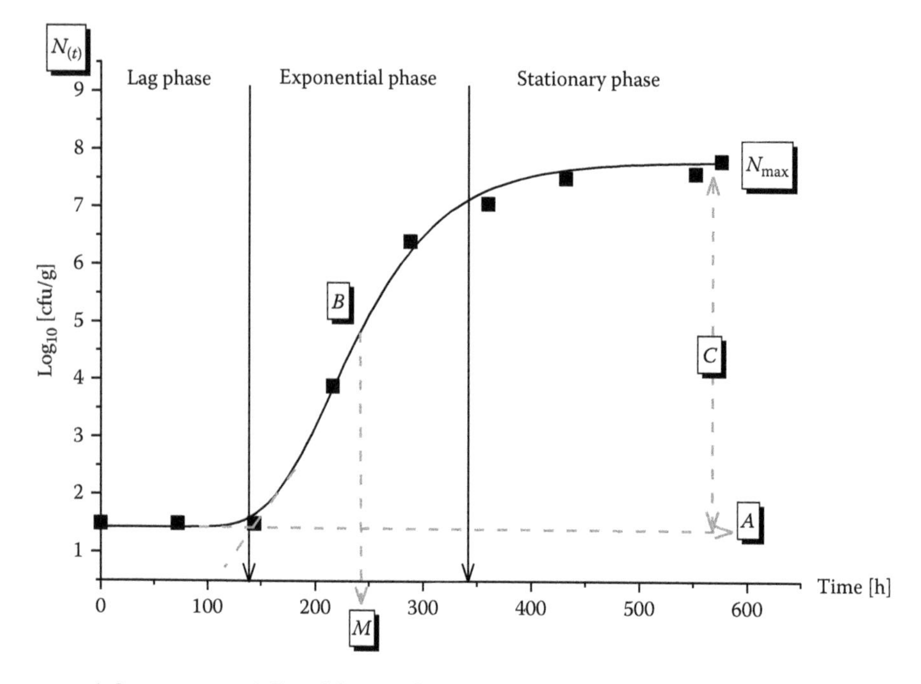

Figure 6.1 Microbiological growth curve and relevant parameters for the determination of shelf life.

6.2.1.1 *Lag Phase (Adjustment Phase)*

During the lag phase, the microorganisms adapt to the environmental conditions. At this time, the cells are actively metabolizing; they are synthesizing ribosomal RNA and subsequently enzymes, and the cells may be growing in mass and volume in preparation for cell division. There are different definitions of the lag phase. From the aspect of one cell, Pirt (1975) defined the lag phase as the time from inoculation of one cell until the time of the first cell division. Zwietering (2005) defined the lag time as the time "obtained by extrapolating the tangent of the growth curve at the time of fastest growth back to the inoculums level." There are further definitions, but it is generally agreed that within this phase there is no apparent increase in the population. An important factor in the lag phase is the initial concentration N_0 of the microorganisms since high initial counts decrease the duration of shelf life.

For the determination of shelf life, the duration of the lag phase is important. The length of this phase depends on a variety of factors, such as the nutritional and physical composition of the food, environmental factors, initial viable count, physical state of the cells related to the history, and time necessary for the synthesis of essential coenzymes. The prolonging of the length of the lag phase is an important aspect in the development of methods to increase shelf life (Buchanan and Cygnarowicz, 1990). During the lag phase, no sensory changes of color, odor, or texture in food appear that are caused by the metabolism of bacteria.

6.2.1.2 *Exponential Phase*

During the exponential phase, the growth of microorganisms is characterized by regular cell division, so the number of bacteria increases exponentially. The growth rate μ reaches its maximum value during this phase when the slope of the growth curve is at its steepest. The length of generation time, the time required for the cell population to double, and the value of the growth rate are influenced by the composition of the food itself and by the processing factors, the environmental factors, and the genetic characteristics of the organisms. Short generation times of the most relevant spoilage organisms lead to reduced shelf life. During the exponential phase, typical sensory changes in color, odor, and texture appear due to metabolic activity and growth of the microorganisms. The end of shelf life for most perishable products is reached in this phase and is expressed as the defined acceptability limit (Borch et al., 1996; Huis in't Veld, 1996; Nychas et al., 2007).

6.2.1.3 *Stationary Phase*

The rate of cell division decreases during the stationary phase. Some cells are dying, some are still slowly dividing, and some are ceasing to grow. This effect can be caused by the exhaustion of essential nutrients and space as well as by the accumulation of inhibitory end products. The cell count reaches the maximum value during this time N_{max}. The maximum number of the most relevant spoilage organisms in perishable products like fresh meat typically varies between 8 and 10 $\log_{10}$ cfu/g.

For the determination of shelf life based on microbiological growth, the most relevant parameters are the initial count, the length of the lag phase, the maximum specific growth rate, and the maximum population.

6.2.2 Factors Influencing Microbial Growth

As stated, the growth behavior of spoilage organisms in food is influenced by several factors and by the type of organisms that are mostly responsible for the spoilage process. For food products, the most relevant influencing factors can be divided into four groups (Mossel, 1971; Heard, 1999):

- Intrinsic factors are specific to the food itself (e.g., water activity, nutrient content, pH value, structure, redox potential).
- Extrinsic factors are environmental parameters of the food (e.g., storage temperature, gas atmosphere, humidity).

- Processing factors are physical or chemical treatments during the processing of the food (e.g., heat treatment).
- Implicit factors describe synergistic or antagonistic influences that are related to the primary selection of organisms (e.g., specific rates of growth, symbiosis).

Since most of these factors influence each other, the relevance of each factor within the spoilage process of different products is also different and difficult to establish exactly. But, there are some factors that are of vital importance for all products. These include the water activity (a_w), the pH, the nutrient content, the initial bacterial count, the temperature, and the gas atmosphere:

Water activity: The water activity a_w, which is the ratio of the vapor pressure of the food to the vapor pressure of pure water, has an important influence on the growth of the microorganisms since an aqueous phase is a main requirement for their metabolic activity. The optimum a_w for most microorganisms is in the range 0.995 to 0.980 (Christian, 2000). Lowering the a_w (e.g., due to salt, sugar, drying, or curing) inhibits microbiological growth or results in an increased lag phase and a decreased growth rate and thus leads to a longer shelf life. Since microorganisms have different demands with regard to water activity, the composition of the microflora is influenced by the prevalent a_w. With a few exceptions, the bacterial growth continues up to $a_w \geq 0.88$ (Mossel, 1971; Christian, 2000); below this value, fungi are predominant within the flora (Jay et al., 2005). Thus, perishable products, like fresh meat or fish, with an $a_w \geq 0.98$, provide optimal conditions for bacterial growth; hence, these products are characterized by short shelf lifetimes (Bruckner, 2010).

pH: Similar to water activity, the pH can also significantly influence the growth of spoilage organisms, which is caused by changes in the enzyme activity and the transport of nutrients in cells. The optimal pH of most microorganisms is around 7 (6.5–7.5), with the minimum pH value above 4.5 and the maximum below 8 (Lund and Eklund, 2000). These values cannot be seen as precise boundaries since the growth of the microorganisms is always influenced by several factors. In particular, the type of the acid in the food has an effect on the pH minimum (Lund and Eklund, 2000; Jay et al., 2005).

Nutrient content: Microorganisms in foods require water, a source of energy (e.g., amino acids, sugars), and sources of nitrogen, vitamins, and minerals. Generally, perishable foods provide an adequate nutrient content for the growth of typical spoilage organisms. But, the availability and composition of the nutrient content also influence the growth of microorganisms. Further, some products contain special components with antimicrobial activity. Examples of such components are lysozyme in egg whites or lactoferrin in cow's milk. Such substances can inhibit or slow the growth of some microorganisms (Jay et al., 2005).

Initial bacterial count: The initial amount of bacteria in the food has a significant influence on the shelf life and on the growth of the microorganisms. High initial counts are mainly caused by inadequate hygienic conditions during processing and lead to a decrease in shelf life as shown in Figure 6.2c.

Temperature: Temperature conditions during processing, transport, and storage are one of the most important factors that influence the spoilage process. Increasing temperature conditions lead to a decrease in the length of the lag phase and generation times and thus to an increase in the growth rate. They also influence protein

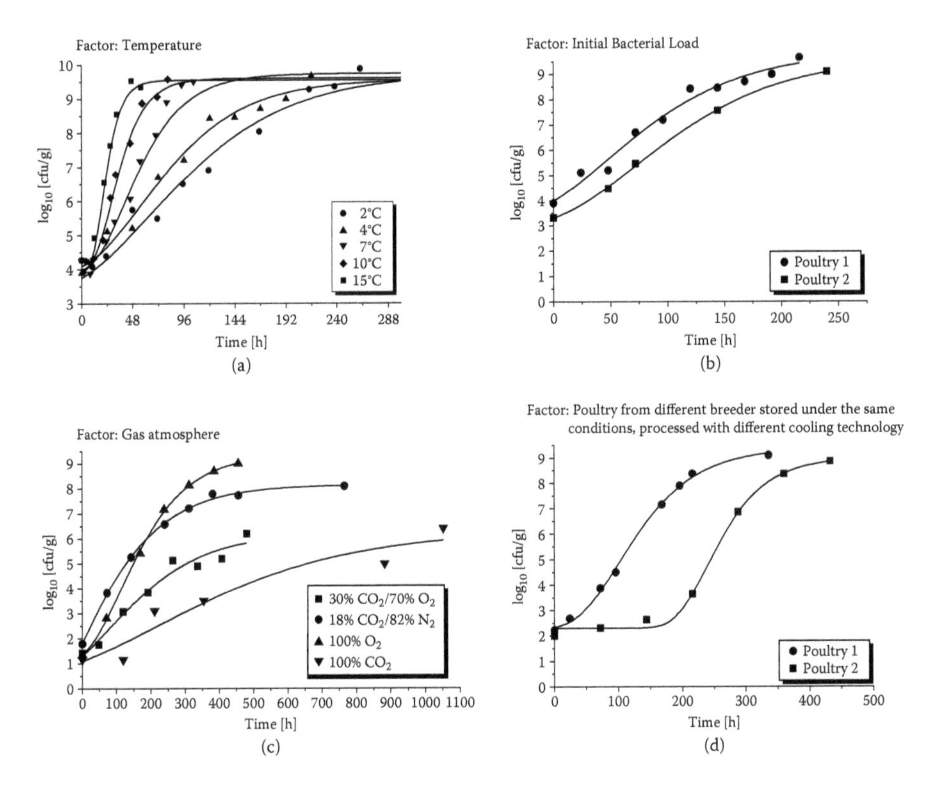

Figure 6.2 Factors influencing microbial growth exemplified for the growth of *Pseudomonas* spp. in poultry: (a) storage temperature (from Bruckner, S. 2010. Predictive shelf life model for the improvement of quality management in meat chains. PhD thesis, Rheinische Friedrich-Wilhelms-Universität Bonn, Germany); (b) initial bacterial load; (c) gas atmospheric composition; (d) different product origin.

synthesis, enzyme activity, solute uptake, and thus the length of shelf life (Herbert and Sutherland, 2000) (Figure 6.2a). As an example, the shelf life of fresh pork is decreased by nearly 25% if the storage temperature is 4°C instead of 2°C (Bruckner, 2010). Also, short temperature abuses can significantly increase the growth rate of SSOs. Shelf life reductions of up to 20% were observed for fresh poultry and fresh pork when the periods with out-of-range temperatures were less than 5% of the total storage time and the temperature shift was 11°C (Bruckner, 2010). Thus, for the determination of shelf life, the temperature dependency of the most relevant spoilage organisms is of vital importance. The activity of various types of spoilage organisms in foods also depends on the temperature since most microorganisms only grow within specific temperature ranges and because they have special cardinal temperatures at which their growth is maximal or minimal. For perishable products, the psychrotrophilic and psychrotrophic bacteria are the most relevant ones since they are also able to grow at low temperatures. Typical microorganisms in these classes are *Pseudomonas* species, *Flavobacterium,* and *Vibrio/Aeromonas* (Dainty and Mackey, 1992; Borch et al., 1996; Huis in't Veld, 1996).

Atmosphere: The growth and survival rates of microorganisms and the shelf life can be influenced by the gas composition in the packaging. For example, by modifying the atmosphere inside a package of fresh meat, the shelf life can be increased up to 50% (Borch et al., 1996). The most commonly used gases in the food industry are oxygen, carbon dioxide, and nitrogen. The three gases have different functions and effects on microbial growth:

- *Oxygen:* Within a defined range, the growth of microorganisms is dependent on the oxygen demand of the organisms. As a consequence, the growth of anaerobic microorganisms can be inhibited by the presence of oxygen, while the growth of microaerophilic and aerobic organisms can be influenced by oxygen, depending on the demand of the specific organism. Due to the different oxygen demands of the microorganisms, the composition of the microflora changes depending on the oxygen level (Phillips, 1996; Molin, 2000). In fresh aerobe-packed poultry, the spoilage flora is dominated by *Pseudomonas* spp. (Figure 6.1c); by reducing the oxygen concentration and increasing the CO_2 level in the package, the spoilage process is mainly dominated by *Brochothrix thermosphacta* and lactic acid bacteria.

  Besides the effect on microbial growth, oxygen helps to maintain the bright red color of certain meats by maintaining the myoglobin in its oxygenated form (oxymyoglobin).
- *Carbon dioxide:* This gas is used because of its bacteriostatic effect, especially against aerobic, Gram-negative, psychrotrophic bacteria. The effect of carbon dioxide is explained by the inhibition of enzyme synthesis, the decrease of the intracellular pH, alteration of cell membrane functions (nutrient uptake, absorption), and changes to the physical–chemical properties of proteins (Farber, 1991; Molin, 2000). These effects lead to an increase in the length of the lag phase and a decrease of the growth rate. Generally, the effect of carbon dioxide depends on the carbon dioxide concentration and on the temperature, as increasing temperature levels reduce the effect of carbon dioxide.
- *Nitrogen:* This inert gas is used as a packaging filler to prevent the package from collapsing because of its high insolubility in water and fat in comparison to carbon dioxide. In addition, nitrogen reduces the growth rate of aerobic bacteria (Farber, 1991; Phillips, 1996; Sandhya, 2009).

Other gases that are currently studied for their possible use in food packaging are noble gases such as argon, helium, xenon, and neon.

6.3 SHELF LIFE DETERMINATION USING MICROBIAL INDICATORS

Different factors and their interactions make the determination of shelf life a complex task. Another difficulty is given by the fact that the acceptability limits for most of the relevant spoilage organisms are not legally defined. As a consequence, shelf life determination in the industry is often subjectively defined. It is based on experiences and on a decision-making process that mostly uses sensory parameters.

A more efficient and effective method to determine the shelf life is to use "predictive microbiology" and "predictive food models" (McMeekin and Olley, 1986;

McMeekin et al., 1993; Baird-Parker and Kilsby, 1987). These models can be applied to calculate the growth of the relevant spoilage organism under certain conditions. With the help of these models and the detected quantity of microorganisms, it is possible to draw conclusions about the existing quality status of the product and its shelf life.

The idea of predicting the growth of a certain spoilage organism to study freshness loss is not new. Scott (1937) presented this concept more than 70 years ago, based on investigations of microbial growth in meat. During recent years, numerous papers have been published that focused on the development of mathematical models to predict the growth of the SSOs in food products, like the work of Gibson et al. (1987); Zwietering et al. (1990); Baranyi et al. (1993); Baranyi and Roberts (1994); Buchanan et al. (1997); and McKellar (1997). Other publications mainly focused on different kinds of food and microorganisms and on different factors that influence microbial growth (Cayré et al., 2003; Koutsoumanis et al., 2006; Mataragas et al., 2006; Gospavic et al., 2008; Kreyenschmidt et al., 2010b). Additional papers have been published about databases and software solutions, the validation of predictive models, new rapid analytical approaches, and the effects of interactions between microorganisms (see also Sections 6.3.7 and 6.4). Reviews of the current situation in predictive microbiology were given by McDonald and Sun (1999); Ross and McMeekin (1995); Whiting (1995); and Ross et al. (1999). Due to the large increases in computer power since the 1990s, the concepts of predictive microbiology have become increasingly realistic for practical use within the food industry. Thus, several software systems and associated databases have been developed to predict the growth of a range of relevant microorganisms under specific conditions (see Section 6.3.7).

Mathematical approaches have been formulated to predict the growth of the spoilage processes based on microbial growth characteristics. The models can be classified in a number of ways (e.g., empirical vs. mechanistic models). Another classification characterizes the models into primary, secondary, and tertiary models, as proposed by Whiting and Buchanan (1993). Since this classification can be applied to most models, it is described next.

Primary models: These are used to describe the growth of the SSOs as a function of time. Such models allow the calculation of relevant growth parameters, like the maximum growth rate, the duration of lag phase, the initial count, and the maximum count. Typical models are the modified Gompertz model, the Baranyi and Roberts model, the three-phase linear model, and the modified Logistic model (Gibson et al., 1987; Zwietering et al., 1990; Baranyi et al., 1993; Baranyi and Roberts, 1994; Buchanan et al., 1997). The modified Logistic model and the modified Gompertz model are comparable in applicability and accuracy and are often used to describe bacterial growth. Both include four parameters to describe the sigmoid growth curve. Next to these models, the Baranyi and Roberts model is also often used (Baranyi and Roberts, 1994). The main difference between this model and other sigmoid curves like Gompertz, Logistic, and so on is that the midphase is close to linear. Another difference is the use of a special parameter, which is a characteristic for the physiological state of the cells in the lag phase. The simplest primary model is the three-phase linear model introduced by Buchanan et al. (1997). The growth rate in the lag phase is assumed to equal zero, while in the exponential phase there is a linear increase of the log value of the bacterial population. It is also assumed that the growth rate is zero during the stationary phase. An overview of typical models is given in Table 6.2.

Secondary models: These models are used to describe the parameters of the primary models that depend on factors like temperature, pH, or water activity (Labuza and Riboh, 1982; Labuza et al., 1992; Bratchell et al., 1990; Ross and McMeekin, 1994; Whiting, 1995; McDonald and Sun, 1999). Most of the developed models calculate a growth factor of a SSO depending on the temperature since this is the main influencing factor for food spoilage (Labuza and Fu, 1993; Zwietering et al., 1991). The most relevant models to describe temperature dependence are the Arrhenius

Table 6.2 Overview of Different Primary and Secondary Models

Primary Model		**Source**
Modified Logistic model	$N(t) = A + \frac{C}{1 + e^{-B\cdot(t-M)}}$	Gibson et al., 1987
Modified Gompertz model	$N(t) = A + C \cdot e^{-e^{-B\cdot(t-M)}}$	Gibson et al., 1987
Baranyi and Roberts model	$N(t) = A + \mu_{max} \circ a(t) + \ln\left[1 + \frac{\exp(\mu_{max} \circ a(t)) - 1}{\exp(N_{max} - A}\right]$ $a(t) = \frac{q(t)}{1 + q(t)}$	Baranyi and Roberts, 1994
Parameter	$N(t)$ = microbial count at any time, A = lower asymptotic line of the growth curve resp. initial bacterial count; N_{max} = maximum population level; C = N_{max} - A, difference between upper and lower asymptotic line; B = relative maximum growth rate at time M; M = time when maximum growth rate is obtained (reversal point) and t is time, m_{max} is maximum specific growth rate; $a(t)$ = adjustment function, which takes into account the lag phase during which the population adapts to the new environment; $q(t)$ = physiological state of the cells at time t	
Secondary Models		
Arrhenius equation	$\ln(B) = \ln F - \left(\frac{E_a}{R \cdot T}\right)$	Arrhenius, 1889
Square root equation	$\sqrt{B} = b * (T - T_0)$	Ratkowsky et al., 1982
Parameter	B = relative growth rate at time M; F = preexponential factor; E_a = activation energy for bacterial growth; R = gas constant; T = absolute temperature; b = is the slope of the regression line; T_0 = theoretical minimum temperature for cell growth	

model, the square root model, or the response surface equation (Arrhenius, 1889; Belehradek, 1930; Ratkowsky et al., 1982; Labuza et al., 1992; Whiting, 1995).

The often-used Arrhenius equation normally covers a limited temperature range from 0°C to 25°C to estimate the microbial growth. Modified versions of the Arrhenius model have been developed to achieve better fits with temperature above or under the aforementioned values (Davey, 1989; Buchanan, 1993; McDonald and Sun, 1999). However, some modified forms have been reported as complex and cumbersome in use (Buchanan, 1993). But, successful applications of the simple Arrhenius model are available for many different meat types and meat products (Giannuzzi et al., 1998; Kreyenschmidt, 2003; Moore and Sheldon, 2003; Mataragas et al., 2006; Kreyenschmidt et al., 2010b).

Besides the Arrhenius model, the square root model is often used. It was first introduced in this field by Ratkowsky et al. (1982) to describe the influence of temperature on growth rate. It is based on a simple linear relationship between the square root of the growth rate and the influencing parameter (e.g., the temperature). Response surface models use standard regression techniques that contain polynomial terms (Whiting, 1995).

There are several other primary and secondary models that have been developed, and already existing ones have been modified or expanded in scope. A detailed overview about existing secondary models was given by Ross and Dalgaard (2004).

Tertiary models: Primary and secondary models can be combined to allow the determination of shelf life under certain environmental conditions. The incorporation of primary and secondary models in "user-friendly" computer software to provide a complete prediction tool is called *tertiary modeling* (Buchanan, 1993; Whiting, 1995). Currently, there is a fairly wide range of different software solutions available to predict the growth of the various spoilage organisms under certain conditions, which are presented in Section 6.3.7.

In the scientific community, these models and software systems like ComBase-PMP (http://www.combase.cc) or the Seafood Spoilage and Safety Predictor (SSP; http://sssp.dtuaqua.dk) are usually used to predict the growth of microorganisms in food. However, the applications of predictive models in the industry differ significantly between countries. The use of predictive food models has been introduced successfully in parts of the Australian food industry. This may well be attributed to the research work and research projects undertaken at the University of Tasmania (McMeekin, 2007). In other countries like Germany, the use of such models is marginal. However, the interest in models to predict the growth of pathogenic bacteria in the European meat industry is increasing due to the Commission Regulation (EC) No. 2073/2005, annex 2.

The absence of the application and implementation of shelf life models in different industries may be caused by a number of different factors, such as a lack of knowledge and experience regarding the application of shelf life models or a lack of confidence in the general approximations of the bacterial growth curves with mathematical models. Many companies do not see the benefits of using predictive shelf life models. Especially, they do not recognize financial benefits due to more accurate predictions of the remaining shelf life during the whole food supply chain, which could also support the optimization of the storage management and waste reduction.

An overview is now given of how the shelf life of a certain product can be defined to a high degree of precision using scientifically based procedures.

Generally, shelf life models can be used for different purposes:

1. For scientific analysis
2. To determine the shelf life of new products or after a substitution of ingredients at the product development stage
3. To predict the shelf life of already-existing products or as a supporting tool for the hazard analysis and critical control point (HACCP) systems
4. To calculate the remaining shelf life during different stages of the chain (e.g., in the incoming or outgoing inspections) and to optimize storage systems

The following explanations mainly concentrate on purposes 2, 3, and 4.

6.3.1 Definition of Product and Process Characteristics

For the successful determination of shelf life, access to relevant literature databases, microbiological testing facilities, and standardized methods for microbiological tests are necessary. Before starting any experiments, the first step should be the exact definition of the aim of the investigation. This can be the determination of a shelf life under constant conditions, the investigation of the effect of one or more parameters on shelf life, or just the validation of an already-existing model (see also Chapter 9).

Furthermore, the precise formulation of the product and its business and production processes are also important requirements for the determination of shelf life. The identification of the parameters that have an influence on the shelf life is of high relevance (see also Section 6.2.2). This should lead to a detailed description of the product and of the process characteristics, and it should include:

- Nutrient composition and ingredients of the product (fat content, salt concentration, lactate concentration)
- Physicochemical properties (pH, a_w)
- Processing technologies and relevant parameters (cooling technology)
- Microbial initial range of the SSO
- Legal microbial requirements (spoilage and pathogenic bacteria) and product requirements
- Environmental conditions during and after processing (packaging material, atmosphere composition, active packaging, temperature conditions)

Since some factors that influence the growth behavior of the SSOs and thus the shelf life can change during the life cycle of the product, it is advisable to define ranges and time spans in which these factors may vary.

Figure 6.3 shows a sample document exemplifying such a process overview for fresh poultry fillets. This document can be modified appropriately depending on the specific product and the processing steps that are undertaken.

Product:	Origin:		Product weight:		
Nutrient:					
Fat content:	Salt content:		Lactate:		
	Step in the Chain				
	Slaughter	Processing	Packaging	Transport	Retailer
Step duration [h] (from – to – mean)[a]					
pH after each step in the chain					
Water activity at T = °C					
Temperature range (from – to – mean)					
Cooling method					
Packaging (material, gas atmosphere, size of the boxes)					
Microbiological load of the spoilage bacteria (lowest, highest, and mean value)					

Figure 6.3 Example of process description for fresh poultry fillet. [a] Specifies the usual step duration in terms of shortest time, longest time, and mean time.

6.3.2 Literature Review

Many papers have been published about the spoilage process of different perishable products. Especially in the fresh meat and fish area, numerous studies have been published about the growth of typical spoilage organisms like *Pseudomonas* spp. (Neumeyer et al., 1997; Koutsoumanis et al., 2006; Gospavic et al., 2008; Bruckner, 2010). In addition, a number of databases and software solutions are already available to predict the growth of selected organisms under different conditions and which can be used to predict shelf life. These research papers, databases, and software solutions provide detailed information about the spoilage processes in perishable products, about the relevant factors influencing the shelf life of certain products, and about the characteristics of the relevant spoilage organisms (see Section 6.3.7). However, it has to be mentioned that due to the complexity of the spoilage processes, microorganisms can behave differently in one specific product type, even if the environmental conditions are the same. Figure 6.2d shows the growth of *Pseudomonas* spp. in fresh poultry (both unsalted) from different breeders and processed with different cooling technologies at 4°C. Such differences can be caused by different processing technologies (e.g., the cooling technology, which can affect the physiological state of the organisms), by differences in the feeding or breeding of the poultry, or by drip losses.

Therefore, literature and the extensive information within databases may help to answer the questions posed in Section 6.3.1.

6.3.3 Shelf Life Studies

The procedures used for microbiological investigations and for the sampling approaches are not explained in this section. Several publications and guidelines describe the approach required in detail, such as International Organization for Standardization (ISO) standards or the laboratory methods published by the Food and Drug Administration (FDA). This section gives a short overview of how experiments have to be planned and conducted to carry out shelf life studies.

For the exact definition of shelf life, a good experimental design is a key requirement. Mistakes made at the stage of the experimental design or during data collection cannot be corrected afterwards and result in a poor-quality prediction. Even if the experiment is planned and conducted in a nearly optimal way, the data will inevitably be affected by errors caused by the fact that (a) living systems are being investigated that have normal biological variations; and (b) current methods to investigate microorganisms are not accurate enough (Van Boekel and Zwietering, 2007). Therefore, since it is not possible to create an optimal experiment, the aim should be to get as close as possible to a virtually perfect design.

The experimental setup and the investigated parameters depend on the product and process characteristics defined in Sections 6.3.1 and 6.3.2. If not enough information is available about the SSOs under the relevant conditions, a preliminary test should be conducted. This test should examine the most relevant organisms and sensory parameters under defined environmental conditions. In the preliminary test, the relevant microorganisms should be investigated directly just after product manufacture. Later, three to five further samples should be investigated before the end of the shelf life. This approach will allow the identification of the most relevant spoilage organisms. The growth of the SSO will then be investigated in detail.

For the determination of shelf life, the growth of the main spoilage organisms will be investigated as a function of time, under defined environmental conditions that are typical for the real process. Sensory analysis should be conducted in parallel to test the most relevant sensory changes (to this purpose, see also Chapters 5 and 7).

All other parameters that may be relevant for the product and the determination of the shelf life, like fat oxidation or histamine formation, should also be analyzed during the storage test.

If the shelf life needs only to be evaluated for a defined temperature and no prediction of the temperature dependency is required, the temperature has to be chosen among those of normal practice. If the temperature dependency is to be modeled, the experiments should be conducted using those temperature ranges that have been defined to describe the real scenarios. If the temperature conditions are usually between 2°C and 20°C, the experiments should be conducted using at least five different isothermal conditions. If the impacts of new ingredients or a new gas composition should be tested, the tests should be conducted in parallel, and the samples should be taken from the same batch. An example of this approach is given in Case Study 9.1 of Chapter 9. This is particularly relevant if the initial contamination of the food shows significant variations, as is often found with fresh meat, fish products, and fresh-cut fruits and vegetables. The number of samples that should be taken depends on the product characteristics, the variations of the initial quality or quality parameters, and the environmental conditions. To be able to develop good functions to predict usable growth rates of the SSOs in food, 10 sampling points during storage are recommended. The frequency of sampling is important and depends on the expected shelf life and on the storage temperature. For instance, for products with an expected shelf life of 4 days at 2°C, samples have to be taken more often than once a day, whereas products that are less perishable need to be tested only once a day or even less. If the storage temperature will be increased, samples have to be taken more often since the shelf life usually decreases with increasing temperatures. The data should be recorded intensively, in particular at the ends of the lag phases, at the ends of the exponential phases, and during the stationary phases to define N_{max}.

Another important point for a good determination of the shelf life is the number of experiments that will be repeated to be able to investigate the variances because the microflora and the initial number of microorganisms of perishable products vary between batches and during the production processes. Generally, the larger the number of repeated experiments, the more precise is the prediction of shelf life. Since the sampling is time and cost effective, it is recommended to use a statistical sampling plan. If the influence of more than one factor on shelf life has to be analyzed (e.g., the effects of different gas mixtures, composed of CO_2, N_2, and O_2) or if the analysis of only a single factor is complex, it is advisable that the experiment should be set up using an appropriate statistical design to reduce the number of experiments and to obtain the maximum information with the least effort. An example of this approach is given in Case Study 9.8 in Chapter 9. For detailed information, refer to the work of Antony (2003); Rasch (2004); Van Boekel and Zwietering (2007); and Montgomery (2009).

6.3.4 Shelf Life Modeling

After processing the data, the mathematical models need to be adjusted. In the following, a brief explanation is given about the adaption of the models to realistic scenarios; this means how to obtain the values for all parameters of the models for specific products within specific environments. Here, the procedure to calculate the bacterial growth and to describe the main parameters are introduced using the modified Logistic model and the modified Gompertz function as the primary model, and how the temperature dependency for microbiological growth can be modeled by using the Arrhenius model as a secondary model is presented. It is also explained in Section 6.3.6 how both models can be combined to predict shelf life under dynamic temperature conditions.

As mentioned in Section 6.2.1, the growth of most SSOs on meat can be described by sigmoid functions. To use these functions, all necessary parameters have to be determined. In the case of the modified Logistic or modified Gompertz function, these parameters are *A, M, B,* and *C* (see Equations 6.1 and 6.2).

For the modified Logistic function,

$$N(t) = A + \frac{C}{1 + e^{-B(t-M)}} \tag{6.1}$$

For the modified Gompertz function,

$$N(t) = A + C \cdot e^{-e^{-B(t-M)}} \tag{6.2}$$

where

- A = Number of bacteria at the beginning (in case of the Logistic function, it is approximately the number of bacteria at the beginning)
- M = Inflection point
- B = Growth rate at the inflection point
- C = Difference between the maximum and minimum number of bacteria

Using these equations, the absolute maximal growth rate μ_{max} can be estimated by the modified Gompertz function:

$$\mu_{\max} = \frac{BC}{4} \tag{6.3}$$

or it can be estimated by using the modified Logistic model:

$$\mu_{\max} = \frac{BC}{e} \tag{6.4}$$

In both cases, the absolute maximal growth rate m_{max} is the maximal slope of the curves. Its values are evaluated from the first derivative of the growth curves *N(t)* by calculating the value for $t = M$, which is the inflection point.

$$\mu_{\max} = \frac{d}{dt} N(t_{=M}) \tag{6.5}$$

By defining the lag phase as the time, where the tangent at the inflection point crosses the line that describes the minimal bacterial count *A*, the duration of the lag phase (*LPD*) computes with the mentioned parameters using the modified Gompertz model to

$$LPD = M - \frac{1}{B} \tag{6.6}$$

and it computes using the Logistic model to

$$LPD = M - \frac{2}{B} \tag{6.7}$$

In the following, the procedure of shelf life determination by modeling the bacterial growth is described as exemplified with the help of the modified Gompertz and modified Logistic function without loss of generality to use other functions (see Table 6.2).

In Equations 6.1 and 6.2, parameters *M* and *B* are functions of different intrinsic and environmental factors like the temperature, but parameters *C* and *A* may be assumed to be constant (Kreyenschmidt et al., 2010b; Bruckner, 2010). Therefore, the techniques to estimate *C* and *A* can be as elementary as calculating an arithmetic mean of all collected data or as complicated as using the most sophisticated "curve-fitting" software, or *C* and *A* may be estimated "just by looking" at the data. But, what really matters for the estimation of *C* and *A* is the fact that the number of cell counts *A* is highly dependent on the history of the food (e.g., whether it was processed under good hygienic conditions). What also matters is the knowledge that there is in general a simple relationship between factors *A* and the shelf life: The higher the *A*, the smaller is the shelf life if the product and all processes remain constant. So, the distributions of all measured numbers *A* should be investigated and analyzed to define *A* for the model by considering the purposes for which the shelf life modeling is done, such as whether it is to provide estimations for the shelf life from a company point of view or to calculate an accurate value for the estimated minimal bacterial count from a scientific point of view.

The same considerations should be taken into account for the estimation of *C*, regarding what finally is the difference between the number of bacteria in the measured stationary phases and the initial count of the bacteria *A*.

To obtain appropriate values for each temperature level for the growth rates *B* and the inflection points *M*, special software applications that read the measured data as input (e.g., Origin, R, or even Excel) will help to fit the modified Gompertz function or the modified Logistic function to the measured data. Depending on the software used, the user needs to have some programming and mathematical skills. If the fitting parameters are given by self-programmed algorithms or just by the software used, indicators for the quality of the fits can be the coefficient of determination R^2, the root mean square error *RMSE,* or other factors like a bias factor or an accuracy factor, which often is used to describe the fits of the secondary models (Mataragas et al., 2006).

Good-fitting parameters can also be obtained by doing "trial-and-error plots" without using sophisticated software and without requiring strong mathematical or programming skills. The following step-by-step process describes how to fit mathematical functions to the measured data of the SSOs as functions of time for each measured temperature with the help of trial-and-error plots. A simple Excel sheet helps to execute these steps and to display the results:

1. The decadic logarithms of the measured data (e.g., colony-forming units per gram [cfu/g]) should be calculated for each measuring point. From a mathematical perspective, all other logarithms would also be viable, but most authors use the decadic logarithms. If there are more measuring points that express the same (e.g., within an experiment that investigates more samples of the same batch of meat within the same environmental conditions), an average value of all measurements should be the value for further analysis. In Figure 6.4, column C shows the average of the measured number of colony-forming units per gram at each measuring time. Column D shows the decadic logarithm of the measured number of colony-forming units per gram.
2. Since each measurement and each experiment has its specific measurement accuracy, this specific accuracy should be taken into account. One possibility to do so is by adding quantified statistical and systematical errors to each average of the

Microsoft Excel - ManualFit.xls

Datei Bearbeiten Ansicht Einfügen Format Extras Daten Fenster ? Frage hier eingeben

SUMME =$C$3+($C$5/(1+EXP(-1*$C$6*(B9-$C$4))))

	p1	p2	p3	p4
A	0.90	0.90	1.00	1.00
M	110.00	120.00	130.00	140.00
C	7.00	8.00	8.00	7.50
B	0.04	0.04	0.05	0.06

t[h]	average of messured data MD(t)	log(MD)	estimated error [%]	p1	p2	p3	p4
1		=$C$3+($C$5/(1+EXP(-1*$C$6*(B9-$C$4))))					1.00
2	8.3	0.92	90	0.99	0.97	1.01	1.00
3				1.00	0.97	1.01	1.00
4				1.00	0.98	1.01	1.00
5				1.00	0.98	1.02	1.00
6				1.01	0.98	1.02	1.00
7				1.01	0.99	1.02	1.00
8	4.2	0.63	70	1.02	0.99	1.02	1.00
9				1.02	0.99	1.02	1.00
10				1.03	1.00	1.02	1.00
11				1.03	1.00	1.02	1.00
12				1.04	1.01	1.02	1.00
13				1.04	1.01	1.02	1.00
14				1.05	1.01	1.02	1.00
15				1.05	1.02	1.03	1.00
16				1.06	1.02	1.03	1.00
17				1.07	1.03	1.03	1.00
18				1.07	1.03	1.03	1.00
19	9.1	0.96	90	1.08	1.04	1.03	1.01
20				1.09	1.04	1.03	1.01
21				1.09	1.05	1.03	1.01
22				1.10	1.06	1.04	1.01
23				1.11	1.06	1.04	1.01
24				1.12	1.07	1.04	1.01

DataFit

Bearbeiten

Figure 6.4 Spreadsheet to create manual fits to measured data. See also text and Figure 6.5.

measuring points. Figure 6.4 shows such an estimated error in column E. When the data are displayed within coordinate systems, the errors are usually displayed as error bars, as shown in Figure 6.5.

3. Next to the measured data, theoretical values need to be calculated using the formulas of the fitting functions (e.g., using the modified Gompertz function or the modified Logistic function) so that the measured data as well as the theoretical values may be plotted within the same coordinate system. In Figure 6.4, the formula of the Gompertz function is entered in each cell in columns F, G, H, and I below p1 to p4 in such a way that the parameters *A, M, B,* and *C* are used as variables. The same procedure may be done with the modified Logistic function. The specific values of the parameters *A, M, B,* and *C* may be entered in the table of the Excel spreadsheet shown in Figure 6.4.
4. When the preparation of the data is done, the decadic logarithm of the measured data in column D as well as the calculated values in columns F, G, H, and I can be plotted versus time in column B. By changing the parameters p1–p4 in the table in a trial-and-error manner, a set of parameters can be found so that the function fits the data sufficiently.

The graph in Figure 6.5 shows an example of decadic logarithms of measured data of SSOs and four different curves that were fitted manually to the data as described. It can be seen that the dotted lines do not fit the data, but the solid lines do.

Even though fitted curves obtained using sophisticated software or algorithms usually match better to the data, the use of manually fitted curves may be more suitable for the desired purposes. Generally, those who manually iterate the parameters based on his or her knowledge may well choose a curve that is closer to reality and more suitable for the specified needs. Sophisticated software and algorithms

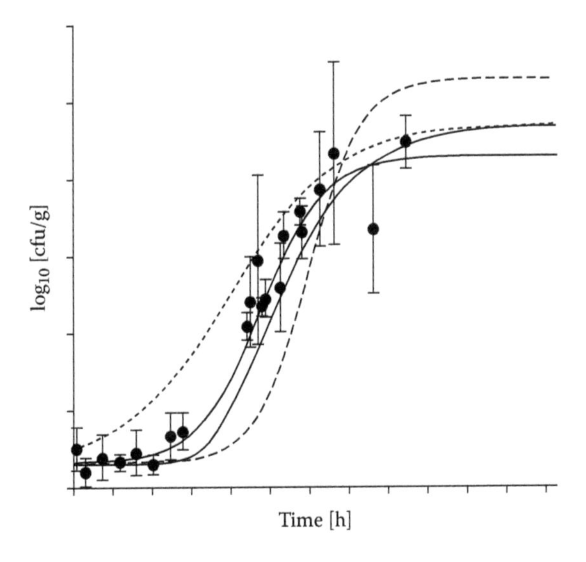

Figure 6.5 Gompertz functions are fitted to data that show the growth of specific spoilage organisms.

would always choose those parameters A, M, B, or C that result in the best R^2 or in other optimized decision-making factors that can be used by the algorithms without requiring additional knowledge of experts.

However, once all parameters A, M, B, and C are evaluated for the measured temperatures, the growth of bacteria can be predicted for each measured temperature using Equations 6.1 and 6.2 of the models described.

When the growth of bacteria shall be modeled for other temperatures than the measured one, it is necessary to know the dependence of parameters B and M on temperature. Therefore, parameter B can be described using the Arrhenius model in most cases, and it can be directly estimated from the so-called Arrhenius plot. Here, the Arrhenius equation will be rewritten in logarithmic units to obtain a linear equation with the transformed variable $1/T$ (see Equations 6.8 and 6.9).

The Arrhenius equation:

$$B = F \cdot e^{-\frac{E_a}{RT}} \tag{6.8}$$

The Arrhenius in logarithmic units:

$$\ln B = \ln F - \frac{E_a}{R} \cdot \frac{1}{T} \tag{6.9}$$

where

B = Growth rate
T = Temperature in degrees of Kelvin
E_a = Activation energy
R = Gas constant
F = Constant

Equation 6.9 shows a linear equation if ln B is plotted versus $1/T$. The slope of this Arrhenius plot is $-E_a/R$, and the axes intercept ln F. To be able to fit a straight line to the data to obtain ln F and E_a/R, the temperature has to be plotted in Kelvin, as shown in Figure 6.6.

Consequently, the apparent activation energy E_a is the most important parameter to describe the bacterial growth rate as a function of temperature.

The function to describe the inflection point M as a function of temperature may also be estimated using an Arrhenius plot. Depending on specific products, processing technologies, and environmental situations, linear equations or a square root model can also be useful to model the data. Table 6.2 shows some selected functions to use as primary- or secondary-level models. For each specific case, the most appropriate function needs to be chosen, for example, with a best-practice test and by considering the specific purposes envisaged for the model.

Figure 6.7 summarizes one possible set of mathematical functions to describe the bacterial growth $N(t)_T$ for different, but constant, temperatures. The bacterial growth on specific foods may be calculated with these functions for given

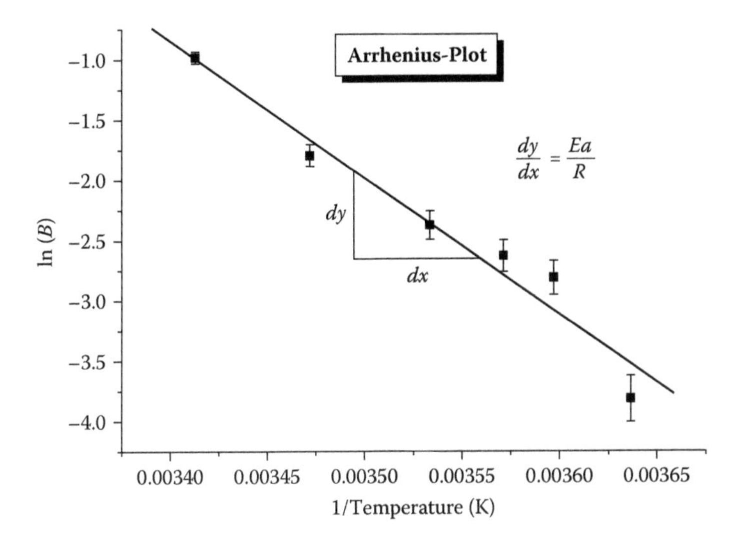

Figure 6.6 The Arrhenius plot and relevant parameters. (From Kreyenschmidt, J. 2003. Modellierung des Frischeverlustes von Fleisch sowie des Entfärbeprozesses von Temperatur-Zeit-Integratoren zur Festlegung von Anforderungsprofilen für die produktbegleitende Temperaturüberwachung. PhD thesis, Rheinische Friedrich-Wilhelms-Universität Bonn, Germany. Bergen/Dumme, Germany: Agrimedia.)

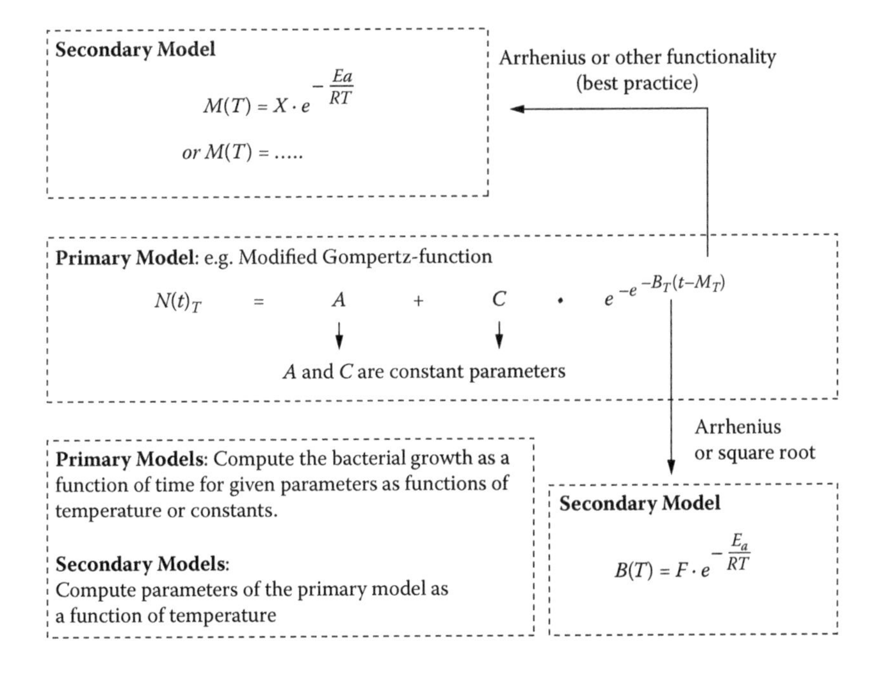

Figure 6.7 Examples of formulas that can be used as primary and secondary models.

constant temperatures using the modified Gompertz function. Alternatively, the bacterial growth could be modeled using the modified Logistic model or another suitable model if the parameters were estimated using this model to fit the measured data.

Hence, since those models are phenomenological descriptions of bacterial growth, the calculated shelf life should be validated with additional measurements when the initial measurements have poor statistical values. It is imperative to mention that these functions are only reliable within the given temperature ranges, where the application of Arrhenius equation is compatible. Even though these temperature ranges cover most of the temperatures that the products are exposed to during processing, transport, and storage, the models should be validated at least for the highest and lowest temperatures. As long as the acceptable limits of bacteria numbers of the foods are known, the shelf life for given temperatures can be calculated using Equation 6.10 or 6.11.

For calculation of the shelf life SL using the modified Gompertz model,

$$SL = -\frac{\ln(-\ln(\frac{N_t - A}{C}))}{B} + M \tag{6.10}$$

For calculation of the shelf life SL using the modified Logistic model,

$$SL = -\frac{\ln(\frac{C}{N_t - A} - 1)}{B} + M \tag{6.11}$$

N_t in both equations is the assumed acceptability value for the number of bacteria. The other parameters were described previously. Some suggestions regarding how to estimate N_t, are given in Section 6.3.5. How to estimate the growth of bacteria under dynamic temperature conditions is described in Section 6.3.6.

6.3.5 Determination of the Acceptability Limit

To determine shelf life of perishable foods, the acceptability limit of the SSO is of high relevance. For most pathogenic bacteria, regulations exist specifying the maximum level allowed in the respective products (e.g., EC 2073/2005). Such criteria unfortunately seldom exist for spoilage bacteria. In the literature, several acceptability limits have been published for perishable products like poultry and cooked ham; for fruits and vegetables, fewer limits of SSOs are published. Although these specific values exist in the literature for several products of the meat industry, these limits vary and are not always useful for all practices. For cooked ham, for example, some authors defined the spoilage level as when the lactic acid bacteria reach a value of 7 $\log_{10}$ cfu/g (Ruiz-Capillas et al., 2007;

Slongo et al., 2009; Kreyenschmidt et al., 2010b). Other authors recommended a level of 6 $\log_{10}$ cfu/g (Vasilopoulos et al., 2008) or 8.3 $\log_{10}$ cfu/g (Mataragas et al., 2006). One reason for the different spoilage levels relevant to the same kind of product can be explained by the presence of different species. In the specific case mentioned, different species of lactic acid bacteria with different metabolic activities could be present. Also, the presence of other spoilage organisms can lead to these differences. The different flora may be caused by different cooking methods, by differing product compositions, by temperature conditions, or by hygiene conditions (Samelis et al., 2000). Generally, for most chilled products of animal origin, for fresh-cut fruits and vegetables, the spoilage levels differ between 6 and 8 $\log_{10}$ cfu/g. Thus, the definition of an appropriate acceptability limit is of vital importance for the prediction of the exact shelf life.

If no regulations exist, the acceptability limit can be defined based on the literature, on company internal decisions, or on industry group recommendations. If generally agreed values are not available, a level has to be defined for the product. For perishable products, the acceptability limit for the SSO is defined at that time when the product is spoiled from a sensory point of view (see also Chapter 3). Depending on the product, a margin should be included. This margin depends specifically on the variation of the initial status of the product. High variations in the initial microbial load lead to higher margins than small variations. In addition, customer–supplier relationships and agreements can influence the magnitude of this margin. Thus for constant environmental conditions, the shelf life can be defined based on the acceptability limits.

6.3.6 Expansion of Shelf Life Models for the Consideration of Dynamic Temperature Conditions

To predict microbiological growth rates under dynamic temperature conditions, a combination of the primary and secondary models can also be used (Labuza and Fu, 1993; Dalgaard et al., 1997, 2002; Koutsoumanis and Nychas, 2000; Koutsoumanis, 2001).

A method is shown next that iterates Equations 6.1 or 6.2 for short time intervals in which the temperatures are assumed to be constant (Kreyenschmidt et al., 2010b; Bruckner, 2010). All dynamic temperature conditions can be assumed to consist of a finite number of short time intervals with constant temperatures within each interval. Depending on the size of each interval, this approach will be more or less precise since the reality would consist of an infinite number of intervals with an infinitesimal size each.

The considerations are shown using the modified Gompertz function and the modified Logistic function as representative functions for the primary models. The parameters are obtained using secondary-level models.

Assuming that the growth of bacteria would be calculated for each interval using the modified Gompertz equation or the modified Logistic equation, by strictly using the formulas as introduced in Section 6.3.4, a number of errors would occur:

- First, the initial number N_0 of bacteria, respectively, A, is obviously wrong for all intervals other than the first one. In all intervals, when the bacteria are not in the lag phase, its count will be enhanced.
- Second, the assumed lag phase due to the sigmoid curves of the modified Gompertz as well as of the modified logistic function would also be wrong for all intervals other than the first one. This can easily be explained by considering the fact that in reality there are a number of intervals within the exponential growth phase of the bacteria, without any lag phase at all. But, the strict use of the formulas, to estimate the development within those intervals, would start with a lag phase.
- Third, the number of bacteria would decrease from one interval to another if the temperature decreases from one interval to another. In reality, only the growth rate would decrease and not the number of the bacteria.

These errors need to be eliminated for a phenomenological but acceptable description of the growth of the bacteria under dynamic temperature conditions.

It seems to be an obvious process to start remedying these errors using corrected values for $A = A(i - 1)$, which would be the number of bacteria at the end of the previous interval if the calculated interval is not the initial one. This would correct two of three errors mentioned, but it would not correct the second error; instead, another error occurs: This procedure would shift the maximum bacterial count $(A + C)$ since A would change at the beginning of each interval. But, the maximum number of bacteria $(A + C)$ and the minimum number of bacteria (A) are assumed to be constant for all temperatures (see Section 6.3.4).

Another attempt to rectify the errors was validated for many conditions (Kreyenschmidt et al., 2010b; Bruckner, 2010). All the mentioned errors can be corrected by fitting the inflection point for each interval so that a smooth curve will be achieved. This leads to the following procedure:

The growth within the first interval is calculated as it is described in Section 6.3.4, and the growth rate within all other intervals is computed by calculating an accurate growth rate B, for example, using the Arrhenius equation. The inflection point M, for each interval, will then be obtained by fitting the inflection point (shifting it to the left or to the right) so that the bacterial count of the previous interval $A(i - 1)$ equals the number at the next interval $A(i)$. This requirement is given by Equation 6.12, where $N(t)$ is written as $N(i - 1)$, and this expresses the number of bacteria at the end of the previous interval. All other parameters are explained in Section 6.3.4.

$$M = \frac{\ln\left(-\ln\left(\frac{N(i-1)-A}{C}\right)\right)}{B_{(T)}} + t \tag{6.12}$$

The same procedure can be used with the modified logistic function, which leads to Equation 6.13:

$$M = \frac{\ln\left(\frac{C}{N(i-1)-A} - 1\right)}{B_{(T)}} + t \tag{6.13}$$

So, the prediction of bacterial growth as a function of dynamic temperatures can be simulated as growth curves within small intervals with constant temperatures each. To do so, routines within software development environments should be programmed. Finally, it has to be emphasized that all models are phenomenological models, so validations are mandatory.

An example is shown next based on data published by Bruckner (2010). The product type was fresh pork loin that was purchased from a local butcher (Bruckner, 2010). In this study, *Pseudomonas* spp. was identified as the SSO.

By using these data, in a first step, the growth rate B of *Pseudomonas* spp. on fresh pork loin, the inflection point M, the initial bacterial count A, as well as the number of bacteria in the saturation phase $A + C$ have to be determined as described in Section 6.3.4 (see also Figure 6.5). In the study of Bruckner (2010), the software Origin 8G (OriginLab Corporation, Northampton, MA, USA) was used to fit the measured growth curves to Gompertz functions to obtain the parameters A, M, B, and C for each constant temperature. Once these parameters are known for the representative temperatures, in this study 2°C, 4°C, 7°C, 10°C, and 15°C, the parameters for the secondary models have to be determined. Therefore, the logarithms of the growth rates B have been plotted versus the reciprocals of the temperatures in units of Kelvin to obtain the growth rates for this product as a function of temperature according to Equation 6.8 (see Figure 6.8).

Figure 6.8 shows an Arrhenius plot. The data are fitted to a linear function with the software Origin. Within the picture, the fitting parameters are the necessary factors $\ln F = 26.00229$ and $E_a/R = 8334.49041$ of Equation 6.9. Since the reciprocals of the temperatures (the X axis of Figure 6.8) are very small numbers, which are displayed with five decimals, and since the measured values $1/T$ are far away from zero (zero would correlate to an infinite high temperature) and the slope of the curve is very high (>8,000), it is necessary to calculate with many decimals for $\ln F$ and E_a/R to obtain reasonable results. In this study, different functionalities have been tested to obtain the inflection point M as a function of temperature. It could be proven that an easy linear function of the temperature matched sufficiently. Figure 6.9 shows the data, a regression line, and its regression parameters.

Next to the parameters to calculate M as a function of temperature, also R^2 is displayed to show the goodness of the fit. So, all parameters A and C as well as the functionality $M(T)$ and $B(T)$ are defined to calculate iteratively the bacterial growth for dynamic temperature conditions. As mentioned, the bacterial count will be calculated with the Gompertz function for each interval with a constant temperature using A and C as constants and $B(T)$ and $M(T)$ as functions of the temperature. To match the bacterial counts at the ends of the intervals with the bacterial counts at

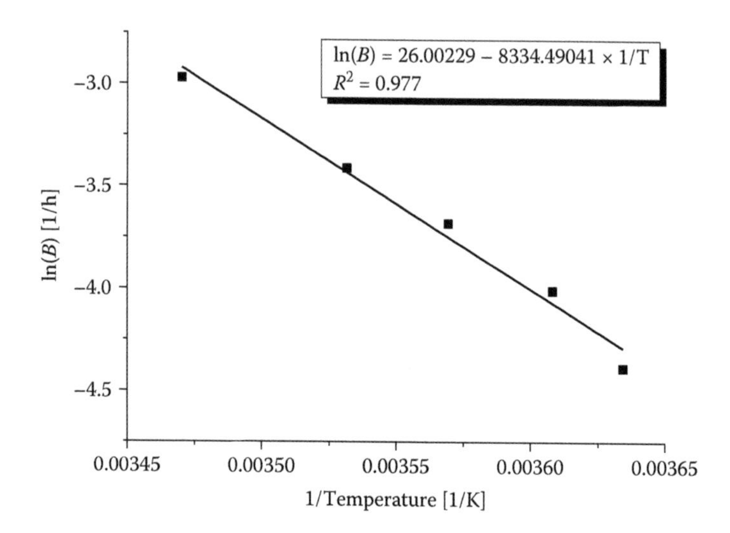

Figure 6.8 Arrhenius plot. Plotted data are measured growth rates of *Pseudomonas* spp. on fresh pork loins for different temperatures between 2°C and 15°C. (From Bruckner, S. 2010. Predictive shelf life model for the improvement of quality management in meat chains. PhD thesis, Rheinische Friedrich-Wilhelms-Universität Bonn, Germany.)

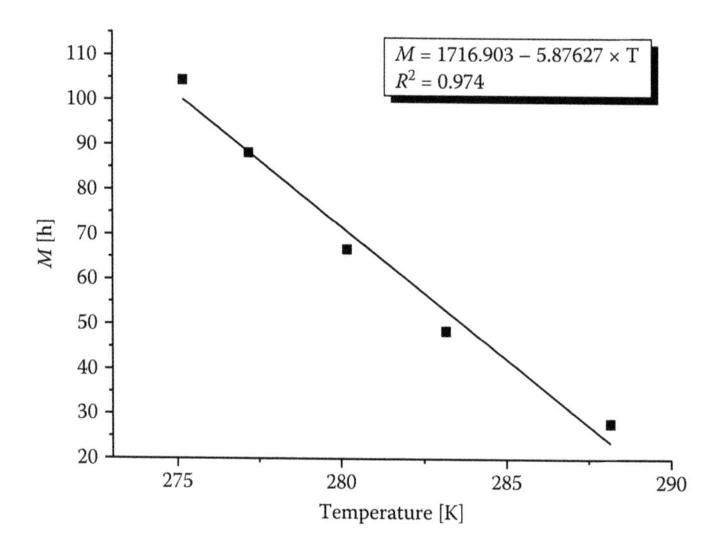

Figure 6.9 Plotted data are measured inflection points of growth curves of *Pseudomonas* spp. on fresh pork loins versus the measured temperatures between 2°C and 15°C, expressed in Kelvin. (From Bruckner, S. 2010. Predictive shelf life model for the improvement of quality management in meat chains. PhD thesis, Rheinische Friedrich-Wilhelms-Universität Bonn, Germany.)

the beginnings of the following intervals with different temperatures, the bacterial counts will be calculated using $M(T)$ of Equation 6.12 for each interval other than the first one. $M(T)$ for the first interval will be calculated using a linear functionality, as shown in Figure 6.9.

Figures 6.10 to 6.14 show simulated scenarios of growth curves of *Pseudomonas* spp. in an Excel program in which the mentioned functionalities are programmed within macros. The following variables have been manually entered in the Excel program to simulate different cold chains: 12 intervals with the temperature for each interval, the initial bacterial count, the final bacterial count, two parameters each for two secondary functionalities (these are the fitting parameters of Figures 6.8 and 6.9), and a bacterial count that is the threshold for being fresh or, in other words, the spoilage level. The automatically generated output after running the macros is a diagram that depicts the bacterial growth in log units of colony-forming units per gram as a function of time as well as the spoilage level. Next to the diagram, the following values are calculated and displayed: the shelf life, the estimated lag phase (calculated using Equation 6.3), and the maximal bacterial growth rate (calculated with Equation 6.6). Both values are calculated using the parameters M and B of the first interval. Therefore, they will not really be reliable if dynamic temperature conditions are simulated.

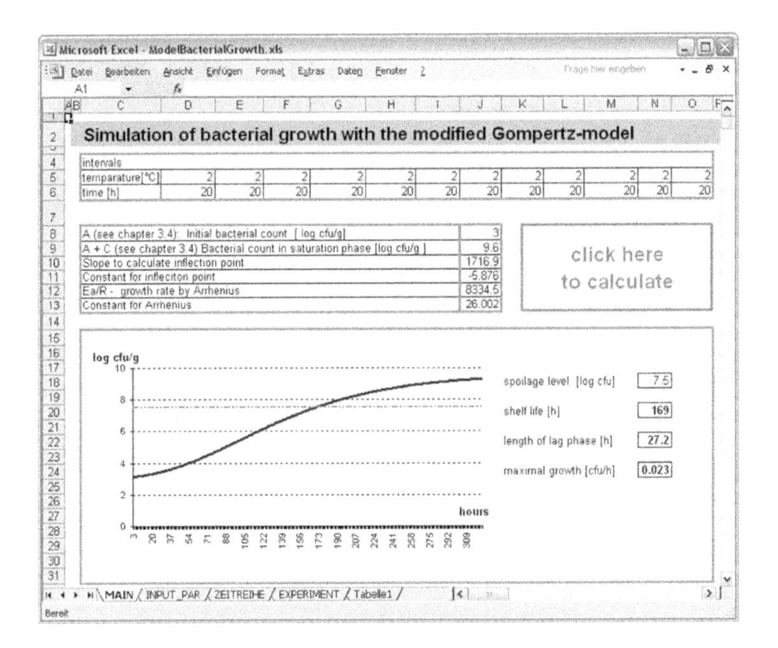

Figure 6.10 This is an easy program to use to simulate bacterial growth. The user has entered the specific values of Figures 6.8 and 6.9 of pork loin from the study by Bruckner (2010): an initial bacterial count of 3 $\log_{10}$ cfu/g (equal to 1,000 cfu/g), a saturation level of 9.6 $\log_{10}$ cfu/g, a cold chain with a constant temperature of 2°C and an acceptability limit of 7.5 $\log_{10}$ cfu/g. The shelf life of the product was simulated to be 169 hours for the entered parameters.

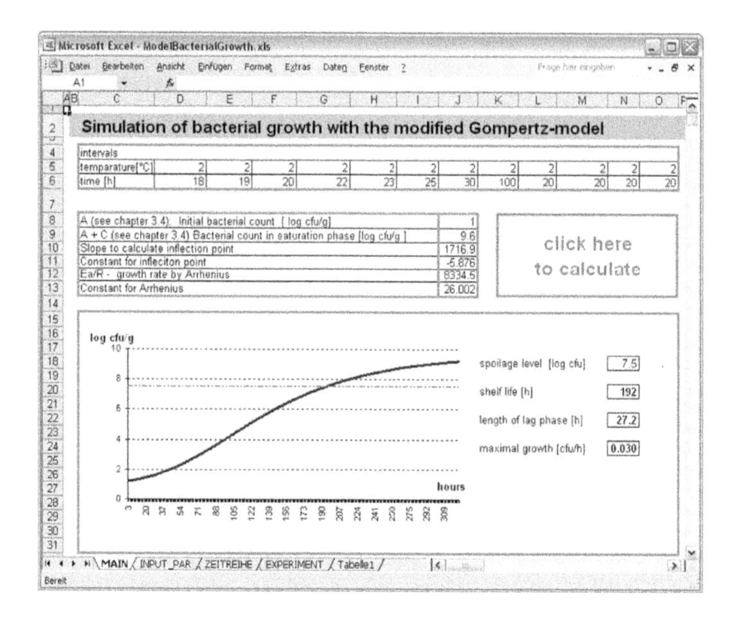

Figure 6.11 A similar input to Figure 6.10, except that the user has entered a reduced initial count of 1 $\log_{10}$ cfu/g, which could have been reached by improved hygienic conditions. The reduced initial count prolongs the shelf life from 169 hours to 192 hours—in other words, an extra day of extended shelf life.

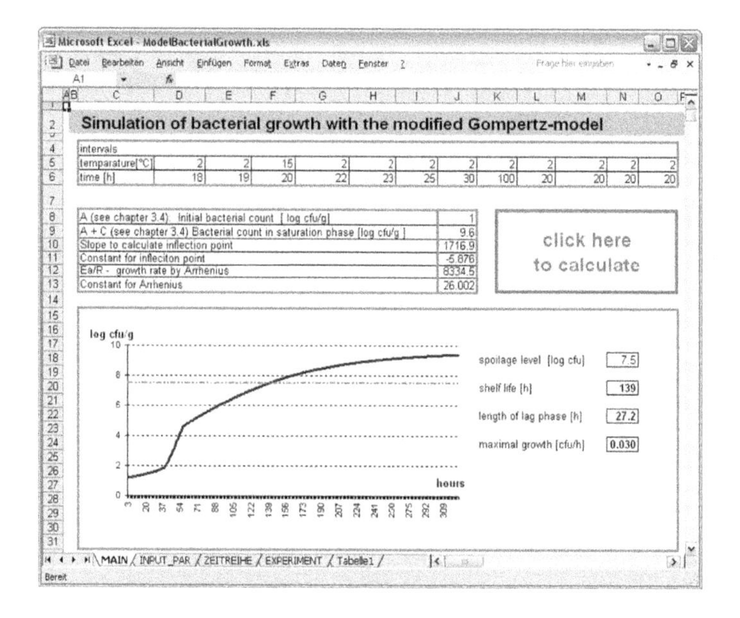

Figure 6.12 A similar input to Figure 6.11. Hypothetically, the user assumes good hygienic conditions in the beginning of the chain, but the cold chain was interrupted after 37 hours for 20 hours, which reduces the shelf life by approximately 50 hours.

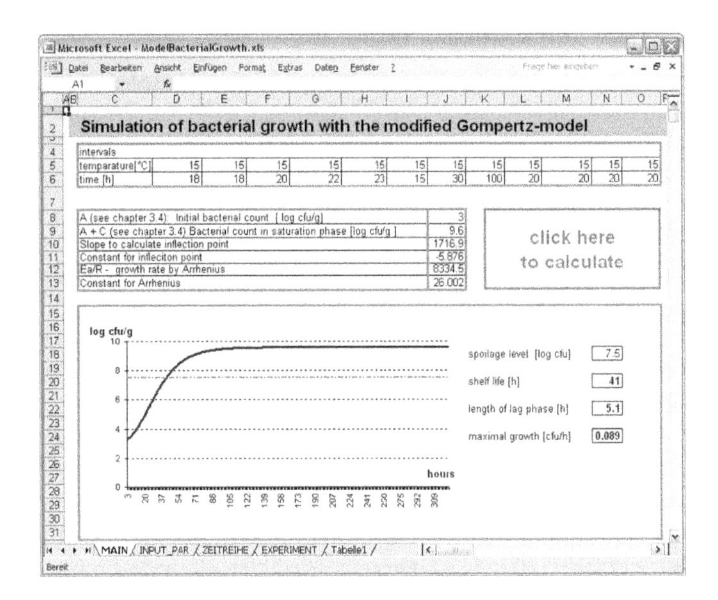

Figure 6.13 A simulation for the same pork loin as in Figure 6.11 but under poor cooling and poor hygienic conditions. The lag phase is reduced to 5 hours and the shelf life to 41 hours.

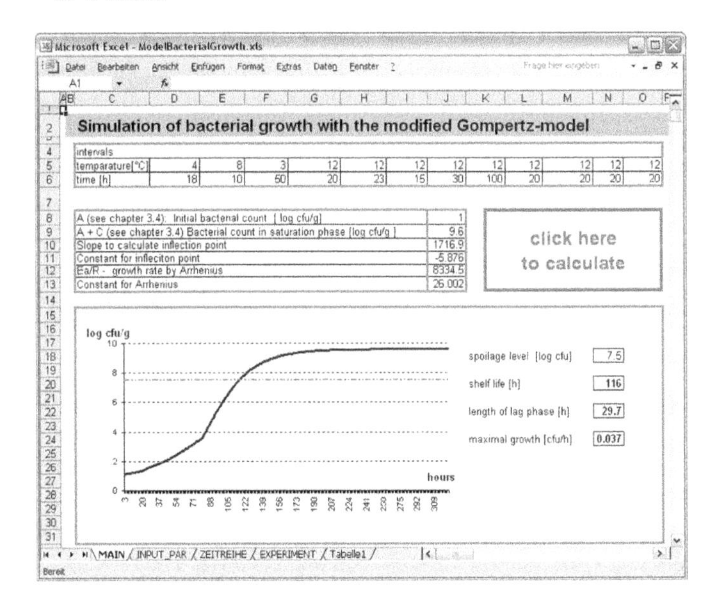

Figure 6.14 This simulation displays a cold chain that may have a great "market share." The steps show different times and temperatures due to different parties within the chain. The first step could be a producer, the second step a transport company, the third step a retailer, and the fourth step the consumer. Even though the initial bacterial count was set to 1 $\log_{10}$ cfu/g, which corresponds to good hygienic conditions, the shelf life is reduced to 116 hours due to temperature abuses.

6.3.7 Database and Software Tools

There are several models with related databases available to compare and to predict bacterial growth under certain conditions. Such systems give answers to various questions with regard to the growth of SSOs, pathogenic organisms, and estimation of shelf life.

Predictions are usually shown as a growth curve in the form of time series and using the most relevant parameters like lag time, growth rate, and maximum bacterial count. Most of the available software systems focus on the growth and inactivation of pathogenic bacteria. There are also a few software systems with related databases available that deal with the growth of spoilage bacteria under certain environmental conditions. In these cases, the data are mostly based on microbial growth in media within a certain range of influencing parameters like pH, a_w, temperature, and NaCl. Some of the databases and predictive models are available free of charge, such as ComBase and the SSSP.

ComBase-PMP (http://www.combase.cc) is a huge database with a modeling program that combines the U.K. Growth Predictor (http://www.ifr.ac.uk/safety/GrowthPredictor) and USDA (U.S. Department of Agriculture) Pathogen Modeling Program (PMP) (Baranyi and Tamplin, 2004; McMeekin et al., 2006a). The initiative is driven by different public research organizations with the objective of making microbial growth data and predictive tools on microbial responses to food environments freely available. Currently, more than 50,000 data records are available for different products, different organisms, and different influencing factors. The growth predictor allows the selection of relevant influencing factors on the growth of certain organisms; typical factors are a_w, pH, NaCl, and temperature or other factors such as CO_2, nitrite, and lactic acid or acetic acid. In addition, the initial count and a value for the physiological state of the cells can be added. The growth prediction is based on the model of Baranyi and Roberts (1994). Growth data are based on culture media under well-controlled environmental conditions. The prediction software DMFit is included in ComBase. This software allows the user to fit bacterial growth with two types of growth models: (a) the model of Baranyi and Roberts and (b) the trilinear, biphasic, or linear models (Baranyi and Roberts, 1994).

Another software package is the SSSP (http://sssp.dtuaqua.dk) developed by the Danish Food Institute. With regard to spoilage, the software can be used to predict the shelf life or the growth of the relevant spoilage organisms in selected fresh and preserved seafood. Special functions allow the comparison between shelf life predicted by the SSSP and observed shelf life and the changing of parameter values to make the model appropriate for various food and bacteria. The tool also delivers the possibility to upload temperature profiles and to predict shelf life. For more information, refer to the work of Dalgaard et al. (2002).

The Sym'Previus Software (http://www.symprevius.org) includes a database with growth and inactivation responses of different microorganisms in foods. It includes predictive models for growth and inactivation of pathogenic bacteria and some spoilage microorganisms (Leporq et al., 2003; Thuault and Couvert, 2008).

Forecast is a shelf life prediction service for chilled foods like fresh meat and fish and was developed by Campden BRI (http://www.campden.co.uk/news/mar09.pdf). For the prediction, the following parameters are used: temperature, NaCl (%aq), a_w, and pH. Spoilage predictions, which are available, are based as an example of the growth of *Pseudomonas* spp., Enterobacteriaceae, lactic acid bacteria, and yeasts. The relevant parameters of the food have to be sent to Campden BRI, and it will provide the relevant growth curves and parameters.

6.4 APPLICATION OF SHELF LIFE MODELS

Key objectives in cold chain management are to optimize food quality and safety, to optimize storage management, and to minimize food waste. Models that allow the calculation of the remaining shelf life in each step of the chain can deliver an important contribution to support these tasks (Kreyenschmidt, 2008).

To calculate the remaining shelf life, the current count of the SSOs has to be known in each step of the supply chain. Since the determinations of the microbial counts must be verified within the short time period of a few hours, modern monitoring technologies can be used instead of using microbiological enumeration techniques, which are time and cost intensive.

One approach is to determine the actual status of special microorganisms or a metabolic end product through rapid methods of food freshness analysis. These include technologies such as the impedance technique, infrared spectroscopy, proton transfer reaction mass spectrometry, or diffuse reflectance spectroscopy (Lindinger et al., 1998; Kreyenschmidt, 2003; Mayr et al., 2003; Lin et al., 2004).

Another possibility to estimate the remaining shelf life is based on the temperature history of the product since temperature is the most relevant influencing factor on shelf life after the food production stage (McMeekin and Olley, 1986; McMeekin, 2004; McMeekin et al., 2006b; Fu and Labuza, 1993; Labuza and Fu, 1993; Labuza, 2006; Kreyenschmidt, 2003, 2008).

Calculations of the remaining shelf life, based on the temperature history, require comprehensive monitoring of product temperature along the entire supply chain. Until now, temperature monitoring in various types of chains has been conducted in several ways; for example, random measurements of the product's core temperature are frequently taken at the incoming inspections. During transportation, the environmental temperature is recorded, which usually does not deliver information about product temperature. Besides that, data analysis, storage methods, and the availabilities of data accesses differ from company to company (Raab et al., 2008). These factors have to be considered if predictive food models and temperature-monitoring systems will be combined to support cold chain management.

However, huge progress has been achieved by the development and the use of wireless sensor networks combined with temperature sensors within food supply chains. Also, conventional thermometry devices and electronic data loggers have been improved and become cheaper. In addition, the use of time-temperature indicators (TTIs) to control temperature conditions in food supply chains from production to consumption has been discussed extensively in recent years.

Several of these new developments with regard to their use in combination with predictive food models have been introduced and discussed at the International Cold-Chain Management Workshops in Bonn, Germany, in relation to monitoring and controlling the cold chain from production to consumption. More information about the results of the workshop are available online (http://www.ccm.uni-bonn.de). The most frequently presented solutions were based on RFID technology and various TTIs.

It is generally agreed that the use of TTIs is currently the only feasible and cost-effective solution that permits an adequate temperature control for a single product from production to consumption and thus along the entire cold chain. At present, there are several different TTI systems available. The principles on which most of the different indicators have been developed are based on enzymatic, chemical, mechanical, electrochemical, or microbiological reactions, which result in a color change of the label depending on time and temperature (Taoukis and Labuza, 1989, 1997, 2003; Taoukis et al., 1991; Tsoka et al., 1998; Brody, 2001; Petersen and Kreyenschmidt, 2004; Giannakourou et al., 2005; Kerry et al., 2006; Mehauden et al., 2008; Yan et al., 2008; Kreyenschmidt et al., 2010a; Maschietti, 2010). Each of these different systems has its own characteristics as well as advantages and disadvantages with regard to implementation in different food chains and specific usability, flexibility, and cost. However, for the combination of TTI systems with shelf life models, it is a key requirement that the response of the TTI is similar to the spoilage of the product (Taoukis and Labuza, 1989). This is necessary so that the TTI value can be read (e.g., by a measurement device) and the obtained value used to calculate the remaining shelf life of the product onto which the TTI is attached. A software solution, which demonstrates how such a system could work, was presented by Raab et al. (2010a, 2010b) and is available online (http://www.ccm-network.com).

Temperature-monitoring systems based on wireless technologies, like RFID systems, have emerged due to their capability to identify, categorize, and manage the products and information flows through the whole supply chain. An advantage in comparison to TTIs is the possibility of making a detailed analysis of the digital data and the collection of real-time information. This allows a continuous recalculation of the remaining shelf life during the whole chain. Also, other sensors than those detecting temperature can be integrated in the RFID tags (e.g., to measure the humidity or ethylene levels), which can support a more precise prediction of the shelf life for specific products. An enormous number of systems in this area have been developed. The temperature-monitoring systems are differentiated according to the power supply into active, passive, or semiactive tags; storage volume; reading distance; data analysis methods; and accuracy of the sensor. These characteristics lead to huge variances in the price of these systems. From an economic point of view, there is no system available that can be used on a single package from production to the consumer for cheap food products. But, there are several devices and systems available that are economical when applied to larger batches and that can be used to control the cold chain from the production stage to the retailer with regard to calculating remaining shelf life and the observance of the cold chain at each stage. Information on the different systems can also be found in the proceedings of the International Workshops of Cold Chain Management and online (http://www.ccm.uni-bonn.de) (Kreyenschmidt and Petersen, 2006; Kreyenschmidt, 2008; Kreyenschmidt et al., 2010a).

The development of such systems will be a big step forward in controlling the temperature through the entire supply chain. One needs to consider that even if technological solutions are in existence, the integration of these systems in combination with predictive food models into different cold chains is a great challenge since standardized concepts for their implementation do not exist. To use temperature-monitoring systems to ascertain a precise prediction of the product temperature, a number of important requirements have to be fulfilled (Raab et al., 2008, 2010a):

- Easy adaptability of the system to the heterogeneous structures of international food chains
- Cost-effective and user-friendly implementation of the systems
- Integration of the devices within preexisting or new technologies
- New software or inspection schemes that run in parallel with existing software implementations

Independently of the systems that are used to control the temperature throughout the chain, there are three further important challenges that have to be met in combining and implementing temperature-monitoring devices with shelf life models. First, collaboration of the participants is needed; this also includes the sharing of real-time data with regard to product characteristics and temperature conditions. Second, the people involved have to be aware of the importance of an accurate temperature-monitoring system to obtain as precise predictions as possible. Third, the users have to be aware of the characteristics of data interpretation, or misinterpretation, with regard to the microbiological parameters.

Last, the application of innovative tools like shelf life models to support cold chain management will only happen if the industry will see the economic benefits of the implemented technologies, such as the optimization of storage or delivery management or of waste management.

6.5 FINAL REMARKS

Shelf life determination of perishable foods is still a complex issue since the growth of relevant microorganisms depends on a wide range of different parameters. Small changes in the environment or in the food itself can lead to different shelf life durations. Even though predictive food models are able to support and to simplify the determination of shelf life, the use of these models is still viewed with suspicion by some branches of the food industry, and there is much scope for their successful introduction. Good conclusions on the likely future trends in predictive modeling have been made by McMeekin (2004, 2007) and Legan (2007); both have stated that predictive food models have not reached their full potential in the area of product and process evaluation and decision making, but they also stated that this process is likely to take a long time. The level of implementation of food safety models has increased greatly in recent years. Also, the application of shelf life models at the product development stage has increased steadily, but the use of such models to optimize processes has not yet reached its potential since the benefits of shelf life models are not widely recognized. Besides that, there are still some challenges to be met with regard to the heterogeneity of foods, the prediction of the lag phase, the determination of the initial bacterial counts, and the determination of the count at the incoming and outgoing inspection. However, due to foreseeable developments and improvements in the field of sensor technologies and rapid testing methods, it can be assumed that the implementation process will proceed much faster than similar developments have since the 1980s.

6.6 REFERENCES

Antony, J. 2003. *Design of experiments for engineers and scientists.* Oxford, UK: Elsevier Science and Technology.

Arrhenius, S. 1889. Über die Reaktionsgeschwindigkeit bei der Inversion von Rohrzucker durch Säuren. *Z Phys Chem* 4: 226–248.

Baird-Parker, A. C., and D. C. Kilsby. 1987. Principles of predictive food microbiology. *J Appl Bacteriol Sym Supply* 63: 43–49.

Baranyi, J., and T. A. Roberts. 1994. A dynamic approach to predicting bacterial growth in food. *Int J Food Microbiol* 23: 277–294.

Baranyi, J., Roberts, T. A., and P. McClure. 1993. A nonautonomous differential equation to model bacterial growth. *Food Microbiol* 10: 43–59.

Baranyi, J., and M. Tamplin. 2004. ComBase: A common database on microbial responses to food environments. *J Food Prot* 67(9): 1967–1971.

Barth, M., Hankinson, T. R., Zhuang, H., and F. Breidt. 2009. Microbiological spoilage of fruits and vegetables. In: *Compendium of the microbiological spoilage of fruits and vegetables*, W. H. Sperber and M. P. Doyle, Eds., 135–183. Secaucus, NJ: Springer Science + Business Media.

Belehradek, J. 1930. Temperature coefficients in biology. *Biol Rev Biol Proc Camb Phil Soc* 5(1–2): 30–60.

Bennik M., H. J., Vorstman, W., Smid, E. J., and L. G. M. Gorris. 1998. The influence of oxygen and carbon dioxide on the growth of prevalent Enterobacteriaceae and *Pseudomonas* species isolated from fresh and controlled-atmosphere-stored vegetables. *Food Microbiol* 15: 459–469.

Borch, E., Kant-Muermans, M. L., and Y. Blixt. 1996. Bacterial spoilage of meat and cured meat products. *Int J Food Microbiol* 33: 103–120.

Bratchell, N., McClure, P. J., Kelly, T. M., and T. A. Roberts. 1990. Predicting microbial growth: graphical methods for comparing models. *Int J Food Microbiol* 11(3–4): 279–287.

Brody, A.L. 2001. What's active about Intelligent Packaging? *Food Technol* 55:75–78.

Bruckner, S. 2010. Predictive shelf life model for the improvement of quality management in meat chains. PhD thesis, Rheinische Friedrich- Wilhelms-Universität, Bonn, Germany.

Buchanan, R. L. 1993. Predictive food microbiology. *Trends Food Sci Technol* 4(1): 6–11.

Buchanan, R. L., and M. L. Cygnarowicz. 1990. A mathematical approach toward defining and calculating the duration of the lag phase. *Food Microbiol* 7: 237–240.

Buchanan, R. L., Whiting, R. C., and W. C. Damert. 1997. When is simple good enough: a comparison of the Gompertz, Baranyi, and three-phase linear models for fitting bacterial growth curves. *Food Microbiol* 14(4): 313–326.

Carlin, F., Nguyen-the, C., Hilbert, G., and Y. Chambroy. 1990. Modified atmosphere packaging of fresh ready-to-use grated carrots in polymeric films. *J Food Sci* 55: 1033–1038.

Cayré, M. E., Vignolo, G., and O. Garro. 2003. Modeling lactic acid bacteria growth in vacuum-packaged cooked meat emulsions stored at three temperatures. *Food Microbiol* 20(5): 561–566.

Chen, J. 2002. Microbial enzymes associated with fresh-cut produce. In *Fresh-cut fruits and vegetables*, O. Lamikanra, Ed., 256–273. Boca Raton, FL: CRC Press.

Christian, J. H. B. 2000. Drying and reduction of water activity. In *The microbiological safety and quality of food*, B. M. Lund, T.C. Baird-Parker, and G. W. Gould, Eds., 146–174. Gaithersburg, MD: Aspen.

Coates, K. J., Beattie, J. C., Morgan, I. R., and P. R. Widders. 1995. The contribution of carcass contamination and the boning process to microbial spoilage of aerobically stored pork. *Food Microbiol* 12:49–54.

Dainty, R. H., and B. M. Mackey. 1992. The relationship between the phenotypic properties of bacteria from chill-stored meat and spoilage processes. *J Appl Bact Symp Suppl* 73(21): 103–114.

Dalgaard, P. 1995. Modeling of microbial activity and prediction of shelf-life for packed fresh fish. *Int J Food Microbiol* 26: 305–317.

Dalgaard, P., Buch, P., and S. Silberg. 2002. Seafood Spoilage Predictor—development and distribution of a product specific application software. *Int J Food Microbiol* 73: 343–349.

Dalgaard, P., Gram, L., and H. H. Huss. 1993. Spoilage and shelf-life of cod fillets packed in vacuum or modified atmosphere. *Int J Food Microbiol* 19: 283–294.

Dalgaard, P., Mejlholm, O., and H. H. Huss. 1997. Application of an iterative approach for development of a microbial model predicting the shelf-life of packed fish. *Int J Food Microbiol* 38: 169–179.

Davey, K. R. 1989. A predictive model for combined temperature and water activity on microbial growth during the growth phase. *J Appl Bacterial* 67: 483–488.

Ercolini, D., Russo, F., Torrieri, E., Masi, P., and F. Villani. 2006. Changes in the spoilage-related microbiota of beef during refrigerated storage under different packaging conditions. *Appl Environ Microbiol* 72(7): 4663–4671.

Farber, J. M. 1991. Microbiological aspects of modified-atmosphere packaging—a review. *J Food Prot* 54:58–70.

Fehlhaber, K., Kleer, J., and F. Kley. 2005. *Handbuch Lebensmittelhygiene*. Hamburg: Behr's Verlag.

Fu, B., and T. P. Labuza. 1993. Shelf-life prediction: theory and application. *Food Contr* 4(3): 125–133.

Giannakourou, M. C., Koutsoumanis, K., Nychas, G. J. E., and P. S. Taoukis. 2005. Field evaluation of the application of time temperature integrators for monitoring fish quality in the chill chain. *Int J Food Microbiol* 102: 323–336.

Giannuzzi, L., Pinotti, A., and N. Zaritzky. 1998. Mathematical modeling of microbial growth in packaged refrigerated beef stored at different temperatures. *Int J Food Microbiol* 39(1–2): 101–110.

Gibson, A. M., Bratchell, N., and T. A. Roberts. 1987. The effect of sodium chloride and temperature on rate and extent of growth of *Clostridium botulinum* type A in pasteurized pork slurry. *J Appl Microbiol* 62: 479–490.

Gill, C. O., and K. G. Newton. 1977. The development of aerobic spoilage flora on meat stored at chill temperatures. *J Appl Bacteriol* 43(2): 189–195.

Gospavic, R., Kreyenschmidt, J., Bruckner, S., Popov, V., and N. Haque. 2008. Mathematical modelling for predicting the growth of *Pseudomonas* spp. in poultry under variable temperature conditions. *Int J Food Microbiol* 127(3): 290–297.

Gram, L., and P. Dillard. 2002. Fish spoilage bacteria—problems and solutions. *Curr Opin Biotechnol* 13(3): 262–266.

Gram, L., and H. H. Huss. 1996. Microbiological spoilage of fish and fish products. *Int J Food Microbiol* 33(1): 121–137.

Heard, G.M. 1999. Microbial safety of ready-to-eat salads and minimally processed vegetables and fruits. *Food Aust* 51: 414–420.

Herbert, R. A., and J. P. Sutherland. 2000. Chill storage. In *The microbiological safety and quality of food*, B. M. Lund, T. C. Baird-Parker, and G. W. Gould, Eds., 101–121. Gaithersburg, MD: Aspen.

Hozbor, M. C., Saiz, A. I., Yeannes, M. I., and R. Fritz. 2006. Microbiological changes and its correlation with quality indices during aerobic iced storage of sea salmon (*Pseudopercis semifasciata*). *LWT—Food Sci Technol* 39: 99–104.

Hu, P., Zhou, G., Xu, X., Li, C., and Y. Han. 2009. Characterization of the predominant spoilage bacteria in sliced vacuum-packed cooked ham based on 16S rDNA-DGGE. *Food Control* 20: 99–104.

Huis in't Veld, J. H. H. 1996. Microbial and biochemical spoilage of foods: an overview. *Int J Food Microbiol* 33: 1–18.

Jay, J. M., Loessner, M. J., and D. A. Golden. 2005. *Modern food microbiology*. New York: Springer.

Jay, J. M., Vilai, J. P., and M. E. Hughes. 2003. Profile and activity of the bacterial biota of ground beef held from freshness to spoilage at 5–7°C. *Int J Food Microbiol* 81: 105–111.

Kakiomenou, K., Tassou C., and G.-J. Nychas. 1996. Microbiological, physicochemical and organoleptic changes of shredded carrots stored under modified storage. *Int J Food Sci Technol* 31: 356–366.

Kerry, J., O'Grady, M., and S. Hogan. 2006. Past, current and potential utilization of active and intelligent packaging systems for meat and muscle-based products: a review. *Meat Sci* 74: 113–130.

Koutsoumanis, K., and G.-J. E. Nychas. 2000. Application of a systematic experimental procedure to develop a microbial model for rapid fish shelf life predictions. *Int J Food Microbiol* 60: 171–184.

Koutsoumanis, K., Stamatiou, A., Skandamis, P., and G. J. E. Nychas. 2006. Development of a microbial model for the combined effect of temperature and pH on spoilage of ground meat, and validation of the model under dynamic temperature conditions. *Appl Environ Microbiol* 72(1): 124–134.

Koutsoumanis, K. 2001. Predictive modeling of the shelf life of fish under nonisothermal conditions. *Appl Environ Microbiol* 67(4): 1821–1829.

Kreyenschmidt, J. 2003. Modellierung des Frischeverlustes von Fleisch sowie des Entfärbeprozesses von Temperatur-Zeit-Integratoren zur Festlegung von Anforderungsprofilen für die produktbegleitende Temperaturüberwachung. PhD thesis, Rheinische Friedrich-Wilhelms-Universität Bonn. Bergen/Dumme, Germany: Agrimedia.

Kreyenschmidt, J. 2008. Cold chain-management. *Proceedings of the 3rd International Workshop Cold-Chain-Management*, Bonn, Germany. June 2–3.

Kreyenschmidt, J., Christiansen, H., Hübner, A., Raab, V., and B. Petersen. 2010a. A novel time-temperature-indicator (TTI) system to support cold chain management. *J Food Sci Technol* 208(45): 208–215.

Kreyenschmidt, J., Hübner, A., Beierle, E., Chonsch, L., Scherer, A., and B. Petersen. 2010b. Determination of the shelf life of sliced cooked ham based on the growth of lactic acid bacteria in different steps of the chain. *J Appl Microbiol* 108(2): 510–520.

Kreyenschmidt, J., and B. Petersen. 2006. Cold chain-management. *Proceedings of the 2nd International Workshop Cold-Chain-Management*, Bonn, Germany. May 8–9.

Labuza, T. P. 2006. Time-temperature integrators and the cold chain: what is next? *Proceedings of the 2nd International Workshop Cold-Chain-Management*, Bonn, Germany, 43–51.

Labuza, T. P., and B. Fu. 1993. Growth kinetics for shelf-life prediction: theory and practice. *J Ind Microbiol* 12(3–5): 309–323.

Labuza T. P., Fu, B., and P. S. Taoukis. 1992. Prediction for shelf life and safety of minimally processed CAP/MAP chilled foods. *J Food Prot* 55: 741–750.

Labuza, T. P., and D. Riboh. 1982. Theory and application of Arrhenius kinetics to the prediction of nutrient losses in foods. *Food Technol* 36: 66–74.

Legan, D. 2007. Application of models and other quantitative microbiology tools. In *Modeling microorganisms in food*, S. Brul, M. Zwietering, and S. Van Gerwen, Eds., 82–109. Cambridge, UK: Woodhead.

Leporq, B., Membré, J.-M., Zwietering, M., Dervin, C., Buche, P., and J. P. Guyonnet. 2003. The "Sym'Previus" software, a tool to support decisions to the foodstuff safety. *Int J Food Microbiol* 100(1–3): 231–237.

Lin, M., Al-Holy, M., Mousavi-Hesary, M., Al-Quasiri, H., Cavinato, A. G., and B. A. Rasco. 2004. Rapid and quantitative detection of the microbial spoilage in chicken meat by diffuse reflectance spectroscopy (600–1,100 nm). *Lett Appl Microbiol* 39: 148–155.

Lindinger, W., Hansel, A., and A. Jordan. 1998. On-line monitoring of volatile organic compounds at pptv levels by means of proton-transfer-reaction mass spectrometry (PTR-MS). Medical applications, food control and environmental research. *Int J Mass Spectrom Ion Process* 173: 191–241.

Liu, F., Yang, R.-Q., and Y.-F. Li. 2006. Correlations between growth parameters of spoilage micro-organisms and shelf life of pork stored under air and modified atmosphere at -2, 4 and 10°C. *Food Microbiol* 23: 578–583.

Lund, B. M., and T. Eklund. 2000. Control of pH and use of organic acids. In *The microbiological safety and quality of food*, B. M. Lund, T. C. Baird-Parker, and G. W. Gould, Eds., 175–199. Gaithersburg, MD: Aspen.

Lund, B. M., and Snowdon, A. L. 2000. Fresh and processed fruits. In *The microbiological safety and quality of food*, B. M. Lund, T. C. Baird-Parker, and G. W. Gould, Eds., 738–758. Gaithersburg, MD: Aspen.

Lyhs, U., Korkeala, H., Vandamme, P., and J. Njörkroth. 2001. *Lactobacillus alimentarius*: a specific spoilage organism in marinated herring. *Int J Food Microbiol* 64: 355–360.

Marchetti, R., Casadei, M. A., and M. E. Guerzoni. 1992. Microbial population dynamics in ready-to-eat vegetable salads. *Ital J Food Sci* 2: 97–108.

Maschietti, M. 2010. A new time-temperature indicator based on high-viscosity liquids. *Proceedings of the 4th International Workshop Cold-Chain-Management*, Bonn, Germany, September 27–28. 47–58.

Mataragas, M., Drosinos, E. H., Vaidanis, A., and I. Metaxopoulos. 2006. Development of a predictive model for spoilage of cooked cured meat products and its validation under constant and dynamic temperature storage conditions. *J Food Sci* 71(6): M157–M167.

Mayr, D., Margesin, R., Klingsbichel, E., Hartungen, E., Jenewein, D., Schinner, F., and T. D. Märk. 2003. Rapid detection of meat spoilage by measuring volatile organic compounds by using proton transfer reaction mass spectrometry. *Appl Environ Microbiol* 69(8): 4697–4705.

McDonald, K., and D.-W. Sun. 1999. Predictive food microbiology for the meat industry: a review. *Int J Food Microbiol* 52(1–2): 1–27.

McKellar, R. C. 1997. A heterogeneous population model for the analysis of bacterial growth kinetics. *Int J Food Microbiol* 36: 179–186.

McMeekin, T. A. 2004. An essay on the unrealized potential of predictive microbiology. In *Modeling microbial responses in food*, R. C. McKellar and X. Lu, Eds., 321–335. Boca Raton, FL: CRC Press.

McMeekin, T. A. 2007. Predictive microbiology: quantitative science delivering quantifiable benefits to the meat industry and other food industries. *Meat Sci* 77: 17–27.

McMeekin, T. A., Baranyi, J., Bowman, J., Dalgaard, P., Kirk, M., Ross, T., Schmid, S., and M. Zwietering. 2006a. Information systems in food safety management. *Int J Food Microbiol* 112: 181–194.

McMeekin, T. A., and J. Olley. 1986. Predictive microbiology. *Food Technol Aust* 38(8): 331–334.

McMeekin, T. A., Olley, J., and Ross, T. 1993. *Predictive microbiology: theory and application.* Research Studies Press and John Wiley & Sons, Tauton, UK.

McMeekin, T. A., and T. Ross. 1996. Shelf life prediction: status and future possibilities. *Int J Food Microbiol* 33: 65–83.

McMeekin, T. A., Smale, N., Jenson, I., Ross, T., and D. Tanner. 2006b. Combining microbial growth models with near real time temperature monitoring technologies to estimate the shelf life and safety of foods during processing and distribution. *Proceedings of the 2nd International Workshop Cold-Chain-Management*, Bonn, Germany, May 8–9. 71–78.

Mehauden, K., Bakalis, S., Cox, P., Fryer, P., and M. Simmons. 2008. Use of time temperature integrators for determining process uniformity in agitated vessels. *Innov Food Sci Emerg Technol* 9: 385–395.

Molin G. 2000. Modified atmosphere. In *The microbiological safety and quality of food*, B. M. Lund, T. C. Baird-Parker, and G. W. Gould, Eds., 214–234. Gaithersburg, MD: Aspen.

Montgomery, D. C. 2009. *Design and analysis of experiments,* 7th ed. New York: Wiley.

Moore, C. M., and B. W. Sheldon. 2003. Use of time-temperature integrators and predictive modeling to evaluate microbiological quality loss in poultry products. *J Food Prot* 66(2): 280–286.

Mossel, D. A. A. 1971. Physiological and metabolic attributes of microbial groups associated with foods. *J Appl Bacteriol* 34(1): 95–118.

Neumeyer, K., Ross, T., and T. A. McMeekin. 1997. Development of a predictive model to describe the effects of temperature and water activity on the growth of spoilage pseudomonads. *Int J Food Microbiol* 38(1): 45–54.

Nychas, G. J. E., Dillon, V. M., and R. G. Board. 1988. Glucose, the key substrate in the microbiological changes occurring in meat and certain meat products. *Biotechnol Appl Biochem* 10(3): 203–231.

Nychas, G.-J. E., Marshall, D., and J. Sofos. 2007. Meat poultry and seafood. In *Food microbiology: fundamentals and frontiers,* M. P. Doyle and L. R. Beuchat, Eds., 105–140. Washington, DC: ASM.

Nychas, G. J. E., Skandamis, P. N., Tassouand, C. C., and K. P. Koutsoumanis. 2008. Meat spoilage during distribution. *Meat Sci* 78(12): 77–89.

Olsson, C., Ahrne, S., Petterson, B., and G. Molin. 2003. The bacterial flora of fresh and chill-stored pork: analysis by cloning and sequencing of 16S rRNA genes. *Int J Food Microbiol* 83: 245–252.

Pao, S., and P. D. Petracek. 1997. Shelf life extension of peeled oranges by citric acid treatment. *Food Microbiol* 14: 485–491.

Petersen, B., and J. Kreyenschmidt. 2004. Ein viel versprechendes Hilfsmittel—Einsatzmöglichkeiten von Zeit-Temperatur-Indikatoren zur Überprüfung der Kühlkette. *Fleischwirtschaft* 10:57–59.

Phillips, C. A. 1996. Modified atmosphere packaging and its effects on the microbial quality and safety of produce. *Int J Food Sci Technol* 31: 463–479.

Pirt, S. J. 1975. *Principle of microbe and cell cultivation.* London: Blackwell Scientific.

Pooni, G. S., and G. C. Mead. 1984. Prospective use of temperature function integration for predicting the shelf-life of non-frozen poultry-meat products. *Food Microbiol* 1: 67–78.

Raab V., Bruckner, S., Beierle, E., Kampmann, Y., Petersen, B., and J. Kreyenschmidt. 2008. Generic model for the prediction of remaining shelf life in support of cold chain management in pork and poultry supply chains. *J Chain Network Sci* 8: 59–73.

Raab, V., Ibald, R., Albrecht, A., Petersen, B., and J. Kreyenschmidt. 2010a. Web 2.0 based software solution to support a practical implementation of time-temperature indicators. *Proceedings of the 4th Cold-Chain-Management Workshop*, Bonn, Germany, September 27–28. 57–58.

Raab, V., Petersen, B., and J. Kreyenschmidt. 2010b. Temperature monitoring in meat supply chains. *Br Food J* 113: 1267–1289.

Rasch, M. 2004. Experimental design and data collection. In *Modeling microbial response in food*, R. C. McKellar and X. Lu, Eds., 1–20. Boca Raton, FL: CRC Press.

Ratkowsky, D. A., Olley, J., McMeekin, T. A., and A. Ball. 1982. Relationship between temperature and growth rate of bacterial cultures. *J Bacteriol* 149: 1–5.

Ross, T., and T. A. McMeekin. 1994. Predictive microbiology. *Int J Food Microbiol* 23(3–4): 241–264.

Ross, T., Baranyi, J., and T. A. McMeekin. 1999. Predictive microbiology and food safety. In *Encyclopaedia of food microbiology*, R. Robinson, C. A. Batt, and P. Patel, Eds., 1699–1710. London: Academic Press.

Ross, T., and P. Dalgaard. 2004. Secondary models. In *Modelling microbial responses in food,* R. C. McKellar, and X. Lu, Eds., 63–149. Boca Raton, FL: CRC Press.

Ross, T., and T. A. McMeekin. 1995. Predictive microbiology and HACCP. In *Advances in meat research: HACCP in meat, poultry and fish processing* 10, A. M. Pearson and T. R. Dutson, Eds., 330–357. London: Chapman & Hall.

Rudi, K., Flateland, S. L., Hanssen, J. F., Bengtsson, G., and H. Nissen. 2002. Development and evaluation of a 16S ribosomal DNA array-based approach for describing complex microbial communities in ready-to-eat vegetable salads packed in a modified atmosphere. *Appl Environ Microbiol* 68: 1146–1156.

Ruiz-Capillas, C., Carballo, J., and F. Colmenero. 2007. Biogenic amines in pressurized vacuum-packed cooked sliced ham under different chilled storage conditions. *Meat Sci* 75(3): 397–405.

Samelis, J., Kakouri, A., and J. Rementzis. 2000. Selective effect of the product type and the packaging conditions on the species of lactic acid bacteria dominating the spoilage microbial association of cooked meats at 4°C. *Food Microbiol* 17: 329–340.

Sandhya. 2009. Modified atmosphere packaging of fresh produce: current status and future trends. *LWT—Food Sci Technol* 43(3): 381–392.

Sapers, G. M., Miller, R. L., Pilizota, V., and A. M. Mattrazo. 2001. Antimicrobial treatments for minimally processed cantaloupe melon. *J Food Sci* 66: 345–349.

Scott, W. J. 1937. The growth of microorganisms on ox muscle. II. The influence of temperature. *J Council Sci Indus Res* 10: 338–350.

Slongo, A., Rosenthal, A., Camargo, L., Deliza, R., Mathias, S., and G. de Argao. 2009. Modeling the growth of lactic acid bacteria in sliced ham processed by high hydrostatic pressure. *LWT—Food Sci Technol* 42(1): 303–306.

Taoukis, P. S., Fu, B., and T. P. Labuza. 1991. Time-temperature indicators. *Food Technol* 45(10): 70–82.

Taoukis, P. S., and T. P. Labuza. 1989. Applicability of time-temperature indicators as shelf life monitors of food products. *J Food Sci* 54: 783–788.

Taoukis, P. S., and T. P. Labuza. 1997. Chemical time-temperature-integrators as quality monitors in the chill chain. In *Proceedings of the International Symposium Quimper Froid* '97, Quimper, France, June 16–18. 291–297.

Taoukis, P. S., and T. P. Labuza. 2003. Time-temperature indicators (TTIs). In *Novel food packaging techniques,* R. Ahvenainen, Eds., 103–126. Cambridge, UK: Woodhead.

Thuault, D., and O. Couvert. 2008. SYM'PREVIUS. Predictive microbiology tools for cold chain management. *Proceedings of the 3rd International Workshop Cold-Chain-Management*, Bonn, Germany, June 2–3. 65–71.

Tsoka, S., Taoukis, P. S., Christakopoulos, P., Kekos, D., and B. J. Macris. 1998. Time temperature integration for chilled food shelf life monitoring using enzyme-substrate systems. *Food Biotechnol* 12: 139–155.

Van Boekel, T., and M. Zwietering. 2007. Experimental design, data processing and model fitting in predictive microbiology. In *Modelling microorganisms in food*, S. Brul, M. Zwietering, and S. Van Gerwen, Eds., 82–109. Cambridge, UK: Woodhead.

Vankerschaver, K., Willocx, F., Smout, C., Hendrickx, M., and P. Tobback. 1996. The influence of temperature and gas mixtures on the growth of intrinsic micro-organisms on cut endive: predictive versus actual growth. *Food Microbiol* 13: 427–440.

Vasilopoulos, C., Ravyths, F., De Maere, H., De Mey, E. , Paelinck, H., De Vuyst, L., and F. Leroy. 2008. Evaluation of the spoilage lactic acid bacteria in modified-atmosphere-packaged artisan-type cooked ham using culture dependent and culture-independent approaches. *J Appl Microbiol* 104: 1341–1351.

Whiting, R. C., and R. L. Buchanan. 1993. A classification of models in predictive microbiology—reply. *Food Microbiol* 10: 175–177.

Whiting, R. C. 1995. Microbial modeling in foods. *Crit Rev Food Sci Nutr* 35(6): 467–494.

Yan, S., Huawei, C., Limin, Z., Fazheng, R., Luda, Z., and Z. Hengtao. 2008. Development and characterization of a new amylase type time-temperature-indicator. *Food Control* 19: 315–319.

Zwietering, M. 2005. Temperature effect on bacterial growth rate: quantitative microbiology approach including cardinal values and variability estimates to perform growth simulations on/in food. *Int J Food Microbiol* 100(1–3): 179–186.

Zwietering, M. H., Jongenburger, I., Rombouts, F. M., and K. van't Riet. 1990. Modeling of the bacterial growth curve. *Appl Environ Microbiol* 56(6): 1875–1881.

Zwietering, M. H., Koos, J. T. D., Hasenack, B. E., Wit, J. C. D., and K. van't Riet. 1991. Modeling of bacterial growth as a function of temperature. *Appl Environ Microbiol* 57(4): 1094–1101.

Zwietering, M., and M. J. Nauta. 2007. Predictive models in microbiological risk assessment. In *Modelling microorganisms in food*, S. Brul, S. van Gewen, and M. Zwietering, Eds., 110–125. Cambridge, UK: Woodhead, and Boca Raton, FL: CRC Press.

CHAPTER 7

Modeling Shelf Life Using Survival Analysis Methodologies

Lorena Garitta and Guillermo Hough

CONTENTS

7.1 WHAT IS SURVIVAL ANALYSIS?

Generally, survival analysis is a collection of statistical procedures for data analysis for which the outcome variable of interest is time until an event occurs (Kleinbaum, 1996). The problem of analyzing time-to-event data arises in a number of applied fields, such as medicine, biology, public health, epidemiology, engineering, economics, and demography (Klein and Moeschberger, 1997). Following are a series of examples of how time to an event of interest is considered (Gómez and Langohr, 2002):

- In a clinical trial of a certain medicine, time zero is when patients are randomly allocated to treatments. Time to event is the time until cancer remission or a clinical indicator falls below a certain level (e.g., viral count falls below 500).
- In industrial durability tests, it is of interest to know, for example, how long a car tire lasts. In this case, instead of "time to event," "kilometers to event" are used. That is, instead of recording the time the tire is on the road, the distance in kilometers run by the car until the tire wears out is recorded.
- In a sensory shelf life (SSL) study, time to event would be measured from when the product left the manufacturing plant until it was rejected by a consumer (Hough et al., 2003).

7.2 CENSORING

Time-to-event data present themselves in different ways, which creates special problems in analyzing such data. One feature often present in time-to-event data is known as censoring, which, broadly speaking, occurs when some lifetimes are known to have occurred only within certain intervals. There are three basic categories of censoring: right-, left-, and interval-censored data.

7.2.1 Right Censoring

Subjects are followed until the event of interest occurs. If the event of interest does not occur during the period the subject is under study, this observation is right censored. Continuing the examples, this type of censoring can occur if

At the end of the study:

- A cancer patient is still alive.
- A tire has not worn out.
- A consumer still accepts the sample stored for the maximum time.

In the middle of a study:

- A patient moves and leaves no forwarding address.
- A tire bursts for extraneous reasons.
- A consumer no longer wants to taste samples stored for successive times.

In all these cases, the event had not occurred up to a certain time, and this information is used in modeling the data.

7.2.2 Left Censoring

Left censoring occurs if the subject has already undergone the event of interest before the study begins. Following are two examples of left-censored data:

- In a study on the aroma persistence of a clothes rinse, standard-size hand towels are washed using the rinse, tumble dried, and kept in a cupboard. At different times after the application, respondents sniff a smelling strip with the aroma of the clothes rinse and are then asked if they can definitely detect the aroma on the towel. Suppose that the first test is 24 hours after the application. If a respondent cannot detect the aroma at this first test, then his or her data are left censored. For this respondent, the aroma disappeared sometime between Time = 0 (application) and Time = 24 hours (first test).
- In an SSL study on mayonnaise, it would not be necessary to ask consumers to taste samples after less than 2 months of storage at 25°C. If a consumer rejected a sample with 2 months of storage because the consumer was particularly sensitive to oxidized flavor, those data are left censored. That is, all that is known is that time to rejection for this consumer was somewhere between Time = 0 and Time = 2 months.

7.2.3 Interval Censoring

Interval censoring occurs when all that is known is that the event of interest occurred within a time interval. The following is an example of interval-censored data:

- In SSL tests, interval censoring is likely to occur. In a study on the SSL of cracker-type biscuits, the maximum storage time at 20°C and 60% relative humidity was considered to be 12 months. To know the exact storage time at which a consumer will reject the crackers, samples would have to be taken on a daily or continuous basis. Obviously, this is not possible. Over the 12-month period, suppose that the following storage times were chosen: 0, 3, 6, 8, 10, and 12 months. If a consumer accepted the sample with 6 months of storage and rejected the sample with 8 months of storage, what is known is that the consumer's rejection time was somewhere between 6 and 8 months of storage. These data are thus interval censored.

Actually, both right-censored and left-censored data can be considered as special cases of interval censoring. For an SSL study, with right-censored data the interval is between the last time the consumer accepted the sample and infinity, and with left-censored data the interval is between Time = 0 and the first storage time.

Table 7.1 presents data for six subjects to illustrate the interpretation given to each subject's data:

Table 7.1 Acceptance/Rejection Data for Six Subjects Who Tasted Yogurt Samples with Different Storage Times at 10°C

	Storage Time (days)							
Subject	**0**	**14**	**28**	**42**	**56**	**70**	**84**	**Censoring**
1	Yes	Yes	Yes	Yes	No	No	No	Interval: 42–56
2	Yes	Yes	Yes	Yes	Yes	Yes	Yes	Right: >84
3	Yes	Yes	No	Yes	No	No	No	Interval: 14–56
4	No	No	Yes	Yes	Yes	Yes	No	Not considered
5	Yes	No	Yes	Yes	No	No	No	Left: $\leq$56
6	Yes	No	No	No	No	No	No	Left: $\leq$14

- Subject 1 was as expected in a shelf-life study; that is, the subject accepted the samples up to a certain storage time and then consistently rejected them. The data are interval censored because we do not know at exactly what storage time between 42 and 56 days the consumer would start rejecting the product.
- Subject 2 accepted all samples. Supposedly at a sufficiently long storage time ($T >$ 84 days), the sample would be rejected; thus, the data are right censored.
- Subject 3 was rather inconsistent, rejecting the sample with 28 days of storage, accepting the sample with 42 days of storage, and rejecting the sample from 56 days onward. Censoring could be interpreted in different ways. One possibility would be to consider the data as interval censored between 14 and 28 days, that is, ignoring the subject's answers after the first time the yogurt was rejected. Another possibility, as shown in Table 7.1, is interval censoring between 4 and 56 days. We consider this option as more representative of the subject's data; that is, we assign a wider uncertainty interval regarding the storage time at which this subject rejected the yogurt.
- Subject 4 rejected the fresh sample. This subject was (a) recruited by mistake (i.e., did not like yogurt), or (b) preferred the stored product to the fresh product, or (c) did not understand the task. It would not be reasonable to consider the results of this subject in establishing the shelf life of a product. For example, for consumers who preferred the stored to the fresh product, a company would have to produce a yogurt with a different flavor profile rather than encourage the consumers to consume an aged product.
- Subject 5 was also rather inconsistent, with alternating no and yes answers. This subject's data were considered left censored. Left censoring is a special case of interval censoring with the lower bound equal to Time = 0 (Meeker and Escobar, 1998). But, as the literature and statistical software distinguish it, we have also done so. The left censoring could be considered as $t \leq 14$ days or $t \leq 56$ days. As for Subject 3, a wider interval is recommended as this reflects the true uncertainty of the subject's response.
- Subject 6 was consistent, accepting the sample with 0 days of storage and rejecting from 14 days onward. The subject's data were considered left censored, $t \leq 14$ days.

As mentioned, one of the characteristics of SSL data is the presence of interval-censored data. Not all statistical software has the necessary procedures to deal with this type of censoring. Commercial software such as TIBCO Spotfire S+ (TIBCO, Inc., Seattle, WA) and SAS (SAS Institute, Inc., Cary, NC) have interval-censoring

Table 7.2 Data in Table 7.1 Transformed According to Procedure for Interval-Censoring Calculations

Row Number	Consumer	Low Time (days)	High Time (days)	Censoring	Code
1	1	42	56	Interval	3
2	2	84	84	Right	0
3	3	14	56	Interval	3
4	5	56	56	Left	2
5	6	14	14	Left	2

procedures. R, a free-access statistical package that can be downloaded (http://www.r-project.org) does have a procedure for interval-censoring calculations.

Table 7.2 presents the data from consumers of Table 7.1 in a form ready to be used in the statistical software mentioned. From the left, the first column shows row number. Note that there are five rows, not six; this is because one consumer rejected the fresh sample, as can be seen in Table 7.1. The second column indicates the consumer numbers corresponding to Table 7.1. The third and fourth columns are the low and high time intervals corresponding to each consumer's rejection time; for right- and left-censored data, the corresponding time is repeated. The fifth column indicates the types of censoring corresponding to each consumer, and the sixth column is the code used by software to interpret each type of censoring: 2, 0, and 3 for right-, left-, and interval-censored data, respectively.

7.3 SURVIVAL AND FAILURE FUNCTIONS

Let T be the time of occurrence of event e. Event e could be death, appearance of a tumor, giving up smoking, end of itching symptoms, or a projector lamp burning out. For SSL studies, event e is rejection of a stored product by the consumer. T is a random nonnegative variable whose distribution can be characterized by the following functions:

- Survival function $S(t)$
- Failure function (also referred to as the cumulative distribution function) $F(t)$
- Probability density function $f(t)$
- Hazard function $h(t)$

If any of these functions is known, the others can be determined univocally. We define the failure function.

The failure or rejection function (also known as cumulative distribution function of T) is the probability of an individual failing before time t: $F(t) = \text{Prob}\ (T \leq t)$ and is defined for $t \geq 0$. In SSL, the rejection function is the probability of a consumer rejecting a food product stored for less than time t. It can also be interpreted as the proportion of consumers who will reject a food product stored for less than time t. Figure 7.1 shows a typical failure or rejection curve. Its basic properties are as follows:

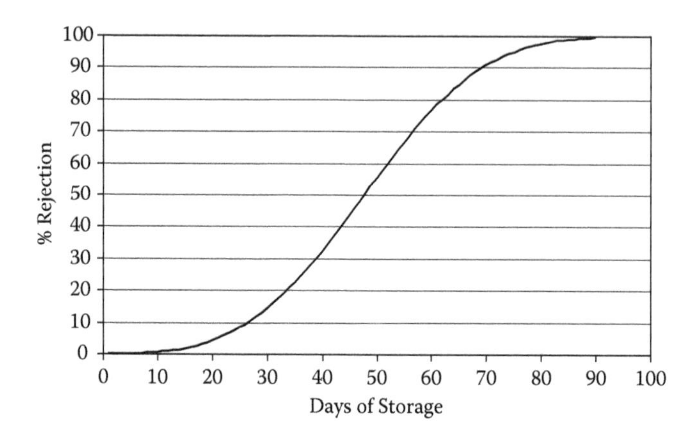

Figure 7.1 Failure or rejection function.

- $F(0) = 0$: The consumer accepts the fresh product.
- $F(\infty) = 1$: The consumer rejects the product stored for prolonged periods.
- $F(t)$ is a monotonously increasing function.
- If T is continuous, $F(t)$ is continuous and strictly increasing.
- $F(t) = 1 - S(t)$.

For full coverage of these variables and functions, refer to the work of Klein and Moeschberger (1997).

7.4 ESTIMATION OF REJECTION FUNCTION

The likelihood function, which is generally used to estimate the rejection function, is the joint probability of the given observations of the n consumers (Klein and Moeschberger, 1997):

$$L = \prod_{i \varepsilon R} (1 - F(r_i)) \prod_{i \varepsilon L} F(l_i)) \prod_{i \varepsilon I} (F(r_i) - F(l_i)) \tag{7.1}$$

where R is the set of right-censored observations, L is the set of left-censored observations, and I is the set of interval-censored observations. Equation 7.1 shows how each type of censoring contributes differently to the likelihood function.

If we can assume an appropriate distribution for the data, the use of parametric models provides adequate estimates of the rejection function and other values of interest. Usually, rejection times are not normally distributed; instead. their distribution is often right skewed. Often, a log-linear model is chosen:

$$Y = \ln(T) = \mu + \sigma W$$

where W is the error term distribution. That is, instead of the rejection time T, its logarithmic transformation is modeled. In the work of Klein and Moeschberger (1997) or Meeker and Escobar (1998), different possible distributions for T were presented, for example, the log-normal or the Weibull distribution. With the former, W is the standard normal distribution; with the latter, W is the smallest extreme value distribution.

If the log-normal distribution is chosen for T, the rejection function is given by

$$F(t) = \Phi\left(\frac{\ln(t) - \mu}{\sigma}\right) \tag{7.2}$$

where $\Phi(\bullet)$ is the standard normal cumulative distribution function, and μ and σ are the parameters of the model.

The rejection function for the Weibull distribution can be expressed as

$$F(t) = 1 - \exp\left[-\exp\left(\frac{\ln(t) - \mu}{\sigma}\right)\right] \tag{7.3}$$

where μ and σ are the parameters of the model.

The parameters of the log-linear model are obtained by maximizing the likelihood function (Equation 7.1). The likelihood function is a mathematical expression that describes the joint probability of obtaining the data actually observed on the subjects in the study as a function of the unknown parameters of the model considered. To estimate μ and σ for the log-normal or the Weibull distribution, we maximize the likelihood function by substituting $F(t)$ in Equation 7.1 by the expressions given in Equations 7.2 or 7.3, respectively.

Once the likelihood function is formed for a given model, specialized software can be used to estimate the parameters (μ and σ) that maximize the likelihood function for the given experimental data. The maximization is obtained by numerically solving the following system of equations using methods like the Newton–Raphson method (Gómez and Langohr, 2002):

$$\frac{\partial \ln L(\mu,\sigma)}{\partial \mu} = 0$$

$$\frac{\partial \ln L(\mu,\sigma)}{\partial \sigma} = 0$$

For more details on likelihood functions, see the work of Klein and Moeschberger (1997) or Meeker and Escobar (1998). In practice, the numerical maximization of the likelihood function is performed with specialized software such as TIBCO Spotfire S+ (TIBCO, Inc., Seattle, WA) or the R Statistical Package, For more details on the use of specialized software, see the work of Garitta et al. (2004a) and Hough (2010).

7.5 DESIGN CONSIDERATIONS

7.5.1 Basic and Reversed Storage Designs

In this section, alternative strategies for storing and retrieving products are presented.

7.5.1.1 Basic Design

The basic design is the first one that comes to mind when thinking about conducting an SSL test. It consists of storing a single batch at the desired environmental conditions (temperature, moisture, etc.) and periodically removing samples from storage and analyzing them. Figure 7.2 shows this design for an SSL study of cookies stored at 20°C. There are two major drawbacks to this design.

One drawback is that for each storage time a corresponding sensory analysis has to be performed. For the example shown in Figure 7.2, and depending on the type of study, this would mean assembling a trained panel, a consumer panel, or both on six separate occasions. Another drawback to the basic design is that trained assessors and/or consumers can become aware that they are participating in an SSL study. This can lead to biased results.

If a fresh sample can be stored without too much difficulty, it is advisable to do so, especially if a trained panel is going to be evaluating stored samples. In this case, it is easier for the trained assessor to compare the stored samples versus the fresh sample.

Ramírez et al. (2001) studied the SSL of sunflower oil at 35°C, 45°C, and 65°C, both in the dark and with 12 hours of daily illumination, using a basic design. A fresh sample was kept at 4°C in the dark. For each storage time, the trained panel compared the stored sample with the fresh sample.

Garitta et al. (2004b) studied the SSL of "dulce de leche," a typical Argentine dairy product prepared from milk that is concentrated by evaporation and sucrose and glucose are added. The product was all from the same batch, packaged in 250-g polystyrene pots, and stored at the following temperatures and times: 25°C for 200 days, 37°C for 122 days, and 45°C for 24 days, using a basic design. Both the 25°C and 37°C samples were removed approximately every 10 days and, for the 45°C temperature, every 4 days, for sensory evaluation. Dulce de leche is stable at room temperature, so some pots were stored at 5°C; it was considered that changes at this temperature were negligible compared with changes at the storage temperatures studied. For each storage time, the trained panel compared the stored sample with the fresh sample kept at 5°C.

7.5.1.2 Reversed Storage Design

The basic idea of the reversed storage design is to have all samples, each with a different storage time, available on the same day. Figure 7.3 shows this design for an

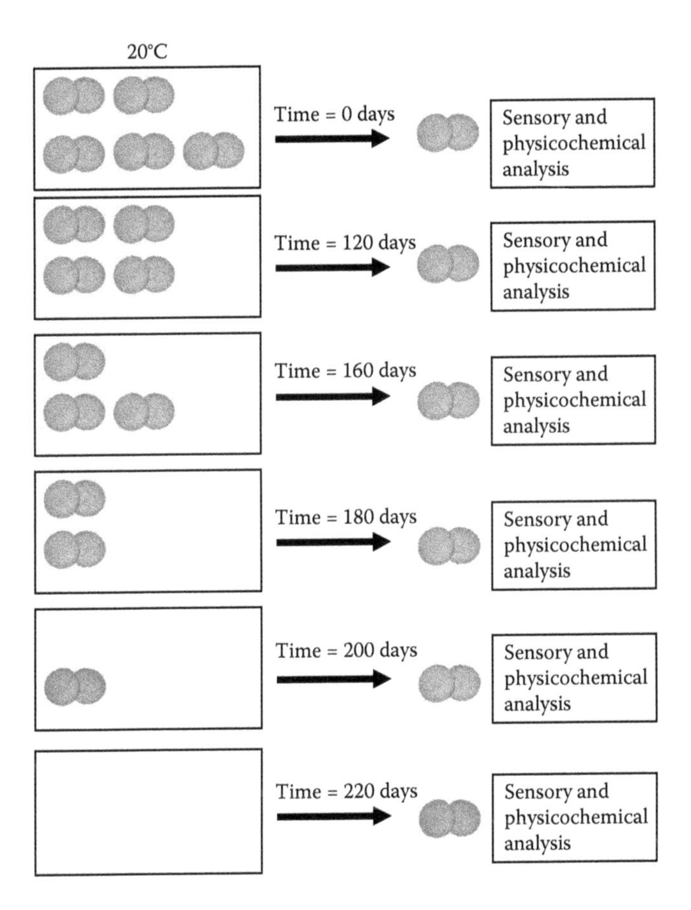

Figure 7.2 Basic design for cookies stored at 20°C.

SSL study of cookies stored at 20°C. It is assumed that changes at 2°C are negligible. As can be seen, the sample that went into 20°C storage at 0 days will have the longest storage time at 20°C: 220 days. The sample that went into 20°C storage at 20 days will have 200 days of storage at the end. The sample that remained at 2°C during the whole period was considered the fresh sample because it was never at 20°C.

The big advantage of the reversed storage design is that all samples are available for evaluation at the same time. If these samples are to be evaluated by consumers, it means recruiting the consumers on a single occasion to evaluate all samples. In the cookies example, this would mean evaluating six samples, which can easily be handled by a consumer with sufficient time between samples. For a trained panel, it is far more efficient to train the participants once and evaluate all the samples than to train the panel repeatedly to evaluate each sample separately as in the basic design.

The reversed storage design is not always possible or convenient. This occurs when there is difficulty in storing samples in conditions for which sensory changes do not occur or occur slowly.

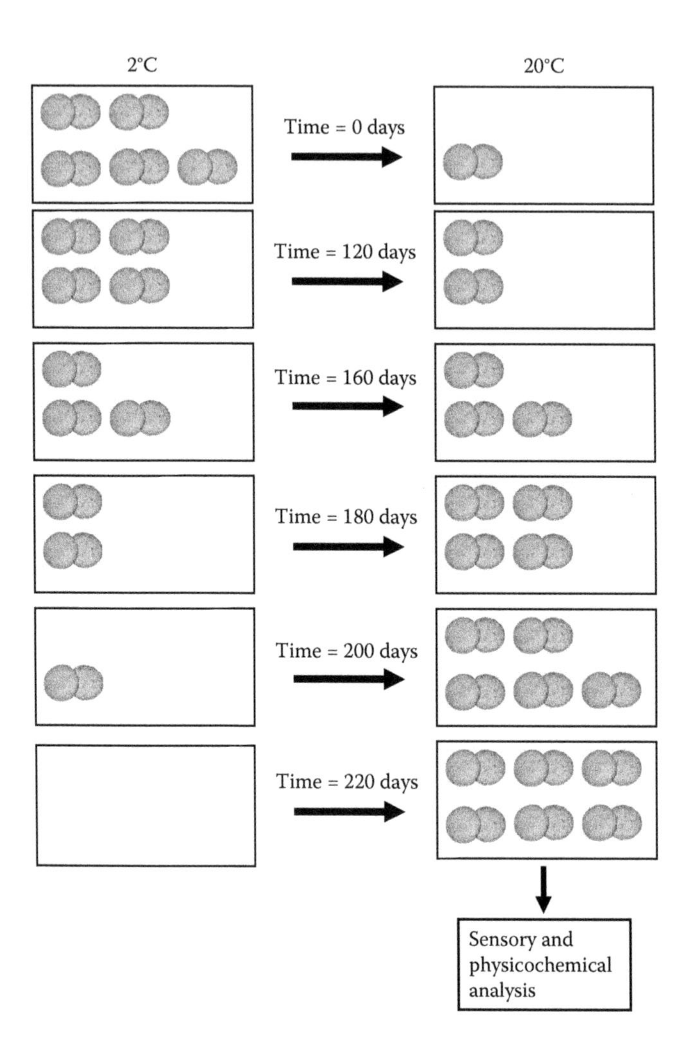

Figure 7.3 Reversed storage design for cookies.

Giménez et al. (2007) applied different approaches to estimate the SSL of brown bread treated with different enzymes. Breads were stored in a temperature-controlled storage room at 20°C for 1, 4, 7, 10, 13, 15, and 17 days. After reaching the desired storage times, breads were frozen at –20°C and stored at –18°C, providing samples with different storage times from one batch. The samples were defrosted at 20°C for 6 hours prior to their evaluation.

Another example of reversed storage design would be potato chips: Samples are stored at 25°C six or seven times. After reaching the desired storage times, potato chips are frozen and stored at -20°C. In this case, a previous test should be performed to ensure that the freezing–defrosting cycle does not change the sensory properties regardless of storage time. To this purpose, samples that have undergone the cycle are compared to samples that have not undergone the cycle by means of a sensory discrimination test, such as triangle or same–different.

7.5.2 Current Status Data

In the examples presented in the preceding section, each consumer tasted the whole set of samples corresponding to the different storage times. Because reversed storage designs were used, it was possible for each consumer to taste all samples in a single session in random order. For the basic design, the same consumers would have to be assembled for each of the storage times; this procedure can be complicated and costly. An alternative methodology that avoids the drawbacks of the basic and reversed storage designs is to have each consumer taste a single sample corresponding to a single storage time. The data thus generated are termed the current status (Shiboski, 1998). In survival analysis, there are three types of censoring: left, interval, and right. For current status data, there is no interval censoring. Statistical software to analyze current status data is the same as that used to analyze interval-censored data. Tables 7.3 and 7.4 show data obtained from current status methodology and their format necessary to be analyzed by statistical software, respectively.

For current status data, each consumer evaluates a single sample, stating whether he or she accepts or rejects the sample. These are simple data to obtain. It could be argued that a relatively large number of consumers has to be recruited (Libertino et al., 2011), but this is made easy due to the simple and quick task that each consumer has to perform. Current status data methodology would be a good solution to having consumers evaluate the samples with different storage times. Simply give each consumer a single sample to take home, consume the product as customary, and then

Table 7.3 Examples of Current Status Data Obtained from 300 Consumers Who Tasted a Product Stored Six Times and Their Censoring

	Storage Time (hours)						
Consumer	**4**	**8**	**12**	**24**	**36**	**48**	**Censoring**
1	Yes	*	*	*	*	*	Right: >4
50	No	*	*	*	*	*	Left: ≤4
51	*	Yes	*	*	*	*	Right: >8
100	*	Yes	*	*	*	*	Right: >8
101	*	*	No	*	*	*	Left: ≤12
150	*	*	Yes	*	*	*	Right: >12
151	*	*	*	No	*	*	Left: ≤24
200	*	*	*	No	*	*	Left: ≤24
201	*	*	*	*	Yes	*	Right: >36
250	*	*	*	*	No	*	Left: ≤36
251	*	*	*	*	*	No	Left: ≤48
300	*	*	*	*	*	No	Left: ≤48

* Storage time not tested by consumers.

Table 7.4 Examples of Current Status Data in Format Necessary for Analysis by Statistical Software

Consumer	Low Time	High Time	Censoring	Code
1	4	4	Right	0
50	4	4	Left	2
51	8	8	Right	0
100	8	8	Right	0
101	12	12	Left	2
150	12	12	Right	0
151	24	24	Left	2
200	24	24	Left	2
201	36	36	Right	0
250	36	36	Left	2
251	48	48	Left	2
300	48	48	Left	2

retrieve the accept or reject response. There are different methods for obtaining the answers: phone, text message, e-mail, or a visit.

Current status data methodology would be convenient for testing an alcoholic beverage such as beer. Having consumers participate in a central location test where they have to taste six or seven beer samples would not be convenient. With this number of samples, the carryover effect could be important, and the alcohol intake could raise safety issues. It would be better for each consumer to receive a single bottle of beer to take home and to state whether the single beer is accepted or rejected.

7.5.3 Storing the Fresh Sample

When conducting an SSL study, it is always of interest to compare samples of different storage with a sample considered fresh. For reversed storage designs (see Section 7.5.1.2), the ability to maintain samples without alterations is central to the design.

For some SSL studies, keeping a fresh sample is relatively simple. Martínez et al. (1998), in their study of SSL of mayonnaise, compared the stored samples with their corresponding controls. The controls belonged to the same batches as the samples under shelf life study but were kept in refrigerated storage at 5°C. It was considered that sensory changes at this temperature were insignificant compared with changes at shelf life study temperatures (>20°C). Ramírez et al. (2001) and Garitta et al. (2004b) showed other examples of uses of fresh samples (see Section 7.5.1.1).

In other cases, keeping a fresh sample is almost impossible. Araneda et al. (2008) applied survival analysis statistics to estimate the SSL of ready-to-eat lettuce stored at 4°C. Due to their "current status data" design, no control sample was necessary. However, if they would have wanted consumers or a trained panel to compare stored samples with fresh, where would they have stored the fresh sample? Freezing a batch of lettuce would not have helped as this process alters the texture considerably. Storing a batch of lettuce just above 0°C to prevent freezing would have retarded deterioration, but this batch would not be completely unaltered in relation to the batch stored at the study temperature of 4°C. One possibility would be to harvest a fresh batch at each storage time, but lettuce, like many other vegetable products, varies from one batch to another. Thus, for a product like lettuce, it is difficult, if not impossible, to have an adequate fresh sample for comparison purposes. When a consumer goes shopping for lettuce, the consumer has an internal standard formed by previous experience of what he or she considers to be fresh lettuce; thus, the lack of a physical sample of fresh product would not be a problem as in the usual shopping experience the consumer does not walk into the shop with a fresh lettuce in hand for comparison purposes. In the case of a trained panel, the lack of a physical standard means they have to rely on the training they received on the sensory characteristics of a fresh sample. Other examples for which a fresh sample is difficult to store are minced meat and fluid milk, among others.

In other SSL studies, keeping a fresh standard has intermediate difficulty. Curia et al. (2005) studied the SSL of commercial yogurt. Bottles (1,000 ml) from different batches were stored at 10°C in such a way as to have samples with different storage times ready on the same day. Storage times at 10°C were 0, 14, 28, 42, 56, 70, and 84 days. All batches were made with the same formulation and were checked to be similar to the previous batch by consensus among three expert assessors.

7.5.4 Number of Consumers

Survival analysis has been applied to estimating the shelf life of foods based on consumers' accepting or rejecting samples stored at different times (Hough et al., 2003, 2006; Gámbaro et al., 2004; Curia et al., 2005; Salvador et al., 2005). The number of consumers used for these estimations has varied between 50 and 80. Although shelf life confidence intervals have been reasonable, there are no systematic criteria in the choice of the number of consumers.

Hough et al. (2007) published an article describing a simulation study to estimate the number of consumers necessary for shelf life estimations based on survival analysis statistics. The results were presented in the form of operational curves: Type IIβ error versus Δt, where Δt is the difference we are willing to accept between the true shelf life and the estimated shelf life. These curves were presented for values of the σ Weibull parameter of 0.17, 0.44, and 0.71 and for supposed shelf life values of 2, 3, and 4 in an adimensional 0–6 time scale.

If a researcher has no previous knowledge of the σ value of the system, an intermediate value of $\sigma = 0.44$ will have to be assumed. Also, he or she can suppose that the shelf life will be close to the middle of the times studied; that is, $t = 3$ in the adimensional time system used by Hough et al. (2007). A reasonable Δt would be 0.5 on the 0–6 adimensional time scale. A type Iα error = 0.05 was considered, and choosing a type IIβ error = 0.2, which corresponds to a power = 0.8, allows reading a value of approximately $N = 120$ from the operational curves. This would be a recommended number of consumers for SSL studies based on survival analysis statistics. In the present chapter, studies performed before Hough et al.'s publication (2007) are presented; thus, different values of N are reported.

7.5.5 Percentage of Rejection

After calculations have been performed, an SSL value has to be recommended. To do this, an adequate percentage rejection has to be adopted. What can be considered adequate? Gacula and Singh (1984) mentioned a nominal shelf life value considering 50% rejection, and Cardelli and Labuza (2001) used this criterion in calculating the shelf life of coffee. Curia et al. (2005) estimated SSL values of yogurt for 25% and 50% rejection probabilities. This means that if a consumer tastes a product with a storage time corresponding to 50% rejection, there is a 50% probability that the consumers will reject the product. This can sound too risky, yet it must be remembered that we are referring to a consumer who tastes the product at the end of its shelf life. Distribution times usually guarantee that the proportion of consumers who taste the product close to the end of its SSL is small. Of this small proportion of consumers, 50% will reject the product, and 50% will accept it.

7.6 APPLYING SURVIVAL ANALYSIS TO ESTIMATE SENSORY SHELF LIFE

To illustrate the methodology to be applied to estimate SSL using survival analysis statistics, data from minced meat and vegetable powdered soup tests are used.

7.6.1 Minced Meat

Minced beef was obtained from boneless arm clod with visible fat removed. Once minced, the whole batch was thoroughly stirred to homogenize. Thirty-gram portions of minced beef were placed in 100-ml clear glass bottles with screw caps. These bottles were frozen to –18°C. The bottles were removed from the freezer at different time intervals so that all samples with different storage times were ready at the same time. The samples were thawed in a refrigerator at 9°C for 2 hours, reaching a final temperature of 4°C; this was considered time $T = 0$ hours.

After thawing for 2 hours, the samples were placed at 2°C for the following times: 0, 1, 2, 4, 6, 8, and 10 days, using a reversed storage design.

Sixty consumers who bought minced meat or prepared food using minced meat at least once a fortnight were recruited from the population of the city of Nueve de Julio, Buenos Aires, Argentina. Each consumer received seven bottles with minced meat (one for each storage time) monadically in random order. For each sample, they had to look at the bottle and answer the question: "Would you normally consume/buy this product? Yes or No?" Their acceptance or rejection of the sample was based exclusively on appearance. Other attributes relevant to the sensory shelf life of minced meat, such as odor or flavor of the cooked product, could have been evaluated, but appearance was chosen for simplicity to illustrate the methodology. The tests were conducted in a sensory laboratory with individual booths with artificial daylight-type illumination, temperature control (between 22°C and 24°C), and air circulation.

Censoring examples: If a consumer accepted all the samples stored less than 2 days and rejected all the samples stored more than 4 days, his or her rejection time was interval censored between 2 days and 4 days. If a consumer accepted all samples tried, then his or her rejection time was larger than the last time observed and was therefore right censored. If a consumer accepted the fresh sample and rejected all the others, this rejection time was left censored. If a consumer rejected the fresh sample, these results were excluded from the study.

Once the censoring for each consumer had been determined, CensorReg procedures from TIBCO Spotfire S+ was used to estimate the μ and σ parameters of the log-normal distribution.

To date, there are no statistical tests to compare the goodness of fit of different parametric models used for interval-censored data. Therefore, visual assessment of how parametric models adjust to the nonparametric estimation is the common practice in choosing the most adequate model.

Figure 7.4 shows how six standard distributions were fitted to the minced meat data. One of these distributions was finally chosen: the log-normal because of the good fit to the data.

The maximum likelihood estimates of the parameters of these models were as follows (see Equation 7.2):

$$\text{Log-normal: } \mu = 1.37, \sigma = 0.71$$

If these parameters are introduced in Equation 7.2, percentage rejection versus storage time can be graphed as shown in Figure 7.5 (e.g., using Excel).

Table 7.5 shows the percentage rejection values for 25% and 50%, together with their confidence intervals and standard errors. Minced meat shelf lives considering a 25% and 50% of rejection were 2 and 4 days approximately, respectively.

To determine the SSL of minced meat, the focus was on the probability of a consumer accepting or rejecting a product after a certain storage time. An important aspect of survival analysis methodology is that experimental sensory work is relatively simple. In this case, 60 consumers each looked at seven bottles with minced meat samples with different storage times, answering yes or no to whether they would consume or buy the samples. This information was sufficient to model the probability of consumers accepting the products with different storage times, and from the model, shelf life estimations were made. There was no necessity to have a trained sensory panel.

Another important aspect is that the information obtained from consumers by this method was directly related to their everyday eating and buying experience. When consumers are confronted with a food product, they either accept or reject it. They do not mentally assign the product a hedonic score of 8 on a 1–9 scale and thus decide the product is acceptable or assign the product a score of 4 and thus decide to reject the product. Survival analysis methodology taps directly into consumer experience.

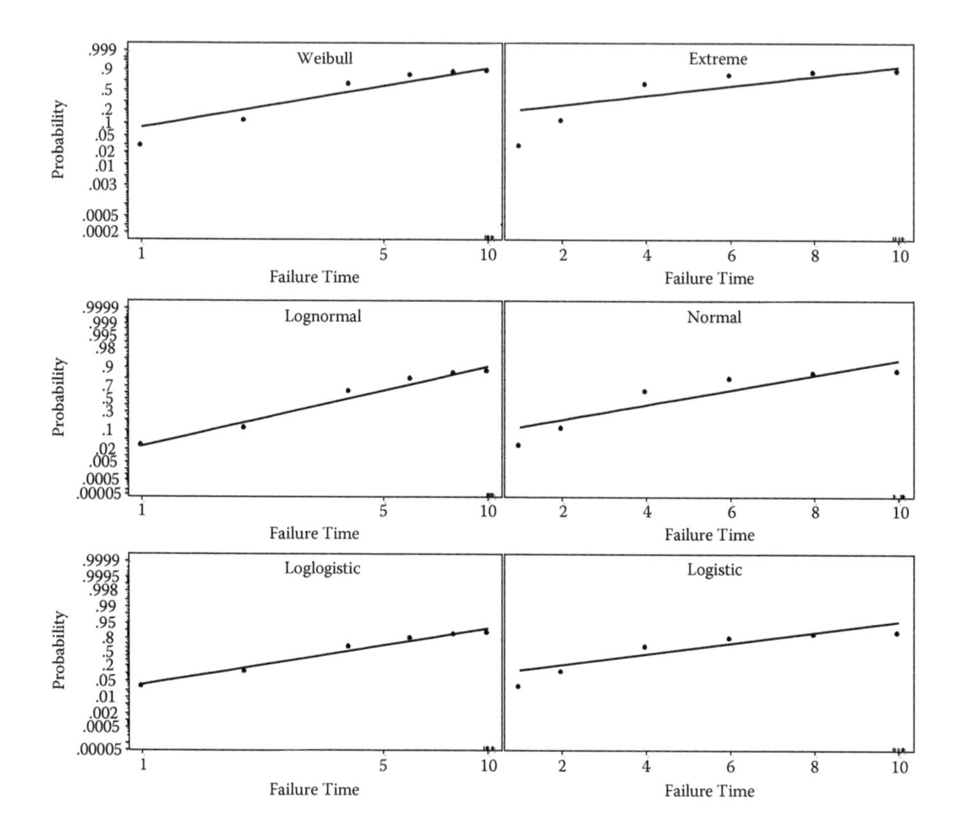

Figure 7.4 Probability of consumer rejecting the minced meat compared with storage time for six distribution models.

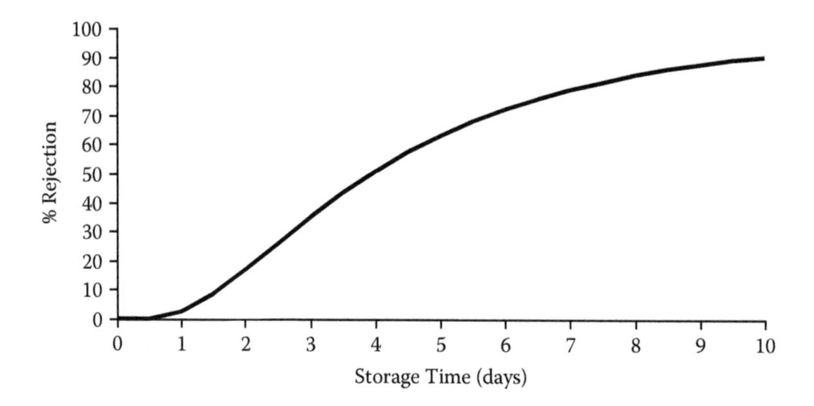

Figure 7.5 Percentage rejection versus storage time corresponding to the minced meat data for the log-normal model.

Table 7.5 Estimated Minced Meat Shelf Life Values Corresponding to 25% and 50% Rejection Probabilities, Together with Lower and Upper 95% Confidence Intervals

Percentage Rejection	Estimated Shelf Life (days)	Lower Confidence Intervals	Upper Confidence Intervals
25	2.44	1.85	3.21
50	3.93	3.13	4.93

7.6.2 Vegetable Powdered Soup

Vegetable powdered soup was stored in a temperature-controlled storage room at 23°C for 3, 5, 8, 10, 12, and 14 months. After reaching the desired storage times, powdered soup was frozen and stored at -20°C, providing samples with different storage times from one batch. Also, a fresh sample was frozen and stored at -20°C corresponding to time = 0. The samples to be tasted were prepared with 16 g of frozen powdered soup and 1 L of boiling water.

Sixty consumers, adult women, who consumed vegetable powdered soup at least once a week, were recruited from the population of the city of Nueve de Julio, Buenos Aires, Argentina. Each consumer received seven soups in Styrofoam cups with lids (one for each storage time) monadically in random order. For each sample, they had to drink and answer the question: "Would you normally consume this product? Yes or No?" The tests were conducted in a sensory laboratory with individual booths with artificial daylight-type illumination, temperature control (between 22°C and 24°C), and air circulation.

Censoring cases: Of the 60 consumers, data from 7 were interval censored, 3 left censored, 35 right censored, and 15 rejected the fresh sample, so their data were not considered. In the local Argentine market, these powdered soups generally reach consumers with over 2 months of storage; thus, they present a certain degree of storage flavors that consumers identify with a normal or fresh sample. When they evaluated a sample with storage time = 0 as in the present study, some consumers found this sample strange and different from their regular powdered soup they were used to eating. This can explain the relatively large number of consumers who rejected the fresh sample.

As mentioned in Section 7.2.3, R software has a procedure for interval censoring calculations, and this statistical software was used in the present study.

To compare which model best fit the data, TIBCO Spotfire S+ offers the possibility of producing a graph that compares the fit of different distributions with the experimental data, and thus a visual comparison defines which is the most adequate distribution. R does not produce this graph; thus, a way to define the best fit is to compare the log-likelihood values; the model that gives the lowest absolute log-likelihood would be the best. In all the data we have processed, the criteria of choosing the model with the lowest log-likelihood coincided with the criteria of visual examination performed with the TIBCO Spotfire S+ software. For the present powdered soup data, the Weibull distribution had the lowest log-likelihood value (see Table 7.6).

Table 7.6 Log-Likelihood Values for Different Models Corresponding to the Vegetable Powdered Soup Data

Model	Log-Likelihood
Logistic	36.3
Gaussian	36.2
Weibull	35.9
Log Logistic	36.0
Log-normal	36.0
Exponential	37.3

Weibull parameters: $\mu = 3.33$, $\sigma = 0.50$ (see Equation 7.3)

Figure 7.6 shows the percentage rejection versus storage time, and Table 7.7 shows the percentage rejection values for 25% and 50%, together with their confidence intervals and standard error of the estimation.

The shelf lives of vegetable powdered soup considering a rejection of 25% and 50% were 15 and 23 months, respectively. For both rejection probabilities, the SSL was greater than maximum storage time. The censored data anticipated this result. There were 35 consumers with right-censored data. That is, more than half the consumers accepted the product with the maximum storage time.

This does not necessarily mean that the study should have been extended beyond 14 months of storage. What the study showed is that, considering a 25% rejection probability, the estimated shelf life was 15 months, 3 more than the regular practice in the country where the study was conducted. This is valuable information.

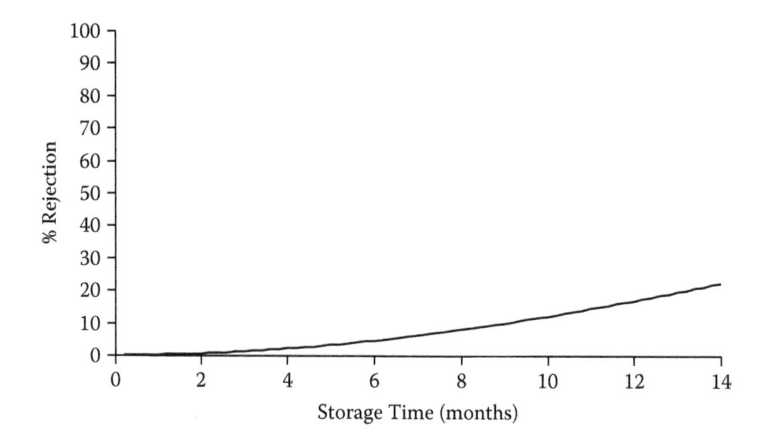

Figure 7.6 Percentage rejection versus storage time corresponding to the vegetable powdered soup data for the Weibull model.

Table 7.7 Estimated Vegetable Powdered Soup Shelf Life Values Corresponding to 25% and 50% Rejection Probabilities, Together with Lower and Upper 95% Confidence Intervals

Percentage Rejection	Estimated Shelf Life (months)	Lower Confidence Interval	Upper Confidence Interval
25	15.01	11.51	19.59
50	23.35	14.96	36.44

As shown for this vegetable soup, estimated sensory shelf lives are not always within the experimental storage times. The advantage of the parametric models is that the SSL of the product can be estimated beyond the last storage time used in the study. However, confidence intervals of these estimations can be large, as shown in the last row of Table 7.7. If all consumers accept the sample stored at the maximum time, calculations are not possible; here again, this information could be valuable.

7.7 PRODUCT MONITORING ON THE MARKET

When collecting a sample from a supermarket shelf, neither the original quality of the production batch nor the storage conditions of the sample are known. Thus, when differences between samples are found, there could be a confusion between production batch, storage conditions, and storage time effects. It could be hypothesized that if quality standards are high and storage conditions are generally uniform, then any differences between collected samples would be due to storage time. Thus, in principle, collecting samples with different "best by" dates from supermarket shelves would be a valid alternative to storing samples under controlled conditions.

Samples of a strawberry fruit drink (SFD) and sweet biscuits with different best by dates were collected from supermarket shelves. The samples were evaluated by consumers using a hedonic scale and an accept or reject decision.

When collecting samples from the shelf with different storage times, it is possible that the sample corresponding to storage time = 0 is not available. In the case of the SFD, the freshest sample that was collected had 8 weeks of storage. Of the 79 consumers who evaluated this sample, 30 rejected it. It was considered that these consumers were particularly sensitive to changes produced during storage of this product; thus, their data were retained for the shelf life estimation. Data from these consumers were considered left censored; that is, their rejection time was between 0 and 8 weeks.

The Weibull model was found to model the data adequately (see Equation 7.3). The estimated parameters were $\mu = 3.42$ and $\sigma = 0.86$.

The estimated shelf life with lower and upper 95% confidence intervals corresponding to a 50% rejection probability was 22 ± 5.6 weeks. The SFD used in the present study had a label "best before" time of 26 weeks (6 months). This was close to the estimation of 22 weeks. It would be advisable for the manufacturers to conduct a controlled shelf life study to confirm the best before date they should be printing on their labels.

Table 7.8 Mean Overall Acceptability and Percentage Rejection for Biscuits Collected with Different Storage Dates

	Digestive Biscuits	
Storage Time (weeks)	Overall Acceptability (1–9 scale)	Percentage Rejection
2	7.1	13
10	6.6	19
14	6.6	19
19	7.0	17
39	6.6	25

Table 7.8 presents overall acceptability and percentage rejection for the biscuits collected from supermarkets with different best before dates. Analysis of variance for the acceptability data resulted in a significant level of 2%. However, these differences between samples did not follow the expected trend in relation to dates. It was not clear which samples were outliers. *T*10 and *T*14 either were from poor batches or suffered mild storage conditions, or *T*19 was from an outstandingly good batch or had suffered mild storage conditions.

It would not be reasonable to submit the data from these samples to shelf life estimations based on methods such as survival analysis or cutoff point (Hough, 2010, Chapters 4 and 6), as the acceptability does not follow the expected trend over time. However, there is valuable shelf life information in the data. The biscuits had a label best before time of 39 weeks. As samples collected over this period had percentage rejections of 25% or less (Table 7.8), the manufacturer can either leave the best before date as it is with the confidence that it is a reasonable shelf life or can initiate a shelf life study under controlled storage conditions to analyze if the best before date can be extended.

Collecting samples from the supermarket shelves can provide approximate shelf life estimates, such was the case of the SFD. For other products, such as the case study of the biscuits, shelf life estimates were not possible. However, the percentage rejection for samples with different storage times can provide information regarding the product's acceptability under realistic storage conditions.

7.8 SHELF LIFE ESTIMATION OBTAINED BY APPLYING SURVIVAL ANALYSIS STATISTICS TO IN-/OUT-OF-SPECIFICATION SCORES

Survival analysis applied to shelf life studies has been presented in this chapter as a methodology with the emphasis placed on the consumer accepting or rejecting a stored product. Consumers are presented with samples that have been stored for different times, and they evaluate if they would normally consume that product or not.

There are situations for which survival analysis with data from consumers is difficult to apply. The foremost difficulty arises when there are several covariates, each with several levels. For example, in a study on the SSL of a salad dressing, the following covariates were of interest:

(a) Two storage temperatures: 4°C and 23°C
(b) Three packaging materials: glass (transparent), PET (polyethylene terephthalate), and PP (polypropylene)

With these two covariates, the total number of treatments was

$$2 \text{ Temperatures} \times 3 \text{ Packaging materials} = 6 \text{ Treatments}$$

For each of these treatments, there were six storage times.

Considering accepted statistical parameters, the recommended number of necessary consumers for survival analysis experiments is 120 (Hough et al., 2007). Thus, for the six treatments, a total of 6 × 120 = 720 consumers would have to be recruited. This is not a reasonable proposal.

Could the same 120 consumers evaluate the six treatments in different sessions? This may be convenient, but not advisable. After the first two or three treatments, the consumers would get wise to the fact that among the samples corresponding to the six storage times of each treatment there were some to be rejected. They also acquire a degree of training in detecting the off-flavors or color changes developed during storage.

There are other cases for which the number of covariates or levels is not high, yet there is difficulty in recruiting consumers.

One alternative to survival analysis using consumer data is the cutoff point methodology described in detail by Hough (2010). Another alternative is the use of a cutoff point with trained panel data to establish when a trained assessor finds a sample out of specification and then applying survival analysis statistics to estimate the shelf life. To illustrate this last alternative, data were taken from a Sensometrics workshop (Garitta and Hough, 2006).

A salad dressing was stored in the dark at two storage temperatures: 4°C and 23°C. The dressing was stored in three different packaging materials: glass, PET, and PP. A panel of 11 trained assessors measured a total of 15 sensory descriptors at six different storage times: 0, 62, 135, 197, 244, and 307 days. Fresh flavor and color intensity were chosen as critical descriptors, and they showed clear changes during storage and can influence consumer acceptability of the product.

To estimate the cutoff point for each of these descriptors, a consumer study would have been necessary. In the absence of consumer data, it was assumed that for color the product was out of specification if the assessor's score was greater than 4 and for fresh flavor if the score was less than 6 on the 1–9 intensity scale. Each assessor evaluated dressings stored in three materials, at two temperatures, by duplicate. Duplicates were averaged for the present study.

Table 7.9 illustrates the treatment of the raw data, for which the variability inherent to the sensory data was taken into account. Assessor 1, when evaluating the dressing stored in glass at 4°C, was absent for the evaluation of storage times 135 and 307 days. These missing data led to a wider censoring interval. Assessor 2, when evaluating glass at 4°C, was not consistent. He evaluated storage times of 62 and 135

Table 7.9 Color Evaluation from Two Trained Assessors

Assessor	Packaging	Temperature (°C)	Storage Times (days)						Censoring
			0	62	135	197	244	307	
1	G	4	3.5	2.9	*	3.4	2.7	*	Right: >244
			In	In		In	In		
1	PP	4	3.9	2.3	*	4.2	4.1	*	Interval: 62–197
			In	In		Out	Out		
2	G	4	3.3	4.8	5	3.2	4.2	4.9	Left: 0–244
			In	Out	Out	In	Out	Out	
2	PET	23	3.3	4.4	5.8	5	5.6	5.6	Interval: 0–62
			In	Out	Out	Out	Out	Out	

Trained assessors evaluated salad dressing stored in glass (G), PET (polyethylene terephthalate), and PP (polypropylene) at 4°C and 23°C for different storage times. The color was considered out of specification when assessed at more than 4 on a 1–9 sensory scale.

* Missing value

Table 7.10 Censored Data Corresponding to the Assessors Shown in Table 7.9: Lower and Upper Time Interval, Type of Censorship, Packaging, and Inverse of Storage Temperature

Assessor	T_{low} (days)	T_{up} (days)	Censor	Packaging	1/Storage Temperature (K)
1	244	244	Right	Glass	0.00361
1	62	197	Interval	PP	0.00361
2	0	244	Left	Glass	0.00361
2	0	62	Left	PET	0.00338

days as out of specification and a storage time of 197 days as in specification; this inconsistency is reflected in a widening of the censoring interval.

Table 7.10 shows the censored data corresponding to assessors from Table 7.9. This table is in the format read by TIBCO Spotfire S+ or the R statistical package. As there were 11 assessors, three packaging materials, and two storage temperatures, the complete table had 11 × 3 × 2= 66 rows.

A log-normal parametric model was fitted to the data. For this model, percentage probability of an assessor finding the color of the dressing out of specification was given by

$$\%Out(T) = \Phi\left(\frac{\ln(T) - \mu}{\sigma}\right) x100 \tag{7.4}$$

Considering an acceleration regression model (Hough et al., 2006) with the inclusion of covariates (Meeker and Escobar, 1998), the value of μ from Equation 7.4 is

$$\mu = \beta_0 + \beta_1 .PET + \beta_2 .PP + \frac{E_a}{R} \cdot \frac{1}{STK} \tag{7.5}$$

where

$\beta_1 =$ 1 if packaging is PET, and = 0 otherwise
$\beta_2 =$ 1 if packaging is PP, and = 0 otherwise
$E_a =$ activation energy
$R =$ gas law constant
$ST =$ storage temperature (K)

A log-likelihood test (Hough, 2010) showed that both packaging and temperature were significant in the model shown in Equation 7.5. In Table 7.11, the estimated parameters of the model are reported.

Figure 7.7 shows the percentage probability of an assessor finding the color of the salad dressing out of specification versus storage time for glass, PET, and PP packaging materials. The packaging effect was clear, with salad dressing in glass having the most prolonged shelf life.

Table 7.11 Parameters of the Log-Normal Model Defined by Equation 7.5 for Color and Fresh Flavor of Salad Dressing Evaluated by a Trained Sensory Panel

Parameters	Color	Fresh Flavor
β_0	-1.53	-13.68
β_1	-0.206	-0.308
β_2	-0.582	-0.661
E_a (cal/mol)	4,090	10,900
σ	0.428	0.700

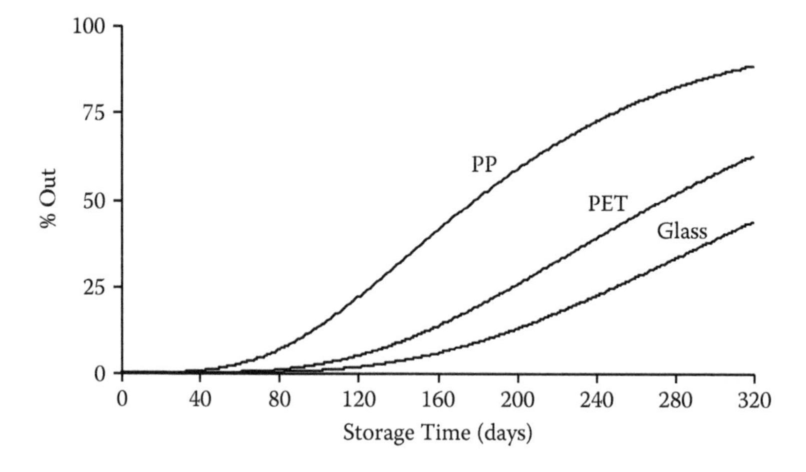

Figure 7.7 Percentage probability of an assessor finding the color of the salad dressing out of specification versus storage time for glass, PET, and PP packaging materials.

Estimated shelf life values for a 25% out-of-specification probability are in Table 7.12. PP packaging would not be advisable as the dressing in this packaging had significantly lower shelf life values than in glass or PET. Whether PET could be used depends on commercial turnover and storage temperature. If a storage temperature other than 4°C or 23°C were to be considered (e.g., 10°C), shelf life values can be estimated using Equations 7.4 and 7.5 and applying parameters from Table 7.9.

The activation energy for the fresh flavor was higher for flavor than for color; this means that with an increase in temperature the rate of fresh flavor loss increased more rapidly than the rate of color gain. The effect is shown in Figure 7.8. At 23°C, out-of-specification probabilities are completely different; however, at 4°C the curves are similar. This can also be observed in the shelf life values in Table 7.12. At higher storage temperatures, loss of fresh flavor is the critical descriptor, while at lower refrigeration temperatures, both flavor and color have to be closely watched. This case study showed how accelerated studies in relatively complex food products, such as a salad dressing, could lead to misleading results if tests are only performed under accelerated conditions.

Table 7.12 Estimated Shelf Lives Corresponding to a 25% Probability of a Trained Assessor Finding a Sample Out of Specification for Salad Dressing Stored at 4°C and 23°C in Glass, PET, and PP

	4°C		23°C	
Packaging	Color (days)	Fresh Flavor (days)	Color (days)	Fresh Flavor (days)
Glass	251	268	144	54
PET	198	190	112	31
PP	127	124	67	14

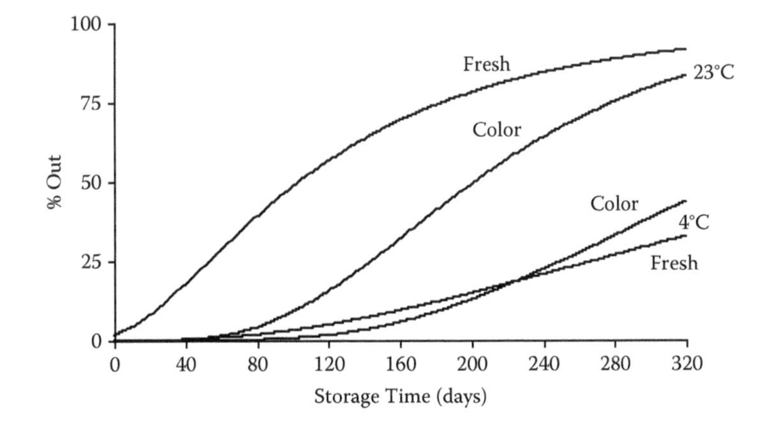

Figure 7.8 Percentage probability of an assessor finding the fresh flavor and color of the salad dressing out of specification versus storage time for salad dressing in a glass package stored at 4°C and 23°C.

It must be remembered that the present application of survival analysis was based on the assumption that for color the product was out of specification if the assessor's score was greater than 4 and for fresh flavor if the score was less than 6 on the 1–9 intensity scale. These values were taken to illustrate the application of survival analysis methodology with data from a trained panel. However, for the method to be valid, cutoff points based on consumer acceptability scores need to be established.

7.9 OTHER APPLICATIONS OF SURVIVAL ANALYSIS

In previous sections, survival analysis was applied to estimate the SSL of food. In these cases, the outcome variable of interest was time until the consumer rejection event occurred. Time can be replaced by other variables in the system under study, for example, concentration of a sensory defect (Hough et al., 2004). In SSL, the rejection function was the probability of a consumer rejecting a food product stored for less than time t. If C is the concentration of sensory defect at which a consumer rejects the sample, the failure function $F(c)$ can be defined as the probability of a consumer (or proportion of consumers) rejecting a food with the level of a sensory defect less than c, that is, $F(c) = P(C < c)$.

In Section 7.2, censoring in SSL studies was discussed. In a study to determine the concentration limit of a sensory defect, samples with different concentrations are presented to consumers. For example, concentrations could be 0, 10, 50, and 100. If a consumer accepts the sample with concentration = 10, and rejects it with concentration = 50, the exact concentration of rejection could be any value between 10 and 50. This is defined as interval censoring. If a consumer rejects the sample with concentration = 10, the rejection concentration is 10 or less, and this is left censoring. If the consumer accepts all concentrations, rejection would occur for a concentration greater than 100, and the consumer's data are right censored.

The model when concentration is the variable of interest is basically the same as the model described in Section 7.4, except that time is replaced by concentration.

To illustrate the methodology to be applied to estimate concentration limits using survival analysis statistics, data from coffee were used. These data were generated together with PhDs Patricia Restrepo and Constanza López from the Universidad Nacional de Colombia.

Coffee is consumed due to its sensory characteristics and as a stimulant. Aroma and flavor can deteriorate during storage. The most notable defect is rancid flavor due to fatty acid oxidations (Clarke and Macrae, 1985).

The aim of the present study was to determine the concentration at which consumers detected deterioration, characterized by oxidized flavor. A stock sample for oxidized defect was prepared as follows: For 2 months, coffee was placed under a glass cover, exposed to sunlight and air bubbling. Coffee treated this way was considered to be 100% oxidized.

The 100% stock sample was mixed with fresh coffee to obtain a series of seven concentrations of oxidized flavor defect: 0%, 10%, 20%, 40%, 60%, 80%, and 100% stock sample. The beverage was prepared with 7 g of coffee and 4.5 g of sugar in 100 ml boiled water. Fifty consumers, between 18 and 40 years old, who drank coffee with sugar more than four times a week, were recruited from the population of the city of Nueve de Julio, Buenos Aires, Argentina. Each consumer received the seven samples of coffee with different degrees of oxidized flavoring in glass cups at a temperature of approximately 70°C. For each sample, they had to answer the question: "Would you normally consume this product? Yes or No?".

The type of censoring is similar to that obtained from SSL data, that is, left-, interval-, and right-censored data, so once the censoring for each consumer had been determined, CensorReg procedures from TIBCO Spotfire S+ or the interval censoring procedure from R could be used to estimate the parameters μ and σ, respectively.

Table 7.13 presents data for 5 of the 50 consumers who tasted the oxidized coffee samples. The format of the raw data is similar to survival analysis SSL data.

Censoring cases: Of the 50 consumers, data for 24 were interval censored, for 19 were left censored, and for 7 were right censored. Log-normal distribution was chosen to model the data. The estimated parameters were $\mu = 3.65$ and $\sigma = 0.98$. These values can be used to plot percentage rejection versus concentration as shown

Table 7.13 Consumer Acceptance/Rejection Data for Samples of Coffee with Different Oxidized Concentrations

	Oxidized Concentration (%)							
Subject	**0**	**10**	**20**	**40**	**60**	**80**	**100**	**Censoring**
1	Yes	No	No	No	No	No	No	Left: ≤10
2	Yes	Yes	No	Yes	Yes	Yes	No	Interval: 10–100
3	Yes	Yes	Yes	Yes	Yes	Yes	Yes	Right: >100
4	Yes	Yes	Yes	Yes	No	No	No	Interval: 40–60
5	Yes	No	Yes	No	No	No	No	Left: ≤40

in Figure 7.9. Table 7.14 shows the percentage rejection values for 10%, 25%, and 50% together with their confidence intervals and standard error of the estimation.

To establish the concentration limit for oxidized flavor, a percentage rejection value has to be chosen. In Section 7.5.5, values of 25% and 50% rejection probability were considered adequate for SSL determinations. This means that if a consumer tastes a product with a storage time corresponding to 50% rejection probability, there is a 50% probability that the consumer will reject the product. As previously discussed, this might appear to be too risky, but it must be pointed out that it refers to a consumer who tastes the product at the end of its shelf life. Distribution times usually guarantee that the proportion of consumers who taste the product close to the end of its SSL is small. Of this small proportion of consumers, 50% will reject the product, and 50% will accept it. Concentration limits of a sensory defect could be used as a quality control specification or used for quality control during product storage. If these are the case, a 50% rejection probability is not acceptable. Thus, it seems reasonable to adopt a 10% rejection probability for these cases or even lower if it is an extremely critical product. For the present case of oxidized flavor, a 10% rejection probability corresponds to a

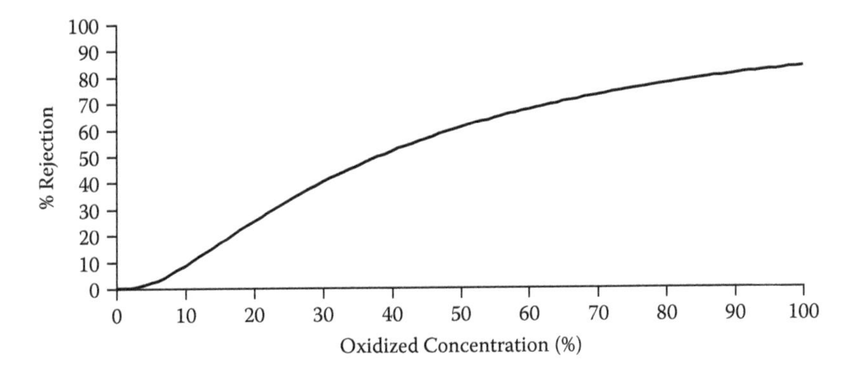

Figure 7.9 Percentage rejection versus oxidized concentration in coffee.

Table 7.14 Estimated Concentration Values Corresponding to 10%, 25%, and 50% Rejection Probabilities, Together with Lower and Upper 95% Confidence Intervals

Percentage Rejection	Estimated Concentration (%)	Lower Confidence Interval	Upper Confidence Interval
10	11.02	6.30	19.26
25	19.94	13.03	30.50
50	38.54	27.64	53.72

concentration limit of 11%, with 6% and 19% lower and upper 95% confidence intervals, respectively.

To determine the concentration limits of sensory defects, the focus has been set on the probability of a consumer rejecting a product with a certain concentration. Survival analysis statistics have been used, replacing the time variable with concentration. It has been shown that different types of censoring have to be considered in determining a concentration limit. An important aspect of this methodology is that experimental sensory work is relatively simple as no trained sensory panel work is necessary. In this work, 50 consumers tasted seven coffee samples with different concentrations of the defect, answering yes or no to whether they would consume the samples. This information was sufficient to model the probability of consumers rejecting the products with different concentrations.

7.10 REFERENCES

Araneda, M., Hough, G., and E. Wittig de Penna. 2008. Current-status survival analysis methodology applied to estimating sensory shelf life of ready-to-eat lettuce (*Lactuta sativa*). *Journal of Sensory Studies* 23: 162–170.

Cardelli, C., and T.P. Labuza. 2001. Application of Weibull hazard analysis to the determination of the shelf life of roasted and ground coffee. *Lebensmittel Wissenschaft und Technologie* 34: 273–278.

Clark, R., and R. Macrae. 1985. *Coffee. Chemistry*. London: Elsevier Applied Science.

Curia, A., Aguerrido, M., Langohr, K., and G. Hough. 2005. Survival analysis applied to sensory shelf life of yogurts. I: Argentine formulations. *Journal of Food Science* 70: S442–S445.

Curia, A., and G. Hough. 2009. Selection of a sensory marker to predict the sensory shelf life of a fluid human milk replacement formula. *Journal of Food Quality* 32: 793–809.

Gacula, M.C., and J. Singh. 1984. Shelf life testing experiments. In *Statistical methods in food and consumer research*, Chapter 8. Orlando, FL: Academic Press.

Gámbaro, A., Giménez, A., Varela, P., Garitta, L., and G. Hough. 2004. Sensory shelf-life estimation of "alfajor" by survival analysis. *Journal of Sensory Studies* 19: 500–509.

Garitta, L., Gómez, G., Hough, G., Langohr, K., and C. Serrat. 2004a. *Estadísticia de Supervivencia Aplicada a la Vida Util Sensorial de Alimentos—Tutorial introductorio y cálculos a realizar utilizando S-PLUS*. Madrid: programa CYTED.

Garitta, L., and G. Hough. 2006. Shelf life estimation obtained by applying survival analysis statistics to in/out of specification scores. In Workshop I—longitudinal sensory data. Eighth Sensometrics Meeting, As, Norway. August.

Garitta, L., Hough, G., and R. Sánchez. 2004b. Sensory shelf life of dulce de leche. *Journal of Dairy Science* 87: 1601–1607.

Giménez, A., Varela, P., Salvador, A., Ares, G., Fiszman, S., and L. Garitta. 2007. Shelf life estimation of brown pan bread: A consumer approach. *Food Quality and Preference* 18: 196–204.

Gómez, G., and K. Langohr. 2002. *Análisis de supervivencia*. Barcelona, Spain: Universitat Politécnica de Catalunya.

Hough, G. 2010. *Sensory shelf life estimation of food products*. Boca Raton, FL: Taylor & Francis.

Hough, G., Calle, M.L., Serrat, C., and A. Curia. 2007. Number of consumers necessary for shelf-life estimations based on survival analysis statistics. *Food Quality and Preference* 18: 771–775.

Hough, G., Garitta, L., and G. Gómez. 2006. Sensory shelf life predictions by survival analysis accelerated storage models. *Food Quality and Preference* 17: 468–473.

Hough, G., Garitta, L., and R. Sánchez. 2004. Determination of consumer acceptance limits to sensory defects using survival analysis. *Food Quality and Preference* 15: 729–734.

Hough, G., Langohr, K., Gómez, G., and A. Curia. 2003. Survival analysis applied to sensory shelf-life of foods. *Journal of Food Science* 68: 359–362.

Klein, J.P., and M.L. Moeschberger. 1997. *Survival analysis, techniques for censored and truncated data*. New York: Springer-Verlag.

Kleinbaum, D.G. 1996. *Survival analysis—a self-learning text*. New York: Springer-Verlag.

Libertino, L., Lopez Osornio, M.M., and Hough, G. 2011. Number of consumers necessary for survival analysis estimations based on each consumer evaluating a single sample. *Food Quality and Preference* 22(1): 24–30.

Martínez, C., Mucci, A., Santa Cruz, M.J., Hough, G., and R. Sanchez. 1998. Influence of temperature, fat content and package material on the sensory shelf life of commercial mayonnaise. *Journal of Sensory Studies* 13: 331–346.

Meeker, W.Q., and L.A. Escobar. 1998. *Statistical methods for reliability data*. New York: Wiley.

Ramírez, G., Hough, G., and A. Contarini. 2001. Influence of temperature and light exposure on sensory shelf life of a commercial sunflower oil. *Journal of Food Quality* 24: 195–204.

Salvador, A., Fiszman, S.M., Curia, A., and G. Hough. 2005. Survival analysis applied to sensory shelf life of yogurts. II: Spanish formulations. *Journal of Food Science* 70: S446–449.

Shiboski, C. 1998. Generalized additive models for current status data. *Lifetime Data Analysis* 4: 29–50.

CHAPTER 8

Packaging–Food Interactions in Shelf Life Modeling

Luciano Piergiovanni and Sara Limbo

CONTENTS

8.1 INTRODUCTION

The shelf life concept is almost always defined and explained with reference to the market and to the commercial life of packaged food products offered by the self-service distribution system (Anonymous 1974, Labuza 1982, Marsh 1986, Lee et al. 2008a). In other words, we seldom talk about the shelf life of unpackaged or raw products and rarely focus attention on shelf life in the home, whereas it is common to ask for a forecast or a measurement of the shelf life of a food product or a beverage with specific reference to its own package and to its own supply chain. In essence, what we always want to know is how long the product, in the form selected and purchased by the final consumer—which form, for the most part, is packaged—will satisfy the consumer's need and expectations.

It is evident that modern packaging is mainly represented by what is called *flexible packaging*, that is, plastic materials, combined in different ways and possibly associated with cellulosic and thin aluminum layers. One peculiarity of flexible packaging is its freedom of shape; the flexible packages are manufactured in many shapes and forms, which can be also different in type, such as basic wrappings or sealed trays, light bottles or multilayer pouches. All forms of flexible packaging are strongly characterized by the specific properties of the constituent materials, which can significantly affect the durability of the packed products and thus their shelf lives.

In contrast with more traditional rigid packaging such as glass bottles and metal cans, flexible packaging materials are permeable to gases and vapors, can be differently transparent to light and ultraviolet (UV) radiations, and, finally, as they are not completely inert, can cause absorption and the release of small molecules, which can also affect the product's commercial life. As a matter of fact, a food or beverage packed in rigid containers can be considered substantially isolated from the external environment, while the same product in a flexible package, due to its inherent properties, is more sensitive to the environmental conditions and to the characteristics of the material. This peculiarity of flexible packaging led in the past to the definition of *shelf life packaging-dependent problems* (Marsh 1986) with the aim of indicating the cases for which shelf life is strongly related to packaging performance. Not only the length of commercial life, but also the prevalent decay process can be significantly influenced by the packaging material properties.

The most obvious example of such a situation may be represented by fresh salad in a plastic pouch. Because the salad has living vegetal tissues able to breathe aerobically as well as anaerobically (Kader et al. 1989), if the product is packaged in a pouch that is highly permeable to oxygen and carbon dioxide, the aerobic respiration will be maintained for a while: The gases permeated will be consumed for energy purposes, and water and carbon dioxide will be produced; the product will decay, fading and consuming all its reservoirs in a given time. On the other hand, if the flexible material of the pouch is a high barrier to the gases, once the oxygen inside is consumed, the carbon dioxide is accumulated in the headspace, and due to the anoxic conditions established, anaerobic respiration will take place. This will lead the vegetal cells to consume other reducing substances present inside, instead of oxygen, producing end products such as acetaldehyde, ethanol, and others, which will seriously affect the sensorial characteristic of the product, determining the shelf life to end in a different and likely shorter time. The flexible packaging properties not only established the shelf life of the salad but also the manner of spoiling.

In this chapter, we refer only to this class of shelf life problems, that is, to shelf life packaging-dependent issues that definitely can be considered the result of food and packaging interactions. All models, forecasting calculi, or shelf life tests for these situations should take into account food vulnerability as well as packaging performance. Implicitly, this chapter refers only to flexible packaging materials.

The first part is dedicated to modeling the exchange of water vapor between the packaged product and the environment, when this transmission phenomenon is modulated by the permeability properties of the material and by the water absorption capability of the food.

The second part is reserved to discussion of the phenomenon of oxygen exchange, assumed to be a prototype of any event of gas exchange, different from water vapor exchange. In a subsequent section, we deal with the migration phenomena from the packaging materials to the contact food or vice versa. These well-known and broadly investigated interaction phenomena are discussed with reference not only to the possible shelf life shortage but also to the perspectives that, by modulating the release of valuable compounds like antioxidants or antimicrobials or scavenging undesired substances, a shelf life extension might be achieved. In the final part of the chapter, the important case of UV and visible light transmission across the packaging is taken into account.

8.2 ASSESSING SHELF LIFE DEPENDENCE ON WATER VAPOR TRANSMISSION ACROSS THE PACKAGING

Many deterioration paths of food quality (chemical, biological, and physical) are dependent on the moisture content in products for their speed and occurrence (Labuza et al. 1970, Eichner 1986). The food moisture content of a packaged product M (g H_2O/g dry matter), and consequently its water activity a_w, can increase or decrease according to the environmental conditions and to the water vapor (WV) resistance of the packaging material used. Even if perfectly sealed and continuous, flexible packaging can permit WV transmission across its intact walls, leading to a given amount of water Q (g) that enters or exits the package. This phenomenon is known as *permeability*, for which many good theoretical descriptions are available in the literature (Barrer 1939, Rogers 1985, Hernandez and Gavara 1999). Like all transport phenomena, WV permeability is modulated by a driving force (see Figure 8.1), corresponding in this case to the difference of water vapor pressure (WVP, bar) between the high- and the low-concentration side of the material. Inside the package, the WVP_{in} is related to the equilibrium between the product and the gas phase at the specific temperature T (°C); outside, it is related to the environmental relative humidity RH (%) and the temperature T (°C) only. Consequently, the outside water vapor pressure WVP_{out} is, or may be considered, generally constant and easy to identify, whereas the WVP_{in} changes according to food characteristics, as discussed in detail here.

The Q transmission is modulated by the package (we could say that the package resists the driving force) according to its wall thickness, its surface area, and the WV diffusivity across the material thickness, which is reduced by all the possible chemical and physical interactions of water molecules with the permeable matrix.

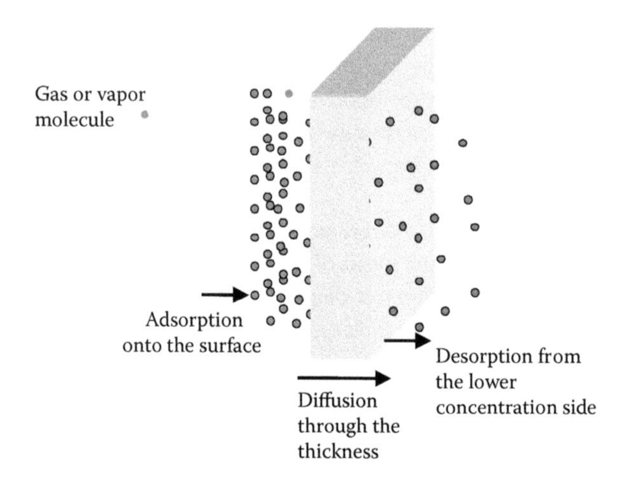

Figure 8.1 The gas and vapor permeation phenomenon.

8.2.1 Water Vapor Permeability of Packaging Materials

These interactions can be very different in strength according to the chemical nature of the material, its structure, and its hydrophilicity. Consequently, several different WV resistances, or WV barriers, are available now as packaging films and sheets, as data in Table 8.1 clearly show. The values presented in Table 8.1 are all values of WV permeability coefficients (KP_{H_2O}) of real packaging flexible materials, and they are expressed using both common engineering units (g µm m^{-2} day^{-1} bar^{-1}) and less common but more appropriate SI units (mol m^{-1} s^{-1} Pa^{-1}).

These values are essential for applying any model for shelf life assessment when WV exchange must be taken into account. However, often the figures available from the testing laboratories or from the material manufacturers are not provided as KP_{H_2O} but as *WVTR* values, that is, not WV permeability coefficients but WV transmission rates, a definitely much more common parameter (American Society for Testing and Materials [ASTM] 2010). It gives values (in g m^{-2} day^{-1}) under specified test conditions, which typically are the so-called tropical conditions: 38°C (100°F) and a gradient of *RH* of 90% across the two sides of the material, which are, de facto, accelerated conditions of the phenomenon driving force. Therefore, it is important to recognize correctly the different parameters available, to know the geometrical dimensions of the true package (permeable surface *A*, m^2 $pack^{-1}$; and thickness *l*, µm) and the driving force under test and real conditions. Only this knowledge will make it possible to transform the technical figures, obtained by the material testing or from the technical sheet provided, into data useful for a shelf life assessment. When making these calculations, the temperature (*T*, °*C*) and *RH* effects must be carefully considered, especially when an accelerated shelf life test (ASLT) is performed, increasing the environmental temperature or making the external humidity significantly different from the internal.

Table 8.1 Water Vapor Permeability Coefficients of Various Polymers (KP_{H2O})

Plastic Polymer	WV Permeability Coefficient[a]	WV Permeability Coefficient[b]
Ethylene vinyl alcohol (EVOH, 44 mol% ethylene)	48,786 to 260,191	3.14E-13 to 1.67E-12
High-density polyethylene (HDPE)	9,757 to 19,514	6.27E-14 to 1.25E-13
Low-density polyethylene (LDPE)	32,524 to 65,048	2.09E-13 to 4.18E-13
Nylon 6 (PA6)	325,238 to 650,477	2.09E-12 to 4.18E-12
Oriented polypropylene (OPP)	6,505 to 16,262	4.18E-14 to 1.05E-13
Plasticized polyvinyl chloride (PVC)	487,858 to 1,300,954	3.14E-12 to 8.37E-12
Polycarbonate (PC)	325,238 to 487,858	2.09E-12 to 4.18E-12
Polyethylene terephthalate (PET)	32,524 to 65,048	2.09E-13 to 4.18E-13
Polystyrene (PS)	227,667 to 325,238	1.46E-12 to 2.09E-12
Polyvinylidene chloride (PVDC)	650 to 19,514	4.18E-15 to 1.25E-13

[a] Units in $g \cdot \mu m \cdot m^{-2} \cdot day^{-1} \cdot bar^{1}$ at 24°C.
[b] Units in $mol \cdot s^{-1} \cdot m^{-1} \cdot Pa^{-1}$ at 24°C.

As is well known, the temperature exponentially affects any transport phenomenon. In particular, above the glass transition temperature, the diffusion coefficient of WV in the flexible packaging materials will be exponentially increased by a linear increase of the environmental temperature. This effect is normally computed by the Arrhenius relationship, and this point is addressed in more detail in the next section, devoted to oxygen transmission. In effect, for the majority of plastic materials used for food packaging applications, which are intrinsically hydrophobic, the temperature effect on the WV diffusion coefficient might not be as high as on the driving force itself (Piergiovanni et al. 1995). A change in temperature definitely changes the *WVP* of pure water, and this effect also is exponential, as the Clausius Clapeyron law states:

$$WVP = WVP_0 \cdot \exp\left[- H_{vap} \cdot \frac{|(T_0 - T)|}{R \cdot T \cdot T_0}\right] \quad (8.1)$$

where WVP_0 (bar) is the water vapor pressure at temperature T_0 (K), ΔH_{vap} is the enthalpy of vaporization or latent heat of vaporization (J mol^{-1}), R is the gas constant (8.314 J mol^{-1} K^{-1}).

The values of *WVP* of pure water at different temperatures are also easily found in tables reported by textbooks and manuals and show that at 38°C (the temperature most used in evaluating the *WVTR*) the water vapor pressure is seven or eight times higher than at refrigerating temperatures, the most common storage temperature of food in flexible packages. The knowledge of *WVP* values of pure water at the temperatures of interest is not sufficient, however; what we also need for correctly establishing the driving force is the relative humidity on the two sides of the walls of the package. Fortunately, *RH* influences the driving forces linearly, setting up which percentage of the pure water vapor pressure is really available in each case. Even if the *RH* gradient normally used in testing the *WVTR* of the packaging materials is quite high (90% usually), the difference of relative humidity in real cases is much lower, often no more than 40% (Piergiovanni et al. 1995). A classical computation that is often required to address a shelf life estimate is to convert a value of *WVTR* collected in tropical conditions $WVTR_{TC}$ in the transmission expected in real conditions $WVTR_{RC}$, that is, under a different driving force, as shown in Equation 8.2:

$$WVTR_{RC} = \frac{WVTR_{TC}}{(WVP_{TC} \cdot RH_{TC})} \cdot (WVP_{RC} \cdot RH_{RC}) \tag{8.2}$$

where the pure water *WVP*, in tropical ($_{TC}$) and in real conditions ($_{RC}$), are values referring to the temperatures of interest, as well as the *RH* values. In Equation 8.2, it is also possible to implement the Clapeyron law, leading to a useful equation (Piergiovanni et al. 1995) that makes it possible to convert values collected in accelerated conditions ($WVTR_{AC}$) to more useful data, pertinent to real conditions, for shelf life assessment (Equation 8.3).

$$WVTR_{RC} = WVP_{AC} \cdot \exp\left[-\frac{H_{vap}}{R} \cdot \left(\frac{1}{T_{RC}} - \frac{1}{T_{AC}}\right)\right] \cdot \left(\frac{RH_{RC}}{RH_{AC}}\right) \tag{8.3}$$

Equations 8.2 and 8.3 for transforming *WVTR* values obtained under different conditions of *T* and *RH* are valid if the effect of temperature is demonstrated to be more substantial for the driving force than for the diffusion of WV as it has been shown only for films with medium and low permeability made of hydrophobic polymers.

The wide use of *WVTR* values instead of KP_{H_2O} for flexible packaging materials is largely due to the peculiar phenomena of WV diffusion into polymeric materials, especially the hydrophilic materials, for which the WV diffusion must often be considered non-Fickian and thus concentration dependent (Lee et al. 2008b). KP_{H_2O} is a temperature-dependent constant that should be independent of the driving force and the thickness of the materials. Since the late 1950s, however, considerable experimental evidence was collected (Rankin et al. 1958, Karel et al. 1959, Woodruff et al. 1972) showing that materials subjected to different *WVP* gradients gave fluxes of WV leading to different values of KP_{H_2O}. In effect, when the WV permeability is high or very high, the relative humidity can also significantly affect the water diffusion coefficient in some polymers. Del Nobile et al. (2002), investigating the water transport properties of cellophane flexible films, even films laminated to a very thin low-density polyethylene (LDPE) layer, confirmed the complexity of the phenomenon of WV permeability in water-sensitive packaging films. The water molecules, acting as a plasticizer and increasing the macromolecular mobility, led to a dependence of both solubility and diffusivity coefficients on the local moisture concentration, which is not easy to take into account. To describe the dependence of diffusion coefficient $D(_{Cw})$ on water molar concentration C_w they have proposed the following empirical expression where *Ai*'s are constants and have to be regarded as fitting parameters:

$$D_{(C_W)} = A_1 - A_2 \cdot \exp\left[-\left(\frac{C_W}{A_3}\right)^{A_4}\right] \tag{8.4}$$

Using Equation 8.4 together with the equation proposed by Flory (1953) to describe water solubilization in the polymer, they were able to demonstrate that the water permeability coefficient depends on WV partial pressures at the upstream and downstream sides of the film. Similar deviations have been noted for possible changes of thickness (Schultz et al. 1949, Rankin et al. 1958, Patel et al. 1964, Banker et al. 1966, Hagenmaier and Shaw 1990); thus, measures performed on different thicknesses of the same films and under the same driving force led to KP_{H_2O}, which increased with film depth. The so-called thickness effects also justified the diffusion of a third parameter for expressing the WV transport phenomena, which is *permeance* or *permeability* (P_{H_2O}, g m^{-2} day^{-1} bar^{-1}), that is, a flux normalized for the surface, the driving force, not for the thickness, which is also common for gas permeability measures (see Section 8.3.1).

All these deviations from a theoretical proportionality of WV transmission (Fickian behavior) have been explained on the basis of interaction phenomena of water with polar groups in the polymeric network and swelling events changing the geometry of the system (Gennadios et al. 1994). Nonetheless, at the beginning of the 1990s, it also became clear that most of the methods and apparatus used for measuring *WVTR* were conceived in such a way that the air gap between the test film surface and the liquid or desiccant, employed to maintain the WVP_{in} constant, offered a not negligible resistance to WV transport, especially in the case of high *WVTR* values. This air resistance led to considerable underestimation of the fluxes ranging between 5% and 46% because of the partial pressure gradient within the "stagnant air layer," as Figure 8.2 shows. As we discuss, the same phenomenon can also affect the shelf life forecast. Because of these achievements, some authors (Biquet and Labuza 1988, Kester and Fennema 1989) proposed using terms such as *effective permeability* or *apparent permeability*; others suggested modification to the procedure of measurements to avoid the stagnant air layer (McHugh et al. 1993) or corrective equations to take into account the measurement errors (Gennadios et al. 1994). Here, the corrective equations proposed by Gennadios et al. (1994) are reported because they are common. Three cases were considered: (a) $WVP_{in} > WVP_{out}$, no *WVP* gradient outside; (b) $WVP_{in} < WVP_{out}$, no *WVP* gradient outside; (c) $WVP_{in} > WVP_{out}$, with a *WVP* gradient outside. With reference to Figure 8.2, the actual value of $WVP_{in,up}$, just below the test film, which is different from the measured WVP_{in}, is obtained using Equation 8.5:

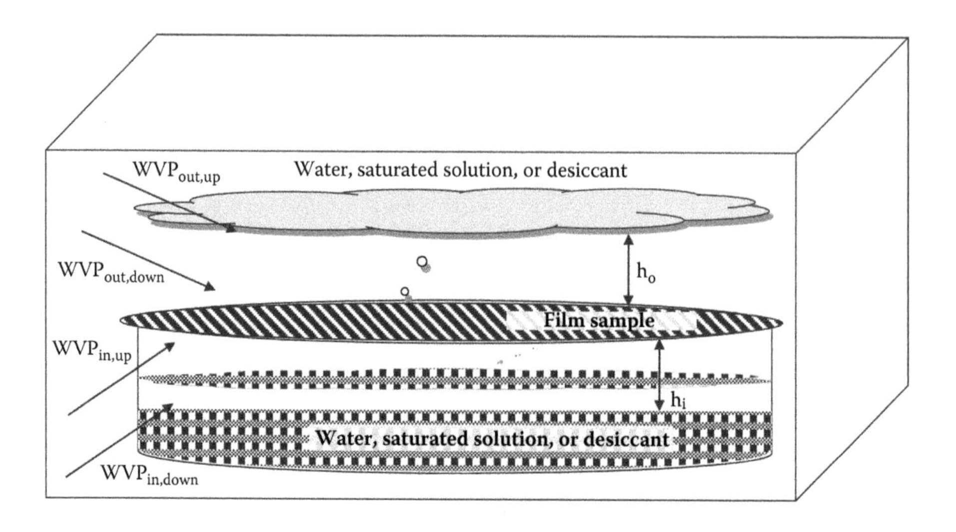

Figure 8.2 Stagnant air layer between the test film and the inner and outer phases and the consequent WVP gradient in the conventional cups for *WVTR* measurements.

$$WVP_{in,up} = P_T - \left(P_T - WVP_{in,down}\right) \cdot \exp \cdot \left(\frac{WVTR_m \cdot h_i}{C_W \cdot D_{H_2O}} \right) \tag{8.5}$$

where P_T is the total atmospheric pressure (Pa), $WVP_{in,down}$ is the water vapor pressure close to the water (or desiccant) inside the measuring cell of Figure 8.2, $WVTR_m$ is the measured WV transmission adjusted (to units mol cm^{-2} s^{-1}) by the factor 6.43 E-11, h_i is the distance (cm) corresponding to the air gap (see Figure 8.2), C_W is the molar WV concentration (estimated from the ideal gas law), and D is the water diffusion coefficient in air (cm^2 s^{-1}), empirically estimated at temperature T and 1 atm (Bretsznajder 1971).

In the first and second cases, the corrected water transmission rate ($WVTR_c$) is evaluated by Equations 8.6 and 8.7, respectively.

$$WVTR_c = WVTR_m \cdot \left[\frac{\left(WVP_{in,down} - WVP_{out,down}\right)}{\left(WVP_{in,up} - WVP_{out,down}\right)} \right] \tag{8.6}$$

$$WVTR_c = WVTR_m \cdot \left[\frac{\left(WVP_{out,down} - WVP_{in,down}\right)}{\left(WV_{out,down} - WVP_{in,up}\right)} \right] \tag{8.7}$$

If a WVP gradient is also assumed outside (the third case), the $WVP_{out,up}$ and $WVP_{out,down}$ are estimated by Equations 8.8 and 8.9, respectively, and the $WVTR_c$ according to Equation 8.10.

$$WVP_{out,up} = P_T - \left(P_T - WVP_{out,down}\right) \cdot \exp \cdot \left(\frac{WVTR_m \cdot h_0}{C_W \cdot D_{H_2O}} \right) \quad (8.8)$$

$$WVP_{out,down} = P_T - \left(P_T - WVP_{out,up}\right) \cdot \exp \cdot \left(\frac{WVTR_m \cdot h_0}{C_W \cdot D_{H_2O}} \right) \quad (8.9)$$

$$WVTR_c = WVTR_m \cdot \left[\frac{\left(WVP_{in,down} - WVP_{out,up}\right)}{\left(WV_{in,up} - WVP_{out,down}\right)} \right] \quad (8.10)$$

See also Figure 8.2 for $WVP_{in/out,up/down}$ meanings.

It is worth mentioning, however, that the most modern and automatic apparatus for determining the WV transmission rates overcome these problems in various ways. As stated, the *WVTR* values are fluxes of *Q*, that is, the amount of water that flows through a unit area per unit of time, under a specified driving force through a given thickness; thus, starting from such information, and proceeding with the shelf life assessment, it will be necessary to transform that value into a more appropriate parameter for the case study. Since we know that the thickness and the permeable surface area will not change during the shelf life, but the driving force will change according to the food moisture absorption or desorption, the most useful figure will be a WV flux normalized per the packaging unit and the driving force (ΦQ_{pack}, g_{H2O} pack^{-1} day^{-1} bar^{-1}). Assuming that the *WVTR* measure has been made on the actual thickness of the material,[1] it will be necessary (as Equation 8.11 shows) to multiply the possibly corrected transmission ($WVTR_c$) by the actual permeable surface of the package (A, m^2 pack^{-1}), and to divide by the real driving force used in the measurement (ΔWVP_{test}, bar):

$$\Phi Q_{pack} = \frac{\left(WVTR_c \cdot A\right)}{WVP_{test}} \quad (8.11)$$

8.2.2 Models to Assess Shelf Life Moisture Dependence

Knowing how much water can enter or exit the package for a driving force unit ΦQ_{pack}, we can proceed with computing the moisture transport phenomena. In doing this, we must now also consider the food behavior and not only the packaging material; that is, it is essential to take into account how the food changes assuming or

[1] If the test has been conducted on a sample of different thickness, it will be mandatory to correct the value, dividing by the thickness of the specimen and multiplying by the actual thickness in the case study.

depleting moisture, thus affecting the dynamics of the driving forces that move the moisture across the package.

In the models currently used to assess the shelf life of moisture-sensitive foods in flexible packaging (Labuza 1982, Fava et al. 2000), it is generally assumed that the Q permeated through the package immediately changes the food moisture content of the product (M), and consequently its water activity (a_w), and at the same time the WVP_{in}, which is believed to be immediately in equilibrium with the RH or a_w of the product. In other words, the driving force will, without any resistance, be progressively reduced because of the transfer of moisture and the sorption/desorption phenomena in the food; thus, the WVP_{in} will approach the WVP_{out}. Actually, this is a useful but not always applicable simplification, and the assumption related to the immediate equilibrium between the product and the WVP_{in} has not yet been verified experimentally in detail. We have recently approached this issue using small humidity sensors and specific acquisition software to continuously record moisture values at different points. A very dry Italian bakery product (*Fetta biscottata*) was chosen for its moisture sensitivity and regular shape and geometry. Different quantities of this product were hermetically packaged in two plastic trays, different for their volume, obtaining several ratios of headspace volume to product weight (cm^3 g^{-1}). During storage in a very moist environment (about 100% RH), the RH values in the product and in the headspace of the packages were measured and recorded every 30 minutes for each combination. The moisture change for a sample with the ratio 4.5 cm^3 g^{-1} (headspace volume to product weight) is shown in Figure 8.3.

For all the samples tested, the difference between the curves recorded was constant in the RH range considered (20–65%) and showed a significant delay of the products in adsorbing the moisture permeated. Experiments conducted until 95% RH was reached inside the packages demonstrated that the two curves joined only after a very long time. The results of these ongoing experiments (Lamiani 2010)

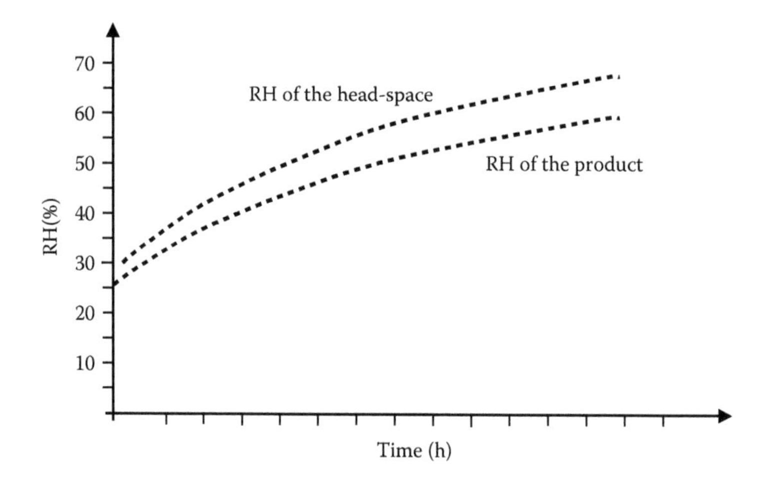

Figure 8.3 Different values of *RH* in different parts of the package during shelf life.

showed that the higher the ratios, the greater the delay of the equilibrium between food and headspace humidity; consequently, the greater is the effect on the shelf life estimate based on the common assumptions. This phenomenon, similar to the stagnant air layer effect discussed regarding *WVTR* measures and little investigated so far, is a further example of interaction between packaging and shelf life forecast and should be taken into account for the most accurate assessments.

Since it is known that the most critical factor for the acceptance of a dry food product can be represented by its moisture content *M*, the shelf life assessment of such products, assuming the common assumptions and the temperature constant as valid, should be monitored by estimating its increase, which can be computed as presented in Equations 8.12 or 8.13:

$$\frac{dM}{dt} = \frac{KP_{\mathrm{H_2O}}}{l} \cdot A \cdot \left(WVP_{out} - WVP_{in}\right) \cdot \left(\frac{1}{W_S}\right) \tag{8.12}$$

$$\frac{dM}{dt} = \Phi Q_{pack} \cdot \left(WVP_{out} - WVP_{in}\right) \cdot \left(\frac{1}{W_S}\right) \tag{8.13}$$

where W_S is the dry weight of the packaged product. Taking into account the relationship between *WVP* and *RH* (conceptually equivalent here to a_w), Equation 8.13 can be written also as in Equation 8.14:

$$\frac{dM}{dt} = \Phi Q_{pack} \cdot WVP_{\mathrm{H_2O}} \cdot \left(RH_{out} - RH_{in}\right) \cdot \left(\frac{1}{W_S}\right) \tag{8.14}$$

As already stressed, we must presume dynamic changes of the RH_{in} inside the package related to the product's characteristics, whereas it is reasonable to consider the outside conditions. Therefore, Equation 8.14 should be better represented by Equation 8.15:

$$\frac{dM}{dt} = \Phi Q_{pack} \cdot WVP_{\mathrm{H_2O}} \cdot \left(RH_{out} - f[M]\right) \cdot \left(\frac{1}{W_S}\right) \tag{8.15}$$

where the change of the internal *RH* is a function of the moisture content of the product.

The relationship between *M* and *RH* (or a_w) of a product is well known as its absorption/desorption isotherm. Therefore, the next step in developing the assessment of the shelf life of the moisture-sensitive product in flexible packaging is to implement a model that describes the isotherm mathematically in Equation 8.15. Many attempts have been made to express the equilibrium relationship in mathematical forms. The most widely used are the equations of Brunauer, Emmett, and Teller (BET), Guggenheim, Anderson, and de Boer (GAB), Halsey, Henderson, Oswin,

Table 8.2 Equations Used to Describe the Absorption and Desorption Isotherms *M* as a Function of a_w

	Equation	Useful a_w range
BET	$\frac{a_w}{(1-a_w)M} = \frac{1}{M_m C_1} + \frac{C_1 - 1}{M_m C_1}$	0.05–0.45
GAB	$\frac{M}{M_m} = \frac{C_1 C_2 a_w}{(1-C_2 a_w)(1-C_2 a_w + C_1 C_2 a_w)}$	0.1–0.9
Oswin	$M = C_1 (\frac{a_w}{1-a_w})^{C_2}$	0.1–0.85
Kuhn	$M = \frac{C_1}{\ln a_w} + C_2$	0.1–0.8
Iglesias and Chirife	$M = C_1 (\frac{a_w}{1-a_w}) + C_2$	0.1–0.6
Linear	$M = C_1 \cdot a_w + C_2$	—

Note: C_1 and C_2: food-related constants; M_m, constant humidity content on monolayer.

Khun, Inglesias, and Chirife (Iglesias and Chirife 1982, Lee et al. 2008c). These equations (see Table 8.2) are used in forecast models to determinate the shelf life of moisture-dependent products and have been implemented in software already commercially available for estimating the shelf life of sensitive food products.

The simplest model that can be used deals with the linearization of the isotherm in the range of interest, as suggested by Labuza et al. in 1972. In this case, the equation of a straight line connecting the initial and the critical values of moisture content M_o, M_c is computed, and an M_e value corresponding to the external (assumed constant) relative humidity is derived. This approximation, however, should be used cautiously because it can lead to major errors in the not rare cases of nonadequate linearity on the isotherm of the *M* values, as Figure 8.4 shows for M_o', M_c'.

Conversely, when appropriate, the linear model can be useful and easy to apply. The linearized isotherm can be described by Equations 8.16 or 8.17.

$$M = C_1 \cdot a_w + C_2 \tag{8.16}$$

$$M = C_1 \cdot \left(\frac{WVP_{in}}{WVP_{H_2O}} \right) + C_2 \tag{8.17}$$

where C_1 is the slope of the linear isotherm, and C_2 is the intercept on the vertical *M* axis in the adsorption or desorption isotherm plot. Equation 8.17 permits computation of whatever value of *WVP* inside the package in the range of interest, for example, the initial value corresponding to M_o, which will be

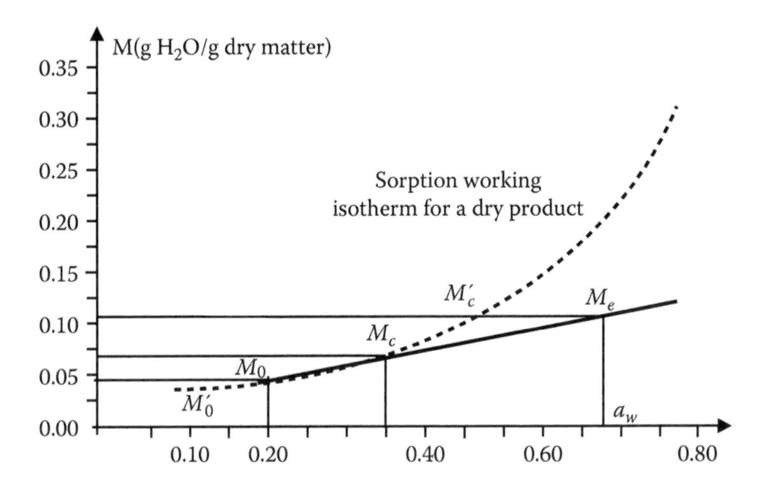

Figure 8.4 Straight-line approximation of moisture sorption isotherm for a dry product.

$$WVP_{in,t0} = \left(\frac{WVP_{H_2O}}{C_1}\right) \cdot (M_0 - C_2) \tag{8.18}$$

and the final value $WVP_{in,eq}$, corresponding to the equilibrium with the external RH_{out}, thus equal to WVP_{out}:

$$WVP_{in,eq} = WVP_{out} = \left(\frac{WVP_{H_2O}}{C_1}\right) \cdot (M_e - C_2) \tag{8.19}$$

Now, the differential Equation 8.13 can be rewritten as

$$\frac{dM}{dt} = \Phi Q_{pack} \cdot \left[\frac{WVP_{H_2O}}{C_1} \cdot (M_e - C_2) - \left(\frac{WVP_{H_2O}}{C_2}\right) \cdot (M - C_2)\right] \cdot \left(\frac{1}{W_S}\right) \tag{8.20}$$

or (and better) as Equation 8.21:

$$\frac{dM}{dt} = \Phi Q_{pack} \cdot \left[\frac{WVP_{H_2O}}{C_1} \cdot (M_e - M)\right] \cdot \left(\frac{1}{W_S}\right) \tag{8.21}$$

Equation 8.21 can be integrated between M_o and M_c and between the corresponding time zero t_0 and shelf life time $t_{s.l}$ as in Equation 8.22:

$$\int_{M_0}^{M_C} \frac{dM}{(M_e - M)} = \Phi Q_{pack} \cdot \frac{1}{W_S} \cdot \left(\frac{WVP_{H_2O}}{C_1} \right) \int_{t_0}^{t_{s.l.}} dt \tag{8.22}$$

Solving Equation 8.22, it is possible to obtain the following formulas useful for assessing shelf life in the cases of moisture gain (Equation 8.23) and moisture loss (Equation 8.24):

$$t_{s.l.} = \frac{\log_n \left[\frac{(M_e - M_0)}{(M_e - M_c)} \right]}{\Phi Q_{pack} \cdot \left(\frac{WVP_{H_2O}}{C_1} \right) \cdot \frac{1}{W_S}} \tag{8.23}$$

$$t_{s.l.} = \frac{\log_n \left[\frac{(M_0 - M_e)}{(M_c - M_e)} \right]}{\Phi Q_{pack} \cdot \left(\frac{WVP_{H_2O}}{C_1} \right) \cdot \frac{1}{W_S}} \tag{8.24}$$

It is worth noting that the variables required to be known for solving the formulas of Equation 8.23 or 8.24 belong to the food product characteristics (C_1, M_o, M_e, M_c, and W_s), to the package ΦQ_{pack}, and to the environment WVP_{H2O}, underlining the close interaction between each in the shelf life assessment and that all of them refer to a specific temperature, assumed to be constant during the shelf life; obviously, a temperature change will modify the *WVTR* and the *M* and a_w relationship. One more aspect that deserves to be mentioned regarding these formulas is related to the possible use of an ASLT. It is actually easy to increase the value of ΦQ_{pack} (e.g., reducing the thickness, increasing the permeable area, or changing the packaging material), and due to its inverse proportionality to the $t_{s.l.}$, a change in ΦQ_{pack} provides a simple way to implement a controllable acceleration of product aging (see Section 8.4).

A particular case of isotherm linearization is represented by the chance of considering the vertical step of the relationship between *M* and a_w as Figure 8.5 shows. In many cases of real shelf life problems, the matter deals with the loss of free moisture (i.e., like weight losses of fresh food or loss of skin turgor in plants and vegetables), therefore an *M* variation that implies only a negligible change in a_w. These conditions correspond to a *Q* transfer that occurs under a constant driving force, owing to a constant difference between WVP_{in} and WVP_{out}.

Assuming the driving force of the moisture transmission across the package is constant, the shelf life assessment can be reduced to a simple ratio of the critical moisture change on the WV flux normalized as per the packaging unit and the driving force of the real conditions, as in Equation 8.25.

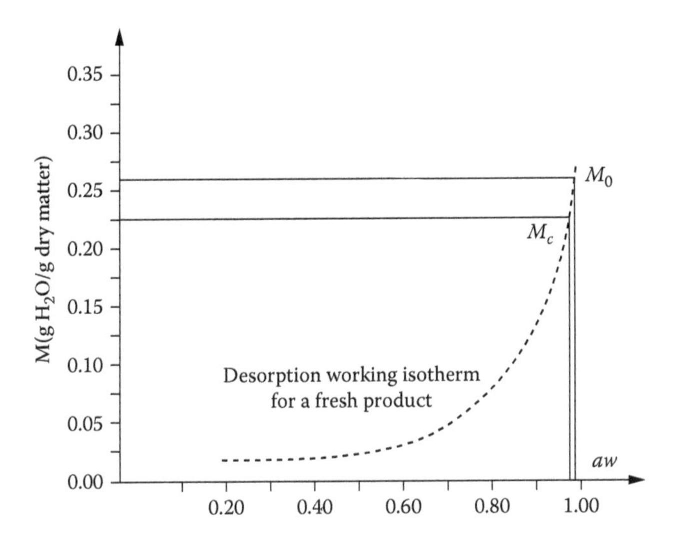

Figure 8.5 Vertical approximation of the moisture desorption isotherm for a fresh product.

$$t_{s.l.} = \frac{(M_i - M_c)}{\Phi Q_{pack} \cdot (WVP_{in} - WVP_{out})} \tag{8.25}$$

8.3 ASSESSING SHELF LIFE DEPENDENCE ON OXYGEN TRANSMISSION ACROSS THE PACKAGING

If the shelf life assessment related to a WV transmission must be described independently and what has been just illustrated can be applied only to a moisture exchange (to no other possible interesting vapors), the models for assessing the shelf life dependent on an oxygen transmission across the package are substantially the same to be used for all gases of interest, such as carbon dioxide, ethylene, or others. However, as is well known (St. Angelo 1996, Kilcast and Subramaniam 2000), the consequences of an oxygen transmission through the package are much more important than those related to other gases; therefore, the topic is here approached with reference to the oxygen exchanges only but can be easily applied to other phenomena involving gas exchange.

The oxygen level inside a package (the internal oxygen partial pressure, P_{pO_2in}) can deeply affect the quality of several different products, influencing microbial development as well as chemical and enzymatic degradation pathways. Despite the huge importance that the oxygen uptake may have on the sensorial and hygienic quality of packaged foods, no adequate knowledge exists regarding these phenomena. The different mechanisms of oxidation (Choe and Man 2006) and the oxygen needs of different microbial strains (Molin 2000, Madigan and Martinko 2006) are, in fact, well

Table 8.3 Oxygen Sensitivity of Different Foods

Food	Maximum Allowable Oxygen (mg kg^{-1})	Food	Approximate Oxygen Uptake Rate in Air (cm^3 h^{-1} kg^{-1})
Beer	1–5	Orange juice	0.35
Wine	1–5	Ketchup	0.20
Fruit juices	10–40	Wheat bran	135–5,000
Soft drinks	10–40	Potato chips	0.90
Dried foods	5–15	Carrots	6
Instant coffee	15–50	Apple (10°C)	4–8
Oil and fats	50–200	Cheese fat (25°C)	19

understood, but the maximum tolerable oxygen levels of different foods are not thoroughly known (Ashley 1985, Brown 1987), and little information exists about the oxygen uptake rates of different food systems and under different circumstances (Galliard 1986, Lee et al. 1996, Piergiovanni et al. 1999, Fonseca et al. 2002). One can only assume that the maximum allowable oxygen amount is generally low (much lower than the critical water amounts), and that the oxygen uptake rates are broadly variable. The data provided in Table 8.3 must be considered only as general indications.

The partial pressure of the internal oxygen changes during the shelf life of a sensitive packaged food, according to the dynamics of two possible concurrent phenomena (i.e., oxygen uptake by the food and the possible inlet of oxygen from the environment into the packaging or, in a few cases, its exit from the package), as regulated by the package oxygen permeability. This situation has been described for many years as the equilibrium between three rates, as shown in Equation 8.26, namely, the rate of oxygen inlet V_1, the rate of oxygen uptake V_2, and the rate of oxygen accumulation in the headspace of the package V_3 (actually the rate of P_{pO_2in} change). Noticeably, the first two rates strongly depend on the oxygen concentration, that is, on the O_2 partial pressure.

$$V_3 = V_1 - V_2 \tag{8.26}$$

If two of the three rates are known, the third one can be easily calculated, and if their dependence on the oxygen concentration is known, the quite simple balance of the rates can be easily represented graphically in a plot of rates versus the internal oxygen partial pressure (see Figure 8.6), which has several useful features.

It clearly shows the dependence of both V_1 and V_2 on the internal oxygen partial pressure, which is linear for the O_2 inlet (with a negative slope, obviously, V_1 is maximum if the P_{pO_2in} is null), whereas the oxygen uptake rate V_2 increases when P_{pO_2in} increases but according to a function similar to that of an asymptote hyperbolic. The plot makes it possible to represent the V_3 evolution, according to Equation 8.26, as the difference between V_1 and V_2, which become negative when V_2 is higher than V_1. Looking at the plot from left to right, that is, starting from the lowest P_{pO_2in} values, it

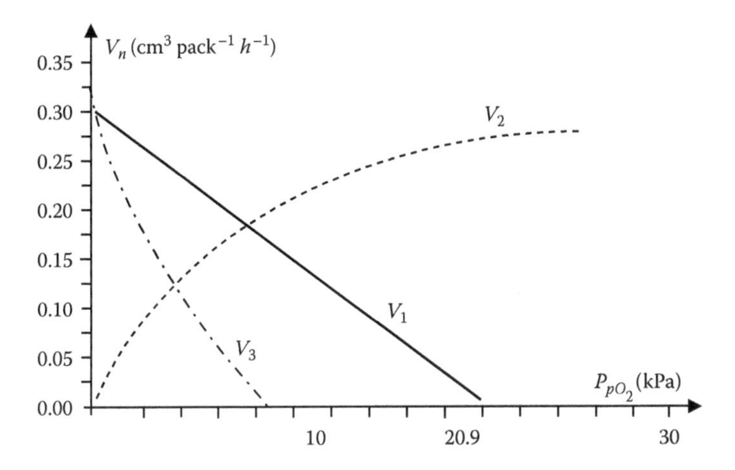

Figure 8.6 The dynamic system of oxygen uptake and oxygen permeation.

shows the advantages of packaging under vacuum or anoxic atmosphere, which lead to the lowest oxygen uptake rates in comparison with the packaging in air (looking at the plot from right to left). Last, but not less relevant, it shows the dynamic system of oxygen uptake and permeation as a special case of food–packaging interactions along the shelf life. Whatever the starting point (at the lowest or at the highest P_{pO_2in}, under vacuum or in air), the evolution inside the packaging will be related to the equilibrium between the food, consuming oxygen, and the package that modulates its inlet.

8.3.1 The Rate of Packaging Oxygen Inlet (V_1)

Considering first V_1 and its dependence on the internal oxygen partial pressure, it is obvious that the rate of oxygen inlet can be written as a first-degree equation as in Equation 8.27:

$$V_1 = a - b \cdot P_{pO_2,in} \tag{8.27}$$

with slope b, which is related to the oxygen permeability of the package, and intercept a, which is the maximum value possible of V_1, corresponding to the absence of oxygen inside the package. V_1 represents the amount of oxygen entering the package in the unit time and its units (as well as for V_2 and V_3) can be assumed to be cm^3 pack^{-1} h^{-1}. This can be easily derived by the oxygen permeability coefficient KP_{O_2} of the packaging material and the geometric features of the package (permeable area and thickness). The oxygen permeability is probably the property most currently measured for various packaging materials. As stated for *water vapor permeability*, the theoretical description of the phenomenon is well understood and easily available in the literature (Barrer 1939, Rogers 1985, Hernandez and Gavara 1999); as for WV permeability, the driving force is the difference (see Figure 8.1) between the high and the low oxygen partial pressure. In this case, we must take into account that the P_{pO2} in the air is constant and is generally assumed as 20.9 kPa or 0.209 bar (conventionally considering the oxygen as 20.9% in volume of air and the total pressure equal to 1 bar); therefore, during shelf life the permeability essentially depends on the P_{pO2in} and its changes. In real life, the oxygen permeability is generally measured, according to well-known standards (ASTM, 2010b) on plain, flat specimens of material, under a driving force of 1 bar of oxygen and a constant temperature, by means of specialized equipment (permeabilimeters) that make it possible to quantify KP_{O_2} (O_2 *permeability coefficient*, cm^3 µm m^{-2} day^{-1} bar^{-1}), P_{O_2} (O_2 *permeance*, cm^3 m^{-2} day^{-1} bar^{-1}, i.e., a flux normalized for the surface, the driving force, not for the thickness), or *OTR* (O_2 *transmission rate*, cm^3 m^{-2} day^{-1}). Some values of KP_{O_2} for common packaging materials are presented in Table 8.4 using both the common engineering units mentioned and the less-common but more appropriate SI units (mol m^{-1} s^{-1} Pa^{-1}). In Table 8.4, other useful parameters are presented: the apparent activation energies of permeability (E_a, kJ), the preexponential factors ($Log_{10} KP_{0,O_2}$) from the Arrhenius equations, as well as the *permeance selectivity* values.

The oxygen diffusion in plastic polymers is strongly affected by the temperature, which exponentially increases the diffusion coefficients, and whose effect, above the polymer glass transition temperature, can be assessed by the Arrhenius law, as Equation 8.28 shows. Therefore, the values proposed in Table 8.4 make it possible to take into account the temperature effect and to estimate the values at whatever temperature for those polymers.

$$KP_{O_2} = KP_{0,O_2} \cdot \exp\left(-\frac{E_a}{R \cdot T}\right) \equiv Log_{10} KP_{O_2} = Log_{10} KP_{0,O_2} - \left(\frac{E_a}{2.3 \cdot R}\right) \cdot \frac{1}{T} \quad (8.28)$$

where KP_{0,O_2} ($cm^3 \cdot \mu m \cdot m^{-2} \cdot day^{-1} \cdot bar^{-1}$) is the oxygen permeability value that we can assume at infinite temperature, where the temperature effect is negligible; E_a is the apparent activation energy of the permeation phenomenon (J mol^{-1}); R is the gas constant (8.314 J K^{-1} mol^{-1}); and T is the temperature considered (K).

The values presented as *permeance selectivity* are both the ratios of carbon dioxide and oxygen permeability and the ratios of oxygen and nitrogen permeability. These values are relatively constant, even when highly permeable or high-barrier materials are considered and, as stated, make it easy to shift from a problem of shelf life depending on oxygen exchange to a different gas-dependent issue, transforming a known oxygen permeability value to an estimate of permeability of different gases.

Once the value of oxygen permeability has been experimentally determined or achieved (e.g., the oxygen permeance P_{O_2}, cm^3 m^{-2} day^{-1} bar^{-1}), it is easy to calculate V_1 from Equation 8.27, with b (cm^3 $pack^{-1}$ h^{-1} bar^{-1}) derived from Equation 8.29:

$$b = P_{O_2} \cdot \frac{1}{24} \cdot A \quad (8.29)$$

Table 8.4 Approximate Gas Permeability Parameters of Various Polymers

Polymer	KP_{O_2} Permeability Coefficient[a]	KP_{O_2} Permeability Coefficient[b]	$Log_{10}(KP_{0,O_2})$ from Arrhenius Equation (units[a])	E_a (kJ mol^{-1})	Selectivity KP_{CO_2}/KP_{O_2}	Selectivity KP_{O_2}/KP_{N_2}
Low-density polyethylene (LDPE)	≃200,000.00	≃1.0334E-09	≃12.94	≃8.07	≃6.5	≃3.9
Unoriented polypropylene (PP)	~60,000.00	≃3.1002E-10	≃13.36	≃8.94	≃2.7	≃4.6
High-density polyethylene (HDPE)	≃25,000.00	≃1.2917E-10	≃11.18	≃6.89	≃3.3	≃3.9
Unplasticized polyvinyl chloride (PVC)	≃4,000.00	≃2.0668E-11	≃4.56	≃0.72	≃8.3	≃3.0
Nylon 6 (PA6)	≃2,500.00	≃1.2917E-11	≃10.86	≃8.16	≃14.2	≃3.8
Polyethylene terephthalate (PET)	≃2,000.00	≃1.0334E-11	≃8.07	≃5.02	≃7.0	≃4.4
Polyvinylidene chloride (PVDC)	≃300.00	≃1.5501E-12	≃13.95	≃12.47	≃5.0	≃5.7

[a] Units in $cm^3 \cdot \mu m \cdot m^{-2} \cdot day^{-1} \cdot bar^{1}$ at 25°C.
[b] Units in mol $s^{-1} \cdot m^{-1} \cdot Pa^{-1}$ at 25°C.

where A is the permeable surface (m^2 $pack^{-1}$) of the package. The intercept a will be obtained by Equation 8.30, with the same units as V_1.

$$a = P_{O_2} \cdot \frac{1}{24} \cdot A \cdot P_{pO_2,out} \tag{8.30}$$

8.3.2 The Rate of Food Oxygen Intake (V_2)

Unlike the oxygen inlet, V_2, which is a food-specific attribute, has been related in different ways to the O_2 partial pressure, according to the different mechanism of oxygen uptake (Lee et al. 1996, Eichner, 1986, Quast et al. 1972); the ones proposed here are the equations that can be used for a chemical (Equation 8.31) and for an enzymatic oxygen consumption (Equation 8.32); even if different, all the equations proposed lead to similar curves in the plot of Figure 8.6.

$$V_2 = \frac{P_{pO_2}}{K_1 + K_2 \times P_{pO_2}} \tag{8.31}$$

$$V_2 = \frac{P_{pO_2} \times V_{\max}}{K_m + P_{pO_2}} \tag{8.32}$$

where K_1 and K_2 are specific constants of the food considered, K_m is the Michaelis constant (bar), and V_{max} is the rate measured in air.

Both Equation 8.31 and Equation 8.32 demonstrate an important point, from a practical point of view, which is also clearly visible from Figure 8.6. At the lowest P_{pO_2in} values (corresponding, for instance, to the shelf life of a product in low oxygen modified atmosphere), the dependence of the oxygen consumption rate V_2 can be approximately represented by a linear relationship and shows a sharp increase for even a small change of oxygen concentration. On the contrary, at the highest P_{pO_2in} values (corresponding, for instance, to packaging in air), it is reasonable to assume a low influence of the oxygen concentration changes on the V_2 that is almost constant. In the range between 15 and 20 kPa of oxygen partial pressure, for the example of Figure 8.6, we can observe an increase of 0.05 cm^3 $pack^{-1}$ h^{-1} of the oxygen uptake, but from 0.00 to 5 kPa, the increase is about 0.15.

The assessment of V_2 is much more complicated than V_1. An investigation of the specific food product must be carried out, measuring the oxygen uptake rates under different oxygen concentrations, to account for the constants of the equations. This can be laborious and requires difficult analyses and special equipment. Broad knowledge of the oxygen uptake rates exists only for the aerobic respiration of vegetable products, for which a considerable amount of research has been done, at different temperatures and also accounting for the possible inhibitory effect that carbon dioxide accumulation can have on the respiration rate (Fonseca et al. 2002).

8.3.3 Models to Assess Shelf Life Oxygen Dependence

Therefore, knowing the V_1 and V_2 for specific packaged food products makes it possible to obtain a representation of how the oxygen concentration inside the package V_3 changes and a graphical route for monitoring the changes in oxygen uptake by the food product (i.e., studying its shelf life). However, the most useful insight provided by Equation 8.26 and its representation in Figure 8.6 is probably the way of categorizing the shelf life problems that can be related to an oxygen exchange into three general cases.

In a first simple case, we can assume that V_1 equals V_2, or this condition is reached soon during the shelf life. The circumstance in which V_3 is null, or can be approximate to zero, corresponds to the intersection point of the V_1 and V_2 curves. This is, for instance, the case of a passive atmosphere modification for a packaged minimally processed vegetable that reached the equilibrium between respiration and permeation. As a consequence of this situation (V_3 null), a further simplification is possible: Considering V_2 as linearly dependent on the oxygen concentration, as Figure 8.7 shows, it is therefore possible to write Equation 8.33:

$$V_2 = b' \cdot P_{pO_2,in} \tag{8.33}$$

where b' is the slope of the linearized function of oxygen uptake rate versus the internal oxygen partial pressure.

On the other hand, Equation 8.34 can represent the oxygen permeation from the outside into the package:

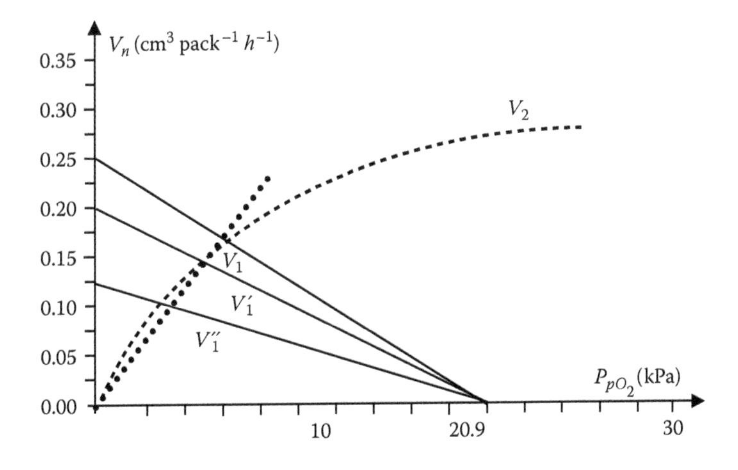

Figure 8.7 Oxygen uptake and oxygen permeation, assuming the same rates.

$$V_1 = P_{O_2} \cdot \frac{1}{24} \cdot A \cdot \left(P_{pO_2,out} - P_{pO_2,in}\right) \tag{8.34}$$

and, because of the assumption of equality, from Equation 8.35,

$$b' \cdot P_{pO_2,in} = P_{O_2} \cdot \frac{1}{24} \cdot A \cdot \left(P_{pO_2,out} - P_{pO_2,in}\right) \tag{8.35}$$

it is possible to derive Equation 8.36:

$$P_{pO_2,in} = \frac{P_{pO_2,out}}{\dfrac{b'}{(P_{O_2}) \times \dfrac{1}{24} A} + 1} \tag{8.36}$$

which leads to the assessment of the constant value of oxygen concentration corresponding to the evenness condition. If we know the constant oxygen partial pressure, it is easy to determine the constant and the equal rates V_1 and V_2. The oxygen consumption will proceed constantly until the critical level is reached and the shelf life expired. This approach has been proposed and used since the 1960s (Jurin and Carel 1963, Labuza and Karel 1967) for the design of different fruit packages. To apply such a simplified approach to the assessment of the shelf life, it is necessary to know the specifications of the package (i.e., the permeability of the three-dimensional package) and the dependence of the product oxygen consumption rate on the P_{pO2}, estimated by the slope b'. However, the final goal will be accomplished only knowing the critical amount of oxygen for the specific packaged product, which divided by the constant rate of O_2 uptake will give the expected shelf life. Unfortunately, often this is the real limitation of this kind of assessment since the information about the oxygen tolerance is unknown or approximate (see Table 8.3). One more interesting point is shown by the Figure 8.7, which again can be seen as a representation of food–packaging interaction behaviors. The more barrier is used for the packaging, the lower is the slope of V_1 and the lower will be the constant oxygen concentration reached at the equilibrium rates, P_{pO_2in}.

The other two prototypic cases are related to the prevalence of V_1 on V_2 or, vice versa, of V_2 on V_1 and are represented in Figures 8.8 and 8.9, respectively.

The situation depicted in Figure 8.8 corresponds to packaging solutions in which a gas-permeable package is used for a food product that has a low or very slow oxygen uptake (e.g., a dry, nonfat product in an inexpensive package). Obviously, depending on these characteristics, and assuming an initial low oxygen content in the packaging, possible oxygen accumulation is feasible in the headspace of the package (i.e., the V_3 can be high and positive). Because the oxygen partial pressure can vary as a balance between permeation and uptake, to assess the attainment of a critical

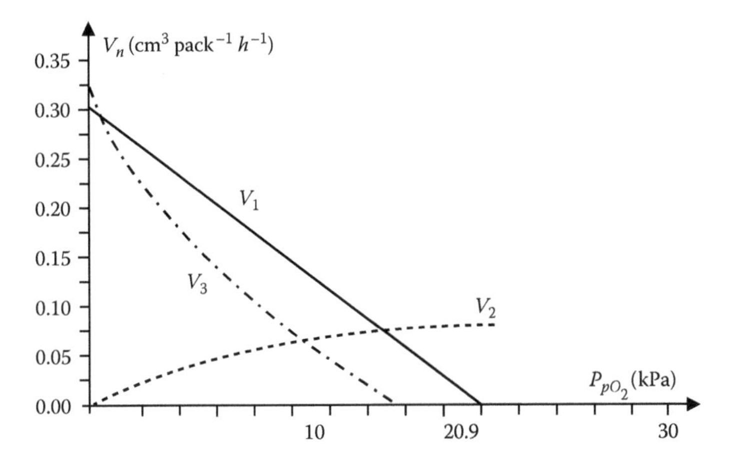

Figure 8.8 Oxygen uptake and oxygen permeation, assuming $V_1 >> V_2$.

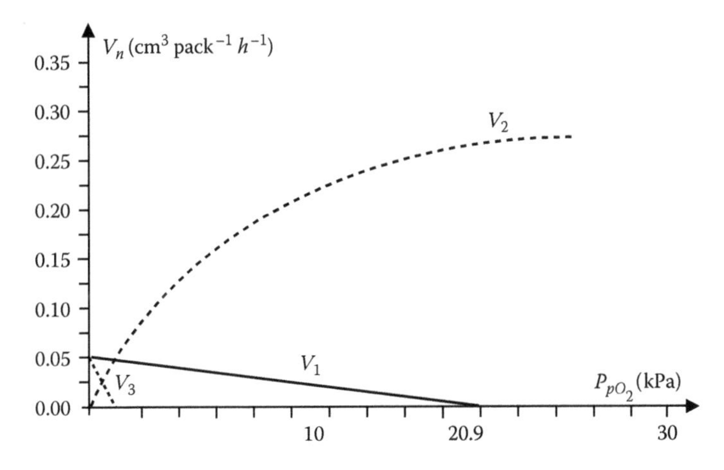

Figure 8.9 Oxygen uptake and oxygen permeation, assuming $V_1 << V_2$.

value, the P_{pO_2in} must be accounted for, and this can be done using the following equation, which assumes V_2 is substantially constant in the range of interest:

$$\frac{dP_{pO_2,in}}{dt} = \frac{\dfrac{KP_{O_2} \cdot A \cdot \left(P_{pO_2,out} - P_{pO_2,in}\right)}{l} - V_2}{UFV} \tag{8.37}$$

where UFV is the unfilled volume (cm^3), that is, the total amount of the gas phase in the package. A further simplification is to completely ignore the contribution of V_2 as in Equation 8.38:

$$\frac{dP_{pO_2,in}}{dt} = \frac{\dfrac{KP_{O_2} \cdot A \cdot \left(P_{pO_2,out} - P_{pO_2,in}\right)}{l}}{UFV} \tag{8.38}$$

which can be integrated (Equation 8.39) in the time range of interest (between t_0 and the time required for reaching a critical concentration inside $t_{s.l.}$) and for the oxygen range (between initial and final oxygen concentration, $P_{pO2,t\,0}$ and $P_{pO2,t\,s.l.}$). The integration leads to the following formula (Equation 8.39), which permits accounting for the shelf life (Lee et al. 2008d):

$$t_{s.l.} = \frac{l \cdot UFV}{KP_{O_2} \cdot A} \cdot \ln\left(\frac{P_{pO_2,out} - P_{pO_2,t0}}{P_{pO_2,out} - P_{pO_2,t\,s.l.}}\right) \tag{8.39}$$

This, apparently quite precise, assessment is definitely an approximation of how long the shelf life of the packaged oxygen-sensitive product is but is quite efficient and easy to apply. We do not use a critical amount of oxygen consumed by the product (not easy to determine) but a critical concentration of oxygen in the surrounding unfilled volume, which is definitely much easier to measure experimentally. Moreover, the shelf life estimates that can be made using this equation, even if not completely accurate, are certainly safe. In fact, the shelf life will, very likely, be underestimated because the critical internal O_2 concentration will be achieved before the maximum tolerable O_2 amount is consumed.

The last of the three prototypic cases is represented in Figure 8.9 and refers to a current common situation. The prevalence of V_2 on V_1 means that we are using a barrier packaging material (too often an excess of gas barrier is provided by the modern food packaging materials) for products that are highly sensitive to the oxygen that is consumed quite rapidly. In these circumstances, it is almost impossible to have oxygen accumulation inside the packaging; therefore, V_3 is extremely low and even negative; that is, a packaging collapse can be observed for fast oxygen consumption, not compensated by an oxygen inlet due to the permeation phenomenon.

The consequence of such a condition is that the low permeation rate acts as the limiting factor in the system; we can assume that as soon as a molecule of oxygen permeates into the package, it will be readily consumed by the product; therefore, we can ignore the V_2 contribution to the dynamic system. The permeation phenomenon, moreover, will proceed under a constant driving force: Because of the negligible oxygen accumulation, the difference between the external and internal oxygen concentration will not change remarkably during the shelf life. This will definitely simplify the problem of shelf life assessment. If we know the critical volume of oxygen $O_{2,max}$ (cm^3 g^{-1}) that will bring the mass W (g) of the packaged product to spoilage, the shelf life will be easily estimated by one of the two forms of Equation 8.40.

$$t_{s.l.} = \frac{O_{2,\max}}{P_{O_2} \cdot A \cdot \frac{1}{24} \cdot \left(P_{pO_2,out} - P_{pO_2,in}\right) \cdot \frac{1}{W}} \quad t_{s.l.} = \frac{O_{2,\max}}{V_1 \cdot \frac{1}{W}} \tag{8.40}$$

8.4 SHELF LIFE ASSESSMENT BY THE SUBSTITUTION METHOD

In the case of assessing the shelf life dependent on both a WV and an O_2 transmission, it is generally assumed that the time available for commercial life is, mainly or exclusively, established by the permeation phenomena. In other words, the shelf life is inversely proportional to the transmission of WV or O_2 into or out of the package, as stated in Equation 8.41, using both formal mathematical symbols and an easy-to-manage formula:

$$t_{s.l.} = \frac{c}{P_{pack}}; \; t_{s.l.} \propto P_{pack}^{-1} \tag{8.41}$$

where P_{pack} is the global packaging permeability (to O_2 or WV), and c is a constant.

This relationship leads to a much more practical and useful equation that deals with the comparison between two specific situations like in Equation 8.42, where it is stated that the product of the shelf life and the packaging permeability is a constant; this means that the shelf life is proportionally shortened if permeability is increased:

$$\frac{t_{s.l.}}{t'_{s.l.}} = \frac{P'_{pack}}{P_{pack}}; \; t_{s.l.} \cdot P_{pack} = t'_{s.l.} \cdot P'_{pack} \tag{8.42}$$

Actually, what often happens is that the shelf life of a product is already known, as are the permeability properties of the packaging used, and the interest is just addressed to a simple substitution thereof; that is, the goal is to shift to a different package, with different and known oxygen or WV permeability and to assess in advance the effects of such a substitution. From Equation 8.42, it is clear that it is possible to determine easily the new shelf life $t'_{s.l.}$ when we know the other three elements of the formula.

The two assumptions of inverse proportionality between shelf life and permeability, and of the uniqueness of the permeation phenomena as the decaying event of the food of interest, are also usefully applicable in ASLTs. Generally, the ASLTs are conceived with an increase of the test temperature in comparison with the common and real temperature to speed up the kinetics of the deterioration mechanisms in the food. As already shown (Equation 8.28), however, the temperature exponentially affects the permeability; therefore, the concurrent exponential effects on food deterioration and on the diffusion processes can be controlled and discriminated with difficulty. In these cases, it seems definitely more appropriate to accelerate product aging, trimming different factors. The permeable surface and the driving force (the partial pressure difference), for instance, are linearly related to the permeation, whereas the thickness is inversely proportional. Consequently, an ASLT can be easily set up, increasing the permeable surface and driving force or reducing the thickness, as well as changing the packaging material to have a different and greater oxygen or WV permeability coefficient. Equation 8.43 provides the basis for interpreting the results of such an accelerated test, assumed for an oxygen-sensitive product, showing the level of shortening obtained changing one or more variables of the permeations phenomenon in the ratio $t_{s.l.}/t'_{s.l.}$

$$\frac{t_{s.l.}}{t'_{s.l.}} = \frac{\left[KP'_{O_2} \cdot \frac{1}{l'} \cdot A' \cdot \left(P_{pO_2,out} - P_{pO_2,in}\right)'\right]}{\left[KP_{O_2} \cdot \frac{1}{l} \cdot A \cdot \left(P_{pO_2,out} - P_{pO_2,in}\right)\right]} \tag{8.43}$$

8.5 ASSESSING SHELF LIFE DEPENDENCE ON A MIGRATION PHENOMENON

Even if in this chapter the "packaging–food interactions" concept has been used in its broader sense, according to several authors (Hotchkiss 1997, Lee et al. 2008e, Duncan and Webster 2009) these words are mainly synonymous with migration phenomena. Migration is a mass transfer from food contact materials into food, or vice versa from food to the packaging, by diffusion processes (Katan 1996). These kinds of interactions between food and packaging, generally considered for safety issues only (the possible migration of harmful substances from plastic materials), can also be detrimental to the sensorial quality of the packaged product, leading to serious reductions in the potential food shelf life. Flavor changes, due to aroma sorption or to the transfer of undesirable odorous substances from packaging materials to foods, for instance, may constitute important mechanisms of sensory deterioration, especially when foods are packaged in plastics. Plastics, in fact, may contain additives, or other small molecules, which are potential migrating substances. At the same time, plastics, as macromolecular matrixes, can easily absorb and dissolve various components of foods and beverages that themselves might act as potential migrants into the contact material and whose loss is unfavorable to commercial life (Hernandez and Gavara 1999, Sheftel 2000).

Nevertheless, these interaction mechanisms have also been seen as possible ways for shelf life extension rather than as a negative event (Zhang et al. 2004, Dainelli et al. 2008). This new vision of the mass transfer phenomena between food and packaging refers to what is generally known as "active packaging." According to E.U. Regulation 1935 (European Parliament 2004, p. 7), "active food contact materials and articles means materials and articles that are intended to extend the shelf-life or to maintain or improve the condition of packaged food. They are designed to deliberately incorporate components that would release or absorb substances into or from the packaged food or the environment surrounding the food." The mass transfers from or into the food, which definitely are packaging–food interaction phenomena, are consequently becoming increasingly important to the shelf life issue, and control release packaging (CRP) has become a target to be achieved and a possible field of research and investigation (LaCoste et al. 2005, Jin et al. 2009, Gemili et al. 2010).

In both cases of looking at migration as an unwanted phenomenon that can reduce shelf life, and as a smart way for shelf life extension, the problem is the full control of the diffusion phenomena that can lead to the capacity to modulate the release of valuable compounds, as well as the ability to monitor the transfer of unwanted migrants. This can only be done by applying the second Fickian laws of diffusion that describe the movement of the migrant in the polymers or in the food and taking into account the concept of partition coefficient $K_{P,F}$, which defines the concentrations of a migrant in two different contacting phases at equilibrium (Hernandez and Gavara 1999, Piringer 2000, Brandsch et al. 2002, Helmroth et al. 2002). With reference to the shelf life issue, the most important model parameter for assessing a migration phenomenon is the diffusion coefficient D_P of the migrant in the plastic, which is seen as the indicator of the rate of the entire phenomenon. In fact, the diffusion coefficient of the migrant in the food D_F is generally neglected and the diffusion assumed to occur without any resistance in the contact food; this happens both for the lack of knowledge about the behavior of the migrants into the different foods and for a prudential "worst-case approach." D_P values are generally low; Table 8.5 presents a selection of diffusion coefficients.

Several authors (Reynier et al. 1999, Pennarun et al. 2004) showed that the D_P can be estimated with prudent excess through calculation, and Equation 8.44 is the one proposed by Piringer and coauthors (Mercea and Piringer 1998, Piringer 2000) that permits overestimation of the diffusion coefficient D_p^* (cm^2 s^{-1}) at any temperature if we know the molecular mass of the migrant and the class of polymer, which establish the value of specific parameters:

$$D_P^* = 10^4 \exp\left(Ap - 0.1351 \cdot M_r^{2/3} + 0.003 \cdot M_r - \frac{10454}{T}\right) \tag{8.44}$$

Table 8.5 Diffusion Coefficients of Various Classes of Substances

Substance	Approximate Coefficient of Diffusion (cm^2 s^{-1})
Gases	10^{-1}
Supercritical fluids	10^{-4}
Liquid, nonviscous	10^{-5}
Viscous liquid	10^{-6}
Rubber-like polymers	10^{-8} to 10^{-12}
Glass-like polymers	10^{-12} to 10^{-18}

In Equation 8.44, M_r is the relative molecular mass of the migrant (Da), and A_p is equal to A'_p - t/T, with T the absolute temperature (K), A'_p a specific polymer "diffusion conductance" parameter, and τ a specific polymer "activation energy." It has been demonstrated that the upper-bound diffusion coefficient D_P*, estimated by Equation 8.44, leads to a worst-case estimation of the migration, which can be useful if an unwanted phenomenon is under consideration. On the contrary, if the effectiveness of active releasing packaging is under examination, the lack of knowledge of the diffusion coefficient into the food D_F as well as an overestimation of D_P (the D_P*) can lead to an incorrect shelf life assessment.

The partition coefficient, the other important parameter that describes the system in which the migration phenomenon takes place, is conventionally defined as the ratio of concentrations of a compound in two immiscible phases at equilibrium (Leo et al. 1971). Partition constant, partition ratio, or distribution ratio are other, even more appropriate, terms used to express the same parameter defined in Equation 8.45 (Wilkinson and McNaught 1997).

$$K_{P,F.} = \frac{C_{P,\infty}}{C_{F,\infty}} \tag{8.45}$$

where $C_{P,\infty}$ is the migrant concentration remaining in the package at equilibrium, and $C_{F,\infty}$ is its concentration achieved in the food (g cm^{-3}).

If the diffusion coefficient establishes, somehow, the rate of the transfer phenomenon, the partition coefficient is an estimate of the extent of the interaction. Figure 8.10 is an effective explanation of the different effects of these two fundamental parameters (Helmroth et al. 2002). In fact, upper curve A is the migration profile obtained for a migrant with established D_P and $K_{P,F}$.

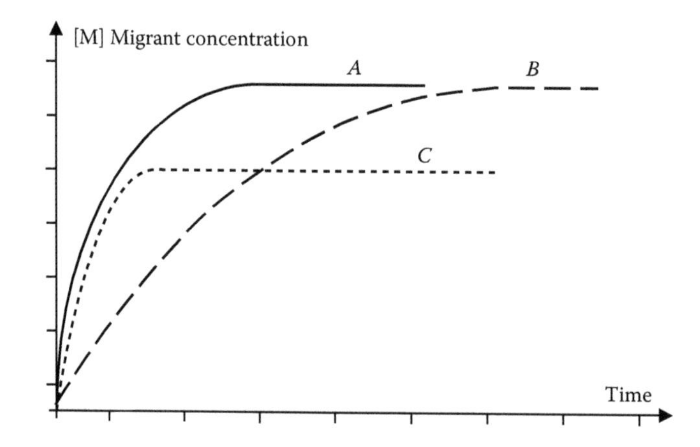

Figure 8.10 The effects of diffusion coefficient and partition ratio on the dynamic of migration: The migrant A has the same diffusion coefficient as migrant C but a higher partition ratio; the migrant B has the same partition as migrant A but a lower diffusion coefficient.

Curve *B* shows what happens for a migrant having a lower D_P and the same $K_{P,F}$ of the migrant represented in curve *A*, that is, the same level of migrant concentration for a longer time. The lowest (*C*), on the contrary, assumes the same D_P as in curve *A* for a migrant with a lower $K_{P,F}$, leading to the same kinetic for a lower level of transfer.

Cautious considerations, however, must be given to all the factors affecting such interactions to minimize undesirable sensorial changes or to manage possible shelf life extension for a controlled-release phenomenon. Variables related to the plastics include the type of polymer and processing method and ratio of mass of polymer to product. Polarity, volume, and contact surface of the food product are also important points, as well as the temperature and the relative humidity of the environment in which the diffusion phenomenon occurs. Diffusion is exponentially promoted by the temperature, while the relative humidity can affect the concentration of volatile substances in the vapor phase as well as the diffusion coefficient in hydrophilic matrixes.

8.5.1 Models to Assess Shelf Life Diffusion Dependence

A common analytical solution for the one-dimensional Fick's second law (Crank 1975, Piringer 2000, Brandsch et al. 2002) for specific initial and boundary conditions is shown in Equation 8.46, assuming a single side contact of infinite slab, no loss of migrant on the external surface, and a homogeneously constant migrant concentration in the plastic layer and a null one in the contact food at time zero.

$$\frac{M_{F,t}}{M_{F,\infty}} = 1 - \sum_{n=1}^{\infty} \frac{2\alpha(1+\alpha)}{1+\alpha+\alpha^2 q_n^2} \exp\left(\frac{-D_P q_n^2 t}{l^2}\right) \tag{8.46}$$

where t is the contact time (s); $M_{F,t}$ is the migrated mass at time t; $M_{F,\infty}$ is the total amount of migrant in food at equilibrium; l is the packaging material thickness (cm); D_P is the diffusion coefficient in the plastic (cm^2 s^{-1}); α is the mass ratio of migrant equilibrated in food to that in packaging material and thus equal to $M_{F,\infty}/M_{P,\infty}$ with $M_{P,\infty}$ the mass of migrant residual in the packaging material at equilibrium; and q_n is the positive root of the transcendent equation of

$\tan(q_n) = -\alpha q_n$; for $\alpha << 1$, $q \simeq n\pi/(1+\alpha)$, and for other a values $q_n \simeq \left[n - \frac{\alpha}{2(1+\alpha)}\right]\pi$

The computation of Equation 8.46 implies knowledge of α, thus the knowledge of the ratio $M_{F,\infty}/M_{P,\infty}$, which is obtainable from the partition ratio and the volumes of food and packaging (V_F and V_P, respectively, cm^3), being equal to $V_F/(K_{P,F} \times V_P)$. As an alternative, the accomplishment of the complete experiment of simulated migration is needed to establish the amounts of the substance migrated in the food and remaining in the plastic.

Therefore, some even more approximated solutions of Fick's diffusion law are often used, adopting a worst-case approach or dealing with a simple analysis of the interaction phenomenon. Prudentially assuming a possible almost-complete transfer of the migrant into the food ($\alpha \simeq \infty$), Equation 8.46 converges to Equation 8.47:

$$\frac{M_{F,t}}{M_{F,\infty}} = 1 - \sum_{n=1}^{\infty} \frac{8}{(2n-1)^2 \pi^2} \exp\left(\frac{-D_P (2n-1)^2 \pi^2 t}{4l^2}\right) \tag{8.47}$$

and to the more simplified forms of Equations 8.48 and 8.49, used for long and short contact times, respectively (assuming the ratio $M_{F,t}/M_{F,\infty}$ not higher than 0.6):

$$\frac{M_{F,t}}{M_{F,\infty}} = 1 - \frac{8}{\pi^2} \exp\left(\frac{-\pi^2 D_P t}{4l^2}\right) \tag{8.48}$$

$$\frac{M_{F,t}}{M_{F,\infty}} = \frac{2}{l} \left(\frac{D_P t}{\pi}\right)^{0.5} \tag{8.49}$$

Actually, if α is very high (this would mean that the migrant is freely soluble in the food to which it is similar, showing an extremely low $K_{P,F}$), $M_{F,\infty}$ can be assumed to be equal to $M_{P,0}$, which is the amount originally present before contact in the packaging material, and Equation 8.49 can be written like Equation 8.50. This is definitely a more practicable equation because $M_{p,0}$, is much easier to find than $M_{F,\infty}$.

$$\frac{M_{F,t}}{M_{P,0}} = \frac{2}{l} \left(\frac{D_P t}{\pi}\right)^{0.5} \tag{8.50}$$

A total transfer of a possible migrant may be desired when $M_{p,0}$ is the amount of a useful substance (e.g., antimicrobials, antioxidants, flavoring substances) specially added to the contact packaging material to provide the right supply to a sensitive product in what is called controlled-release active packaging. Equation 8.50 therefore gives the basis, in an approximate estimate, for assessing the amount of the functional substance released at a specific time $M_{F,t}$.

What is really noticeable from the equations is that the migration phenomenon is proportional to the square root of time, and in fact, the linearity of migration versus $\sqrt{t}$ is widely used for determining the apparent diffusion coefficient of migrants from the slope of linear plots $M_{F,t}/M_{p,0}$ versus $\sqrt{t}$ according to Equation 8.50. Starting from these assumptions and knowing some data about real migration, the very approximated Equation 8.51 can be used to assess the amount of migration:

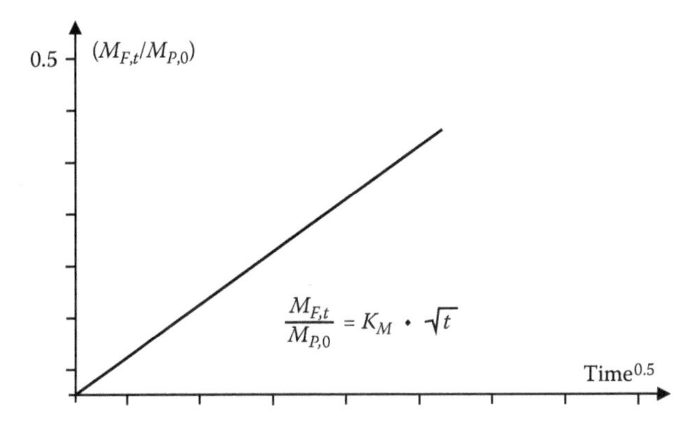

Figure 8.11 The linearity of the migration phenomenon on the square root of contact time.

$$\frac{M_{F,t}}{M_{P,0}} = K_M \cdot \sqrt{t} \tag{8.51}$$

where K_M (the slope of the linear plot) can be defined as an *apparent migration coefficient*, which implies the diffusion coefficient of the migrant in the plastic at the selected temperature and the geometrical features of the system under investigation or as the coefficient of proportionality between the ratio $M_{F,t}/M_{P,0}$ and the square root of the contact time. K_M can be easily calculated by the slope of plots as in Figure 8.11, which shows the dynamics of migration for a limited extent, just for the initial stage of migration or sorption, with $M_{F,t}/M_{P,0}$ generally below 0.6.

The proposed equation, obviously, can also be used to estimate the time t to achieve an effective concentration, positive or negative to the shelf life, as shown in Equation 8.52, which was derived from Equation 8.51:

$$t = \left(\frac{M_{F,t}}{M_{P,0}}\right)^2 \times \frac{1}{K_M^2} \tag{8.52}$$

8.6 ASSESSING SHELF LIFE DEPENDENCE ON LIGHT TRANSMISSION THROUGH THE PACKAGING

During storage, distribution, retail display, and postpurchase, foods are exposed to natural and artificial light sources that may deteriorate quality attributes such as appearance, flavor or fragrance, nutritional integrity, and marketability, depending on the sensitivity of the product and the protective role of the packaging material.

Light and ultraviolet (UV) regions are the parts of the electromagnetic spectrum involved, to different extents, in food degradation during shelf life; according to electromagnetic wave theory, light is a small part of the electromagnetic spectrum, sandwiched between UV and infrared radiation (from about 380 to about 780 nm), while UV radiation, sometimes incorrectly referred to as "UV light," has shorter wavelengths than visible radiation (light) (Taylor 2000).

The effects of exposure to UV and light radiations on the degradation of foods during storage have been widely investigated, especially when lipidic substrates and fatty foods (milk, cheese, butter, yogurt, meat and meat products, oils, etc.) are involved (Lennersten and Lingnert 1998, Bosset et al. 1992, Mortensen et al. 2004, Kalua et al. 2006, Andersen et al. 2005, Piergiovanni and Limbo 2004). More recently, the capability of light in speeding up degradation reactions has been investigated, taking into consideration colored aqueous systems and soft drinks, which are widely consumed all over the world (Jespersen et al. 2005, Limbo et al. 2007, Manzocco et al. 2008).

The effect of these electromagnetic radiations on the susceptible substrates may be attributed to photolytic free-radical autooxidation or photosensitized oxidation. Photolytic free-radical autooxidation is the production of free radicals, primarily from lipids, during exposure to far-UV and high-energy light (Min and Boff 2002).

Photosensitized oxidation begins with the absorption of light by molecules, called photosensitizers, that are able to absorb light to elicit a specific response (Eie 2009). Chlorophylls, carotenoids, flavonoids, anthocyanins, myoglobin, and synthetic colorants are typical photosensitizers present in foods. These components, in fact, are characterized by their conjugated double-bond system. When electromagnetic energy is absorbed, an electron is boosted to a higher level, and the sensitizer becomes unstable. In the presence of oxygen, the excited sensitizer may interact easily with another molecule, producing free radicals or radical ions, resulting in oxygenated products. In the same way, the excited sensitizer may compete with ground-state molecular oxygen, leading to the formation of highly reactive singlet oxygen. The reaction rate of singlet oxygen with some food components is high because singlet oxygen can attack double bonds directly; for example, its reactivity with linoleic acid is about 1,450 times faster than that of ground triplet state oxygen. The consequences of these reactions include fading or discoloration, off-flavor development, and nutritional loss (Min et al. 2002).

The extent of degradation phenomena in foods sensitive to electromagnetic radiations in the UV-visible range is directly related to several factors, including the electromagnetic spectrum and intensity. In fact, the catalytic effects of light are most pronounced for radiations in the lower wavelengths of the visible/UV spectrum where the energy is higher, passing from 150 to 300 kJ mol^{-1} in the visible interval to values close to 590 kJ mol^{-1} in the far UV range.

Lamps used in current retail display cabinets for foods are standardized neither among stores nor within stores in a particular retail food chain (Torri 2007). Halogen lamps and fluorescent tubes (warm white or cool white) are the most widely used sources in retail markets for different kinds of food products (Figure 8.12).

Spectral emission of lamps relating to UV energy output is of public concern, and their obligatory shielding with specific filters can reduce, but not completely eliminate, the energy available to detrimentally affect food product quality. For

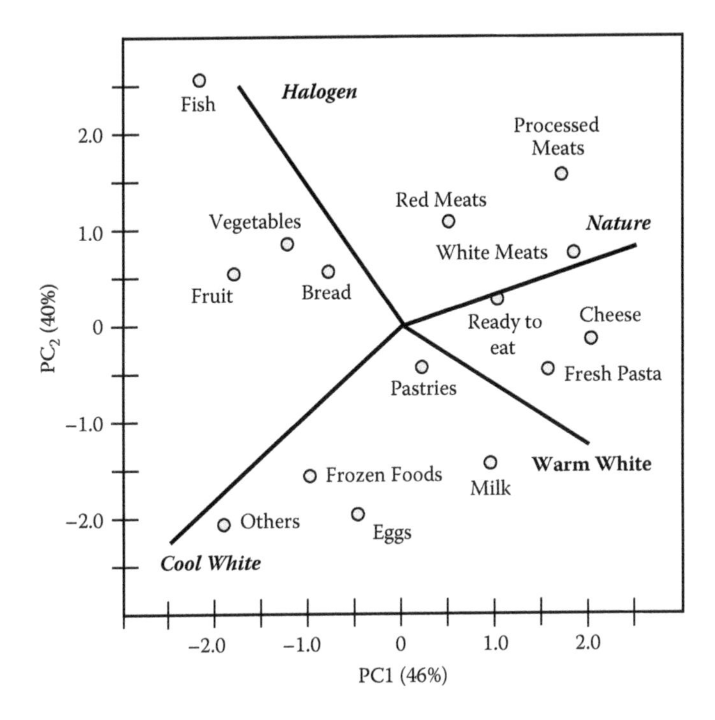

Figure 8.12 Distribution of the lamps in local supermarkets as function of different food products.

example, the UV part of the light spectrum breaks down the sulfur-containing molecules in beer, forming highly odorous mercaptans and other sulfurous compounds (Black 2011).

Considering the visible region, the halogen lamps emit energy in the entire visible region, with low energy in the blue part of the spectrum, while fluorescent tubes are characterized by uncontiguous peaks of energy, with a pattern typical for each type of tube (Eie 2009, Limbo et al. 2007). Moreover, the light energy from the halogen lamp is 2.3 times stronger than that of the fluorescent tube.

It is well known that packaging could play an important role in protecting foods from the detrimental effects of UV and visible radiations (Piergiovanni and Limbo 2004, Mortensen et al. 2004, Karel and Lund 2003). When electromagnetic radiation reaches the surface of a flat material, it can be reflected, transmitted, or absorbed as a function of the intrinsic properties of the material itself (Figure 8.13(a)). Different packaging material categories offer varying degrees of protection against UV radiations and light-induced changes due to differences in reflectance, transmittance, and oxygen permeability determined by the molecular composition of the material (Mortensen et al. 2004). However, when the material has a defined shape for containing the food, the interactions between the electromagnetic radiations, the material, and the food are more complex and depend on optical, geometrical (shape, dimensions, headspace, etc.), gas transmission (especially to oxygen) features of the whole package, and the optical and chemical composition of the foods (Figure 8.13).

So, the correct protection for foods during shelf life can be assessed only if the material itself is able to minimize the effects of UV and light radiations that are considered primary sources of damage exerted on both food and polymeric materials at ambient conditions.

With regard to the UV range, many ingredients like flavor, vitamins, and colorants are sensitive to UV radiations, and significant shelf life improvement can be obtained with higher levels of UV protection. In the same way, most of the polymers used as food packaging material are susceptible to degradation initiated by UV and visible light (Singh and Sharma 2008). Normally, the near-UV radiations (290–400

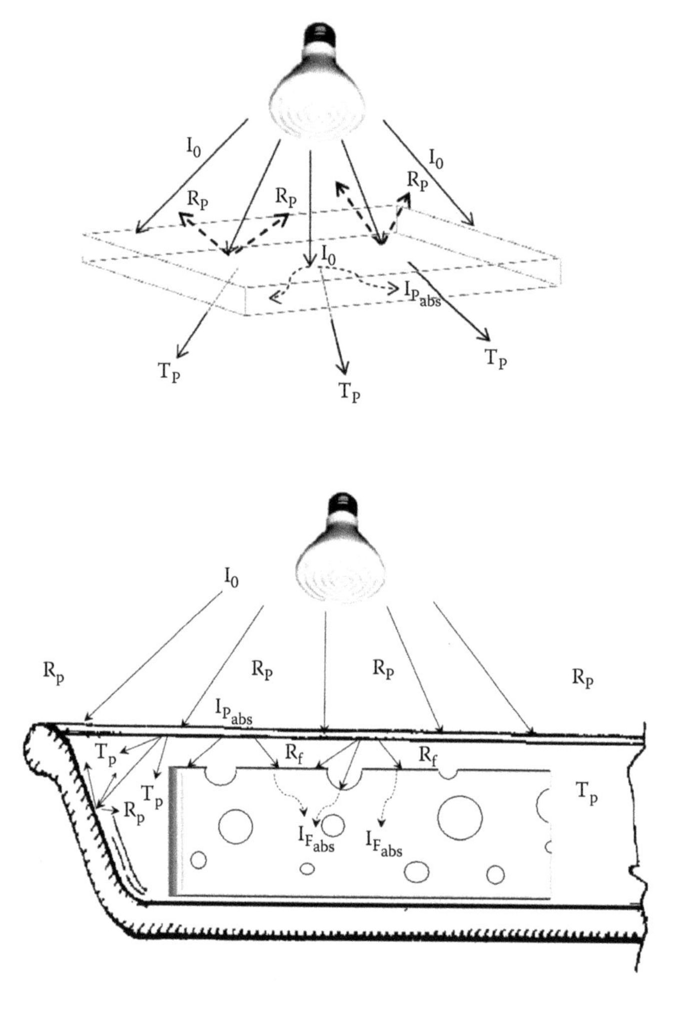

Figure 8.13 Interaction between the electromagnetic radiations emitted by an artificial source and the material: (a) flat film sheet; (b) real finished package.

nm) in sunlight determine the lifetime of polymeric materials in outdoor applications; in several plastic and cellulosic materials, they can lead to changes in color, loss of flexibility, transparency, and enhancement of migration (Lee et al. 2008f). In fact, UV irradiations have sufficient energy to cleave C-C and prolonged or uncontrolled exposures can generate low molecular weight molecules (ester, aldehyde, propyl end groups, etc.) with potential migration capacity into foods during shelf life. Some synthetic polymers (e.g., aromatic polyesters and polyamides) have inherent absorption of UV radiation, causing excitation, radical formation, oxygen addition, and splitting off of small molecules that can alter smell and taste in food and beverage. As mentioned, although the levels of UV radiations emitted from artificial sources are low, they could present a serious problem for various foods during shelf life, especially because the real effects are often underestimated.

To reduce the potential damages induced by UV radiations, UV-absorbing substances may be incorporated into the packaging film (Gugumus 2002). Moreover, novel technologies like UV treatments are specifically used for disinfection and sterilization of the packaging material itself; the 250- to 280-nm range appears effective against microorganisms. In this specific case, the packaging material is involved and exposed to conditions that may alter its structure and consequently its mechanical and mass transfer (barrier and migration) properties. In particular, the main impact of UV radiation on the inertia of plastic films derived from oxidations led to the formation of new low molecular weight substances such as phthalic esters, alkanes, alkenes, ketones, and peroxides coming from the degradation of low-density polyethylene (LDPE) and LDPE additives or carbonyls, carboxylic acids, and hydroperoxides coming from treated polystyrene (PS) and polypropylene (PP) (Guillard et al. 2010).

In the respect that the optimization of the shelf life of a *packaging-dependent* food cannot be separated from the safety of the packaging material itself, the protection ability of the material to UV radiations has to be taken into account in the assessment of the durability of a product.

8.6.1 Models to Assess Shelf Life Lighting Dependence

The prediction of deteriorative changes as a function of intrinsic (product- or material-related) and extrinsic (environmental) factors is an important area of research. However, there is really not much information concerning the modeling of phenomena as far as UV radiations are involved, even if the prediction of material degradation should be an important issue, especially for understanding the impact on the potential material and food shelf life reductions.

Escobar and Meeker (2006) proposed a simple model to assess the degradation of a material at a certain time t taking into consideration the extrinsic factors like the total effective dosage D_{tot}, defined as the cumulative number of photons absorbed into the degrading material able to cause chemical changes. The model is based on the assumption that the radiation effectiveness of a source is equal to the sum of the effectiveness of its spectral components, especially when the range of intensities used in experimentation and actual application is not too broad.

From a mathematic point of view, the total effective absorbed dosage D_{tot} at a certain time t can be expressed by the following formula:

$$D_{tot}(t) = \int_0^t D_{ins}(t)dt \tag{8.53}$$

where D_{ins} is the instantaneous absorbed effective UV dosage at real time t; it can be expressed as

$$D_{ins}(t) = \int_{\lambda_1}^{\lambda_2} D_{ins}(t,\lambda)d\lambda = \int_{\lambda_1}^{\lambda_2} E_o(t,\lambda) \times \{1 - \exp[-A(\lambda)]\} d\lambda \tag{8.54}$$

The terms that appear in Equation 8.54 have the following meaning: The interval λ_1, λ_2 (where λ_1 and λ_2 are respectively the minimum and the maximum wavelength in a specific range) can be taken over the UV-B range (290–320 nm) or UV-C range (200–290 nm) as these are the ranges of wavelengths over which both $\Phi(\lambda)$ and $E_o(t,\lambda)$ are importantly different from zero.

$E_o(t,\lambda)$ is the spectral irradiance (i.e., the power of the electromagnetic radiation considering each frequency in the spectrum separately); the apparent quantum efficiency f is defined here. The E_o term can be measured directly or estimated from literature data depending on the source of the radiation and the wavelength interval.

$\{I - exp[- A(\lambda)]\}$ is the spectral absorbance of the material being exposed to a specific radiation range. In fact, damage is caused only by radiation that is absorbed into the material. As previously, this one also can be measured directly.

Therefore, using Equation 8.55, it is possible to estimate the total effective absorbed dosage D_{tot}:

$$D_{tot}(t) = \int_0^t \int_{\lambda_1}^{\lambda_2} E_0(t,\lambda)\{1 - \exp[-A(\lambda)]\} d\lambda dt \tag{8.55}$$

$D_{tot}(t)$ can also be related to the damage or degradation D_g at time t of the material (experimentally measured, for example, following the Fourier transform infrared [FTIR] evolution of a specific peak or the concentration increase/decrease of a specific substance) by a general function, easily represented in a damage/absorbed dosage plot (Figure 8.14):

$$D_g(t) = f[D_{tot}(t)] \tag{8.56}$$

In this way, only radiation absorbed in the material that causes degradation is actually taken into account. The mathematic solution of Equation 8.56 gives back the

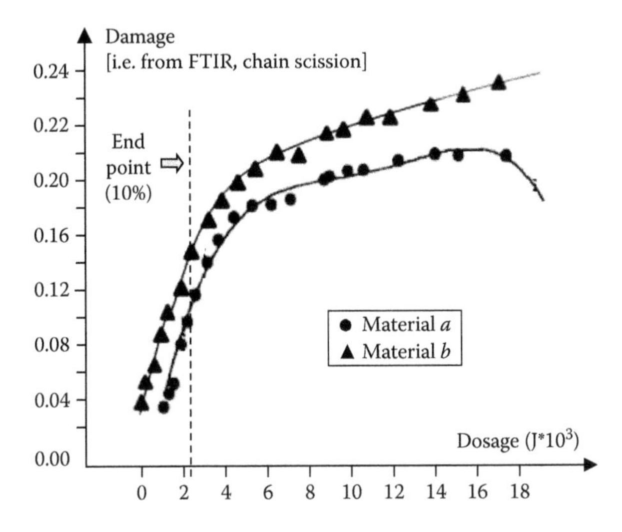

Figure 8.14 Example of a damage-dosage curve during the time of exposure to UV light.

time to reach the failure level of degradation (i.e., a threshold level of an odorous substance), with the advantage of taking into account only radiation really absorbed in the material that causes degradation.

Moreover, from the initial slope of the damage/total absorbed dosage curve (Figure 8.14), using, for example, as the end point the slope at 10% of the total absorbed dosage, it is also possible to estimate the apparent quantum efficiency $\Phi(\lambda)$, which is defined by the rate of a given photophysical or photochemical process divided by the total absorbed radiation (Braslavsky et al. 2011) and that is a useful index for comparing degradation of different polymers to a specific radiation range (Nguyen et al. 2002).

While reduction or elimination of UV energy that reaches food is important and, often, is successfully obtained with the use of specific additives (Salem et al. 2002, Ammala et al. 2002), most food components do remain sensitive over time, albeit to a lesser degree, to light energy in the visible spectral region, especially if oxygen is present in the headspace of the package during storage. Its presence, in fact, is a prerequisite for light-induced oxidation (Singh and Singh 2005), and techniques such as modified atmosphere packaging, active packaging (especially oxygen absorbers), combined with the use of gas barrier materials may in fact reduce the oxygen concentration in the package headspace and, consequently, reduce food degradation. In general, for flexible packaging the adopted solutions able to block radiation in the visible range include the use of colorant additives in the formulation; the use of printed sleeves stuck on bottles, trays, and the like; and the metallization and the cavitation of plastic films. However, for marketing reasons, photosensitive foods are commonly packed in see-through materials and exposed on highly lit shelves, increasing the risk of light-induced oxidation (Monzocco et al. 2008). Although the exposure is commonly short, the lighting environments of retail stores provide energy-inducing photodegradation, particularly evident in the presence of oxygen. The interplay among product, packaging, light, and oxygen is complex, and the degree of food photooxidation also depends on the ability of the package to prevent the radiation from reaching the product. In fact, as represented in Figure 8.13(b), only part of the photons from a light source may be absorbed by the product because, when the incident light is transmitted through the material into the package headspace, a part of the radiation is reflected or absorbed by food, whereas the remaining part can be transmitted, absorbed, or reflected by the internal walls of the package (Limbo et al. 2011).

For these reasons, to assess the shelf life of light-sensitive foods, the material's ability to absorb, transmit, and reflect electromagnetic energy not only in the UV but also in the visible spectrum should be well known and taken into account. These spectra, recorded generally between 380 and 800 nm in wavelength, make it possible to quantify the transparency at any wavelength and consequently to know which proportion of light energy is absorbed by the material (Lee et al. 2008f). In fact, as said previously, when a beam of monochromatic light meets a slab of material along its path, some part of it is reflected, some absorbed, and some transmitted.

The fraction of the incident light transmitted by any given material follows the Lambert-Beer law (Karel and Lund 2003):

$$I = I_0 e^{-kx} \tag{8.57}$$

where I is the intensity of the light transmitted by the packaging material; I_0 is the intensity of the incident light; k is the characteristic constant (absorbance) for the packaging material that depends on the nature of the material and on the wavelength; and x is the thickness of the packaging material.

Assuming that the total transmitted light (I/I_0) is absorbed or reflected by the product and that there is no dissipation phenomenon involving the internal walls of the tridimensional package (through reflection, absorption, or transmission mechanisms), the total amount of light absorbed by a packaged food can be calculated using the following formula (Karel and Lund 2003):

$$I_{abs} = I_0 T_p \frac{1 - R_f}{\left(1 - R_f\right) R_p} \tag{8.58}$$

where I_{abs} is the intensity of the light absorbed by the food; I_0 is the intensity of the incident light; T_p is the fractional transmission by the packaging material; R_p is the fraction reflected by the packaging material; and R_f is the fraction reflected by the food.

This formula can be simplified if the reflection of the packaging material and the food are negligible. Therefore, Equation 8.58 will become Equation 8.59:

$$I_{abs} = I_0 T_p \tag{8.59}$$

When an incident light, at a given wavelength of intensity equal to I_0, passes through a simple *food–package* model like that represented in Figure 8.15, the profile of the light intensity changes as follows:

- $I_i = I_0 e^{-k_p X_p}$ passing through the package thickness X_p. In this case, I_i refers to the intensity of the light at the surface of the food, and k_p is the specific absorbance of the material for any one wavelength;
- $I_x = I_0 e^{-k_f X_f}$ passing from the surface of the food up to a specific plane X_f of the food. In this case I_x refers to the intensity of the light absorbed at a given plane X_i, and k_f is the specific absorbance of the food for any one wavelength.

Combining Equations 8.58 and 8.59, it is possible to write the following equation, which can also be easily converted into the logarithmic form:

$$I_x = I_0 e^{-(k_p X_p + k_f X_f)} \quad \text{or} \quad \ln\frac{I_0}{I_x} = k_p X_p + k_f X_f \tag{8.60}$$

In a shelf life study, the intensity of the light absorbed by the packaged food at time t, which is $I_x(t)$, may be related to the real damage suffered by the food itself. For example, if the concentration of riboflavin is the key attribute to define the shelf life of a food, its decomposition over time has to be measured. In these cases, the quantum yield for the decomposition (or formation) of a substance can be defined as follows (Turhan and Sahbaz 2001):

$$\Phi(\lambda) = \frac{\text{number of molecules decomposed or formed per unit time}}{\text{number of quanta absorbed per unit time}}$$

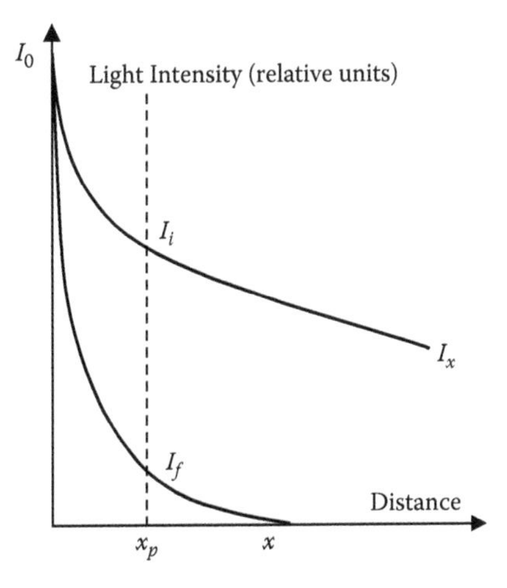

Figure 8.15 Profile of the light intensity as a function of the package thickness and the depth of a solid food.

Because the number of quanta absorbed per unit time is the absorbed intensity, this expression becomes

$$\Phi(\lambda) = \frac{\pm dC / dt}{I_x(t)} \tag{8.61}$$

where $\pm dC/dt$ is the rate of decomposition or formation of a substance in the food.

This last expression would usually be suggested by knowledge of the kinetic model (e.g., linear for zero-order kinetics and exponential for first-order kinetics), although empirical curve fitting may also be adequate.

The model proposed is suitable for estimating the damage that occurs in a packaged solid food under the effect of lighting. The same model probably fails if a packaged liquid food (e.g., a beverage in a bottle) is considered. The reason is that while the light cannot reach the photooxidative substances located in the depth of the bottle, a diffusion process can occur in this liquid product, and an exchange can occur between the photooxidative substance in the surface and that in the interior, thus continuously supplying more substance to the irradiation surface. Similarly, free radicals produced by light at the surface of the liquid can migrate inward and react to damage the molecules in the interior (Karel and Lund 2003). Even in a solid product with limited diffusion, surface reactions often suffice to produce damage resulting in unsalability of the product. This is obviously the case with sensitive colors or when small amounts of potent off-flavors or toxic compounds are produced.

In the evaluation of the degree of protection of plastic materials against light radiations and light-induced changes, the transmittance properties (both total and at specific wavelengths) often play a more important role than the reflection ones, especially if the material is transparent and no superficial treatments like metallization are present.

The use of a spectrophotometer equipped with an integrating sphere is the most common way for evaluating some of these properties. The commercially available spheres are multipurpose instruments designed to be used for both reflectance and absorption/transmittance measurements in the entire UV-visible range. In the applications for packaging materials, the spectrum of the regular (also called specular) transmittance is usually considered (Figure 8.16), that is, the fraction of the incident intensity that does not deviate from the incident direction as light emerges on the other side of the film.

However, for correct interpretation of the optical characteristics of materials, it is useful to consider the different components of transmitted light and not only regular light. The integrating sphere can be used in specific configurations that allow the evaluation of the light that, in passing through a specimen, deviates from the normal incident beam. In fact, as suggested by Roos and Ribbing (1988), the light that passes through a sheet can be divided into different components, each resulting in its own separate contribution to the detector signal. Using specific configurations of the integrating sphere, it is possible to observe that each component differently contributes to the total transmittance, as shown in Figure 8.17 (Torri 2007, Limbo et

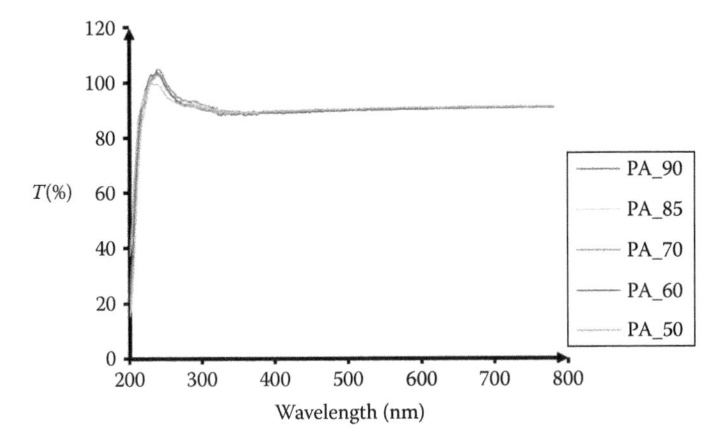

Figure 8.16 Regular transmittance of multilayer materials with different total layers in the UV-visible range.

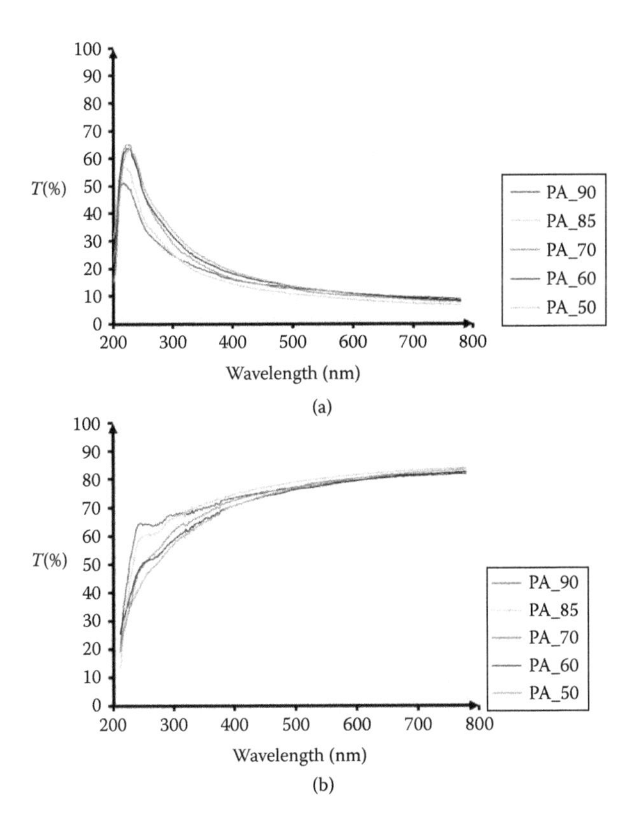

Figure 8.17 Different components of the transmitted light: (a) regular and (b) largely diffused of some plastic materials.

al. 2011). These contributions depend on the scattering angle and mode of operation with the integrating sphere and are important in defining the optical characteristics of a material that is not perfectly regular or perfectly diffuse or, in other words, that is nonideal transmitting.

In designing a food package, it is essential to take into account the regular and diffused components to optimize the food package effectiveness against photooxidation. For example, as reported by Limbo et al. (2011), in tray lidding packaging solutions, where the food is horizontally placed onto a semirigid tray and sealed or wrapped with a transparent film lid, those components play an important role in cheese photooxidation, especially if correlated to other packaging features, like the distance between the product and lid film and the optical characteristics of the tray (color, transmittance spectrum, thickness, etc.). For example, if no headspace exists between the lid and the product, all the transmitted light (regular and diffused) will be absorbed (or partially reflected) by the food. Otherwise, if a certain headspace is maintained, the higher the diffused components of the lid film, the higher is the probability that a part of the photons reaches the internal walls of the tray instead of the product, reducing the risk of photooxidation. In the same way, if the tray is characterized by good absorption of the visible radiations (e.g., if it is black), the light that is scattered by the lid film can be easily absorbed by the internal walls of the tray itself, further reducing the photooxidation phenomena as represented in Figure 8.18 (Limbo et al. 2011).

It is hoped future research will result in predictive safe modeling of photooxidative quality changes based on measurements of early effects. Combining predictive modeling with product and packaging expertise will ultimately prevent light-deteriorated cheeses from reaching consumers (Mortensen et al. 2004).

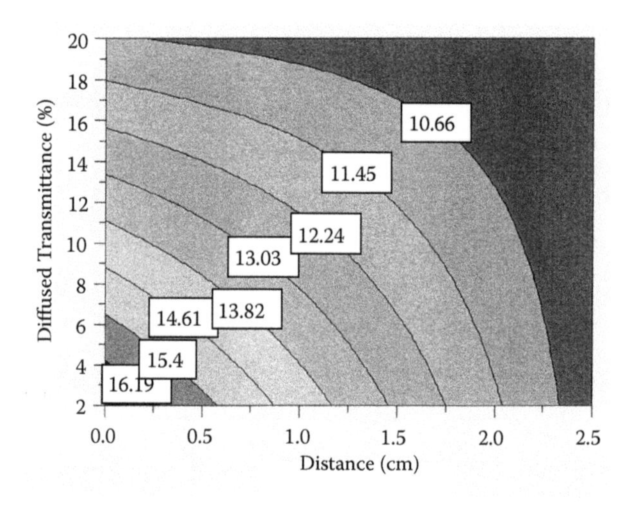

Figure 8.18 Variation of b* color index of sliced cheese during storage packaged in a black tray with a transparent lid. Effect of distance (lid–product) and diffuse transmittance of the film.

8.7 REFERENCES

American Society for Testing and Materials. 2010a. *ASTM E96/E96-10. Standard test method for water vapor transmission of materials.* ASTM International, West Conshohocken, PA.

American Society for Testing and Materials. 2010b. *ASTM D3985-05(2010)e1. Standard test method for oxygen gas transmission rate through plastic film and sheeting using a coulometric sensor.* ASTM International, West Conshohocken, PA.

Ammala, A., Hill, P., Meakin, A.J., Pas, S.J., and T.W. Turney. 2002. Degradation studies of polyolefins incorporating transparent nanoparticulate zinc oxide UV stabilizers. *Journal of Nanoparticle Research* 4: 167–174.

Andersen, C.M., Vishart, M., and Holm, V.K. 2005. Application of fluorescence spectroscopy in the evaluation of light-induced oxidation in cheese. *Journal of Agricultural and Food Chemistry* 53: 9985–9992.

Anonymous. 1974. Shelf life of foods. *Journal of Food Science* 39: 861.

Ashley, R.J. 1985. Permeability and plastics packaging. In: *Polymer permeability*, J. Comyn, Ed., 269–308. London: Elsevier Applied Science.

Banker, G.S., Gore, A.Y., and J. Swarbrick. 1966. Water vapor transmission properties of free polymer films. *Journal of Pharmacy and Pharmacology* 18: 457–466.

Barrer, R. M., and E.K. Rideal. 1939. Permeation, diffusion and solution of gases in organic polymers. *Transactions of the Faraday Society* 35: 628–643.

Biquet, B., and T.P. Labuza. 1988. Evaluation of the moisture permeability characteristics of chocolate films as an edible moisture barrier. *Journal of Food Science* 53: 989–998.

Black, R.G. 2011. Retail beer and wine display lighting. http://www.safespectrum.com/applications_beer_wine.html (accessed July 6, 2011).

Bosset, J.O., Gallmann, P.U., and R. Sieber. 1992. Influence of light transmittance of packaging materials on the shelf-life of milk and dairy products. A review. In *Food packaging and preservation,* M. Mathlouthi, Ed., 222–265. London: Blackie Academic and Professional.

Brandsch, J., Mercea, P., Ruter, M., Tosa, V., and O. Piringer. 2002. Migration modeling as a tool for quality assurance of food packaging. *Food Additives and Contaminants* 19(supplement): 29–41.

Braslavsky, S.E., Braun, A.M., Cassano, A.E., Emeline, A., Litter, M.I., Palmisano, L., Parmon, V.N., and N. Serpone. 2011. Glossary of terms used in photocatalysis and radiation catalysis (IUPAC recommendations 2011). *Pure and Applied Chemistry* 83(4): 931–1014.

Bretsznajder, S. 1971. *Prediction of transport and other physical properties of fluids.* Oxford, UK: Pergamon.

Brown, W.E. 1987. Selecting plastics and composite barrier systems for food packages. In *Food product-package compatibility,* J.I. Gray, B.R. Harte, and J. Miltz, Eds., 200–228. Lancaster, PA: Technomic.

Choe, E., and D.B. Min. 2006. Mechanisms and factors for edible oil oxidation. *Comprehensive Reviews in Food Science and Technology* 5: 169–186.

Crank, J. 1975. *The mathematics of diffusion.* Oxford, UK: Clarendon, pp. 44–68.

Dainelli, D., Gontard, N., Spyropoulos, D., Zondervan-van den Beuken, E., and P. Tobback. 2008. Active and intelligent food packaging: legal aspects and safety concerns. *Trends in Food Science and Technology* 19: S103–S112.

Del Nobile, M.A., Fava, P., and L. Piergiovanni. 2002. Water transport properties of cellophane flexible films intended for food packaging applications. *Journal of Food Engineering* 53: 295–300.

Duncan, S.E., and J.B. Webster. 2009. Sensory impacts of food-packaging interactions. *Advances in Food Nutrition Research* 56: 17–64.

Eichner, K. 1986. The influence of water content and water activity on chemical changes in foods of low moisture content under packaging aspects. In *Food packaging and preservation,* M. Mathlouthi, Ed., 67–92. London: Elsevier Applied Science.

Eie, T. 2009. Light protection from packaging. In *The Wiley encyclopedia of packaging technology*, L. Yam, Ed., 655–659. Hoboken, NJ: Wiley.

Escobar, L.A., and W.Q. Meeker. 2006. A review of accelerated test models. 2006. *Statistical Science* 21(4): 552–577.

European Parliament. 2004. Regulation (EC) No. 1935/2004 on materials and articles intended to come into contact with food and repealing Directives 80/590/EEC and 89/109/EEC. *Official Journal of the European Union*, 13.11.2004, L 338/4–L 338/17.

Fava, P., Limbo, S., and L. Piergiovanni. 2000. La previsione della shelf life di prodotti alimentari sensibili agli scambi di umidità. *Industrie Alimentari* 39: 121–126.

Flory, P.J. 1953. *Principles of polymer chemistry.* Ithaca, NY: Cornell University Press, pp. 495–512.

Fonseca, S.C., Oliveira, F.A.R., and J.K. Brecht. 2002. Modelling respiration rate of fresh fruits and vegetables for modified atmosphere packages: a review. *Journal of Food Engineering* 52(2): 99–119.

Galliard, T. 1986. Oxygen consumption of aqueous suspensions of wheat wholemeal, bran and germ: involvement of lipase and lipoxygenase. *Journal of Cereal Science* 4: 33–50.

Gemili, S., Yemenicioglu, A., and S.A. Altınkaya. 2010. Development of antioxidant food packaging materials with controlled release properties. *Journal of Food Engineering* 96: 325–332.

Gennadios, A., Weller, C.L., and C.H. Gooding. 1994. Measurement errors in water vapor permeability of highly permeable, hydrophilic edible films. *Journal of Food Engineering* 2: 395–409.

Gugumus, F. 2002. Possibilities and limits of synergism with light stabilizers in polyolefins 2. UV absorbers in polyolefins. *Polymer Degradation and Stability* 75: 309–320.

Guillard, V., Mauricio-Iglesias, M., and N. Gontard. 2010. Effect of novel food processing methods on packaging: structure, composition, and migration properties. *Critical Reviews in Food Science and Nutrition* 50: 969–988.

Hagenmaier, R.D., and P.E. Shaw. 1990. Moisture permeability of edible films made with fatty acid and (hydroxypropyl)methylcellulose. *Journal of Agricultural Food Chemistry* 38: 1799–1803.

Helmroth, E., Rijk, R., Dekker, M., and W. Jongen. 2002. Predictive modelling of migration from packaging materials into food products for regulatory purposes. *Trends in Food Science and Technology* 13: 102–109.

Hernandez, R.J., and R. Gavara. 1999. *Plastic packaging. methods for studying mass transfer interactions*. Leatherhead: PIRA International Reviews, pp. 1–42. Leatherhead, UK.

Hotchkiss, J.H. 1997. Food-packaging interactions influencing quality and safety. *Food Additives and Contaminants* 14(6–7): 601–607.

Iglesias, H.A., and J. Chirife. 1982. *Handbook of food isotherms*. New York: Academic Press, pp. 262–319.

Jespersen, L., Strømdahl, L.D., Olsen, K., and L.H. Skibsted. 2005. Heat and light stability of three natural blue colorants for use in confectionery and beverages. *European Food Research Technology* 220: 261–266.

Jin, X., Zumbrunnen, D.A., Balasubramanian, A., and K.L. Yam. 2009. Tailored additive release rates in extruded plastic films produced with smart blending machines. *Journal of Plastic Film Sheet* 25:115–140.

Jurin, V., and M. Karel. 1963. Studies on control of respiration of McIntosh apples by packaging methods. *Journal of Food Science* 17: 104–108.

Kader, A.A., Zagory, D., Kerbel, E.L., and C.Y. Wang. 1989. Modified atmosphere packaging of fruits and vegetables. *Critical Reviews in Food Science and Nutrition* 28: 1–30.

Kalua, C.M., Bedgood, D.R., Bishop, A.G., and Prenzler, P.D. 2006. Discrimination of storage conditions and freshness in virgin olive oil. *Journal of Agricultural and Food Chemistry* 54: 7716–7151.

Karel, M., and D.B. Lund 2003. Protective packaging. In *Physical principles of food preservation*, 2nd ed. 492–569. New York: Dekker.

Karel, M., Proctor, B.E., and G. Wiseman. 1959. Factors affecting water vapor transfer through food packaging films. *Food Technology* 13(l): 69–74.

Katan, L.L. 1996. Introduction. In *Migration from food contact materials*, L.L. Katan, Ed., 1–10. London: Blackie Academic and Professional.

Kester, J.J., and O. Fennema. 1989. An edible film of lipids and cellulose ethers: barrier properties to moisture vapor transmission and structural evaluation. *Journal of Food Science* 54: 1383–1389.

Kilcast, D., and P. Subramaniam. 2000. Fats and oils. In *The stability and shelf life of food*, D. Kilcast and P. Subramaniam, Eds., 3–11. Cambridge, UK: Woodhead.

Labuza, T.P. 1982. *Shelf-life dating of foods*. Westport, CT: Food and Nutrition Press, pp. 1–87, 99–118.

Labuza, T.P., Mizrahi, S., and M. Karel. 1972. Mathematical models for the optimization of flexible film packaging of foods for storage. *Transactions of ASAE (American Society of Agricultural Engineers)* 15: 150–155.

Labuza, T.P., Tannenbaum, S.R., and M. Karel. 1970. Water activity and stability of low-moisture and intermediate moisture foods. *Food Technology* 24(5): 35–42.

LaCoste, A., Schaich, K.M., Zumbrunnen, D., and K.L. Yam. 2005. Advancing controller release packaging through smart blending. *Packaging Technology and Science* 18: 77–87.

Lamiani, P. 2010. Innovative approaches and instruments in modeling and monitoring the shelf life of packaged perishable foods. In *Proceedings of the 15th workshop on the developments in the Italian PhD research on food science and technology*, Portici, Italy, September 15–17, 2010, pp. 257–258.

Lee, D.S., Song, Y., and K.L. Yam. 1996. Application of an enzyme kinetics based respiration model to permeable system experiment of fresh produce. *Journal of Food Engineering* 27(3): 297–310.

Lee, D.S., Yam, K.L., and L. Piergiovanni. 2008a. Shelf life of packaged food products. In *Food packaging science and technology*, D.S. Lee, Ed., 480–506. Boca Raton, FL: Taylor & Francis.

Lee, D.S., Yam, K.L., and L. Piergiovanni. 2008b. Permeation of gas and vapor. In *Food packaging science and technology*, D.S. Lee, Ed., 82–97. Boca Raton, FL: Taylor & Francis.

Lee, D.S., Yam, K.L., and L. Piergiovanni. 2008c. Shelf life of packaged food products. In *Food packaging science and technology*, D.S. Lee, Ed., 505–521. Boca Raton, FL: Taylor & Francis.

Lee, D.S., Yam, K.L., and L. Piergiovanni. 2008d. Shelf life of packaged food products. In *Food packaging science and technology*, D.S. Lee, Ed., 532–540. Boca Raton, FL: Taylor & Francis.

Lee, D.S., Yam, K.L., and L. Piergiovanni. 2008e. Overview of food packaging systems. In *Food packaging science and technology*, D.S. Lee, Ed., 8–9. Boca Raton, FL: Taylor & Francis.

Lee, D.S., Yam, K.L., and L. Piergiovanni. 2008f. Physical properties of packaging materials. In *Food packaging science and technology*, D.S. Lee, Ed., 43–76. Boca Raton, FL: Taylor & Francis.

Lennersten, M., and H. Lingnert. 1998. Influence of different packaging materials on lipid oxidation in potato crisps exposed to fluorescent light. *Lebensmittel Wiss und Technologie* 31: 162–168.

Leo, A., Hansch, C., and D. Elkins. 1971. Partition coefficients and their uses. *Chemical Review* 71(6): 525–616.

Limbo, S., Torri, L., Adobati, A., and L. Piergiovanni. 2011. Investigation of optical and geometrical package features in photo-oxidation decay of a semi-hard cheese. *Italian Journal of Food Science* special issue, 68–71.

Limbo, S., Torri, L., and L. Piergiovanni. 2007. Light-induced changes in an aqueous b-carotene system stored under halogen and fluorescent lamps, affected by two oxygen partial pressures. *Journal of Agricultural and Food Chemistry* 55: 5238–5245.

Madigan, M.T., and J.M. Martinko. 2006. *Brock, biology of microorganisms*, 11th ed. Upper Saddle River, NJ: Prentice Hall Pearson Education.

Manzocco, L., Kravina, G., Calligaris, S., and M.C. Nicoli. 2008. Shelf life modeling of photosensitive food: the case of colored beverages. *Journal of Agricultural and Food Chemistry* 56: 5158–5164.

Marsh, K.S. 1986 Shelf life. In *The Wiley encyclopedia of packaging technology*, M. Bakker, Ed., 578–582. New York: Wiley.

McHugh, T.H., Avena-Bustillos, R., and J.M. Krochta. 1993. Hydrophilic edible films: modified procedure for water vapor permeability and explanation of thickness effects. *Journal of Food Science* 58(4): 899–903.

Mercea, P., and O. Piringer. 1998. Data bank to validate a mathematical model to estimate the migration of additives and monomers from polyolefins into foodstuffs and food simulants. EU-contract No. ETD/97/501370 (DG3).

Min, D.B., and J.M. Boff. 2002. Chemistry and reaction of singlet oxygen in foods. *Comparative Reviews in Food Science and Food Safety* 1: 58–72.

Molin, G. 2000. Modified atmospheres. In *The microbiological safety and quality of food*, B.M. Lund, T.C. Baird-Parker, and G.W. Gould, Eds., Vol. 1, 214–234. Gaithersburg, MD: Aspen.

Mortensen, G., Bertelsen, B., Mortensen, B.K., and H. Stapelfeldt. 2004. Light-induced changes in packaged cheeses—a review. *International Dairy Journal* 14: 85–102.

Nguyen, T., Martin, J.W., Byrd, E., and E. Embree. 2002. Quantum efficiency of spectral UV on polyurethanes. *Proceedings of 9th International Conference on Durability of Buildings Materials and Components* http://www.irb.fraunhofer.de/CIBlibrary/search-quick-result-list.jsp?A&idSuche=CIB+DC9932 (accessed July 2011).

Patel, M., Patel, J.M., and A.P. Lemberger. 1964. Water vapor permeation of selected cellulose ester films. *Journal of Pharmaceutical Sciences* 53: 286–290.

Pennarun, P.Y., Dole, P., and A. Feigenbaum. 2004. Overestimated diffusion coefficients for the prediction of worst case migration from PET: application to recycled PET and to functional barriers assessment. *Packaging Technology and Science* 17: 307–320.

Piergiovanni, L., Fava, P., and S. Ceriani. 1999. A simplified procedure to determine the respiration rate of minimally processed vegetables in flexible permeable packaging. *Italian Journal of Food Science* 11(2): 99.

Piergiovanni, L., Fava, P., and A. Siciliano. 1995. A mathematical model for the prediction of water vapour transmission rate at different temperature and relative humidity combinations. *Packaging Technology and Science* 8(2): 73.

Piergiovanni, L., and S. Limbo. 2004. The protective effect of film metallization against oxidative deterioration and discoloration of sensitive foods. *Packaging Technology and Science* 17: 155–164.

Piringer, O. 2000. Transport equations and their solutions. In *Plastic packaging materials for foods*, O. Piringer and A.L. Baner, Eds., 183–219. Weinhem, Germany: Wiley-VCH.

Quast, D.G., Karel, M., and W.M. Land. 1972. Development of a mathematical model for oxidation of potato chips as a function of oxygen pressure, extent of oxidation and equilibrium relative humidity. *Journal of Food Science* 37: 673–678.

Rankin, J.C., Wolff, I.A., Davis, H.A., and C.E. Rist. 1958. Permeability of amylose film to moisture vapor, selected organic vapors, and the common gases. *Industrial and Engineering Chemistry Research Chemical Engineering Data Series* 3, 120–123.

Reynier, A., Dole, P., and A. Feigenbaum. 1999. Prediction of worst case migration: presentation of a rigorous methodology. *Food Additives and Contaminants* 16: 137–152.

Rogers, C.E. 1985. Permeation of gases and vapours in polymers. In *Polymer permeability*, J. Comyn, Ed., 11–73. London: Elsevier Applied Science.

Roos, A., and C. Ribbing. 1988. Interpretation of integrating sphere signal output for non-Lambertian samples. *Applied Optics* 27(18): 3833–3837.

Salem, M.A., Farouk, H., and I. Kashif. 2002. Physicochemical changes in UV-exposed low-density polyethylene films. *Macromolecular Research* 10(3):168–173.

St. Angelo, A.J. 1996. Lipid oxidation in foods. *Critical Reviews in Food Science and Nutrition* 36: 175–224.

Schultz, T.H., Miers, J.C., Owens, H.S., and W.D. Maclay. 1949. Permeability of pectinate films to water vapor. *Journal of Physical and Colloid Chem*istry 53: 1320–1330.

Sheftel, V.O. 2000. *Indirect food additives and polymers. Migration and toxicology*. Boca Raton, FL: Lewis, pp. 1–5.

Singh, B., and N. Sharma. 2008. Mechanistic implications of plastic degradation. *Polymer Degradation and Stability* 93: 561–584.

Singh, R.K., and N. Singh. 2005. Quality of packaged foods. In *Innovations in food packaging*, J.H. Han, Ed., 24–44. London: Elsevier.

Taylor, A.E.F. 2000. Illumination fundamentals. Lighting Research Center. Booklet, Rensselaer Polytechnic Institute. http://www.opticalres.com/lt/illuminationfund.pdf (accessed June 3, 2011).

Torri, L. 2007. Valutazione degli effetti della luce su alimenti fotosensibili e dell'efficacia protettiva di soluzioni di confezionamento. PhD diss., Milan State University, Italy.

Turhan, K.N., and F. Sahbaz. 2001. A simple method for determining light transmittance of polymer films used for packaging foods. *Polymer International* 50: 1138–1142.

Wilkinson, A.M., and A.D. McNaught. 1997. *Partition coefficient. Compendium of Chemical Terminology: IUPAC Recommendations*. Oxford, UK: Blackwell Science.

Woodruff, C.W., Peck, G.E., and G.S. Banke. 1972. Effect of environmental conditions and polymer ratio on water vapor transmission through free plasticized cellulose films. *Journal of Pharmaceutical Science* 61, 1956–1959.

Zhang, Y., Chikindas, M.L., and K.L. Yam. 2004. Effective control of *Listeria monocytogenes* by combination of nisin formulated and slowly released into a broth system. *International Journal of Food Microbiology* 19(1): 15–22.

CHAPTER 9

Case Studies

Monica Anese, Rosalba Lanciotti, Fausto Gardini, and Corrado Lagazio

CONTENTS

9.1 INTRODUCTION

The aim of this chapter is to show, through a number of case studies, practical applications of the methodologies described in the previous chapters for the assessment of the shelf life of foods.

This chapter, which is mainly addressed to operators of the food sector as well as to students in food science and technology, makes no pretense to assess the shelf life of a series of food products. On the contrary, its intention is to find the most suitable strategies and methodologies for shelf life evaluation in relation to food type as well as its process and storage history.

For each case study considered, after a brief presentation of the product in question, the problem most relevant to the product shelf life was outlined. Then, the shelf life study was carried out first by defining the acceptability limit (if pertinent), identifying the indicators of quality loss, and finally assessing the expiration time by instrumental or sensory tools.

Based on this, Chapter 9 comprises a number of case studies dealing with refrigerated, frozen, and ambient stable foods chosen as representatives of food categories generally presenting short, medium, and long shelf lives. As a consequence, this choice let us identify different quality decay indicators and descriptors and carry out shelf life studies by different methodologies described in previous chapters.

In case study 9.1, the shelf life of fresh-cut apple slices packaged under different atmospheres is assessed. The quality decay descriptor is represented by microbial growth, which was predicted by applying mathematical models such as those described in Chapter 6. Once the upper limit of microbial growth was defined (see Chapters 3 and 6), the model was used to predict product expiration date.

Case study 9.2, "Fruit Juice New Process Design and Shelf Life Assessment," deals with the application of a nonconventional technology, an alternative to pasteurization, for extending the shelf life of fruit juices. After having identified the best processing conditions to inactivate the spoilage agent, the microbial growth was followed during refrigerated and ambient storage. The fruit juice shelf life was then estimated at the time at which the microbial count overcame the defined acceptability limit.

Case study 9.3 describes a shelf life study applied to a newly formulated product, a frozen ready-to-eat pasta meal. Due to the product complexity and the absence of experience, an accelerated stability test (see Chapter 5) was applied to identify the earliest quality decay event. Afterward, the product acceptability limit was estimated using a sensory test, which was carried out by applying the methodology of survival analysis described in Chapter 7. The index best describing the quality loss was then followed over time under actual storage conditions, and the length of time needed to reach the acceptability limit was estimated.

Case study 9.4, "Shelf Life Estimation of Pasteurized Orange Juice," is a simple one, aimed to predict the shelf life of an ambient stable product, pasteurized orange juice, based on vitamin C degradation. In this case study, the acceptability limit, which is not compulsory, was chosen based on the E.U. and U.S. recommended daily allowances for ascorbic acid.

In case study 9.5, "Shelf Life Prediction of Gorgonzola Cheese Breadsticks," the end of the shelf life of another microbiologically stable food is considered. The shelf life testing was carried out under accelerated conditions by applying survival analysis. Such analysis allowed simultaneous assessment of the acceptability limit and the shelf life of the product. Also, to overcome the problems (costs, consumer panel assembly, data elaboration and interpretation) generally encountered when applying sensory analysis, instrumental testing was carried out and a mathematical model accounting for consumer acceptability developed.

Case study 9.6 focuses on the assessment of the secondary shelf life of ground roasted coffee. It described the methodology that can be applied to estimate the shelf life of the product under usage. Here, the application of a sensory test allowed simultaneous assessment of the acceptability limit and shelf life. As in case study 9.5, a relationship between sensory acceptability and changes in the quality descriptor determined by instrumental analysis was established.

The last two case studies aim not only to predict the shelf life of a food product but also to show how predictive methods can represent "problem-solving" tools. For example, case study 9.7 was aimed to develop a predictive model to verify the empirical rule, widely used at the industrial level to estimate the shelf life of frozen foods (here, shrimp), stating that the quality changes that happen in a storage month at -18°C correspond to those that happen in 1 week at -7°C. Therefore, the aim of the study was to find a relationship between the shelf life at a given temperature (i.e., -7°C) and the shelf life at a reference temperature (i.e., -18°C). Case study 9.8, "Shelf Life Assessment of Fruit-Based Noncarbonated Soft Drinks," dealt with beverages stabilized by the addition of nonconventional natural preservatives. The aim was to evaluate stabilization conditions that have to be applied to meet the producer's shelf life requirement. To this purpose, an experimental design (not described in Chapter 6) was applied. Although not easy, as it requires statistical competence for application, this methodology allowed simultaneous study of the main and interactive effects of the different variables (preservative type and concentration, thermal treatment, inoculum concentration) on the growth of spoilage microorganisms. This case study offered an example of how the application of the experimental design appeared well suited for application in shelf life studies as it allowed reduction of the number of experiments, thus saving work and time.

The case studies presented here were inspired by research carried out at the authors' research centers or drawn from the literature.

9.2 CASE STUDY 9.1. SHELF LIFE ASSESSMENT OF FRESH-CUT APPLE SLICES PACKAGED UNDER DIFFERENT ATMOSPHERES

9.2.1 Product Presentation

The fresh-cut fruit and vegetable sector is a rapidly growing market (Buta et al., 1999). The fruits and vegetables in question are presented for sale conveniently washed, peeled, cored, or sliced in prepacked packages. They include ready-to-eat salads, which account for most of these sales. The success of fresh-cut fruits and vegetables lies in their high convenience attribute, the right size package, and consistent fresh-like quality and taste. The shelf life generally attributed to the fresh-cut products ranges from a few days to 2 weeks.

9.2.2 Problem

As it is well known, processing operations such as peeling, slicing, and shredding can speed up microbial and enzymatic spoilage due to destruction of plant cells, making fresh-cut fruits and vegetables more perishable than the original raw materials (Lanciotti et al., 1999). Shelf life extension of these products is generally achieved by applying minimal processing (i.e., minimizing mechanical injuries, adopting proper hygienic practices, and maintaining the chill chain). However, a critical aspect relevant to fresh-cut products is represented by their actual shelf life, which is often lower than that estimated by the company. The correct prediction of the end of the shelf life of these products is therefore of great importance. In fact, consumers are generally aware that the product should be used rapidly after purchase and well before the expiring date reported on the label, which is generally a few days.

This case study dealt with the shelf life assessment of fresh-cut apple slices packaged under vacuum or ordinary or modified atmosphere supplemented or without hexanal during refrigerated storage.

9.2.3 Shelf Life Study

9.2.3.1 *Identification of the Early Quality Decay Event and Critical Descriptor*

The destruction of tissue and the subsequent release of nutrients due to the technological operations enhance the growth of naturally occurring microorganisms. The low pH of most fruits restricts the microflora to acid-tolerant microorganisms. Due to the chemical and physical characteristics of the products, yeasts, molds, and lactic acid bacteria are favored as spoilage agents. Coliforms are part of the natural microflora of fruits and processing lines (Beuchat, 2002). A variety of pathogenic microorganisms such as *Listeria monocytogenes*, *Salmonella* and *Shigella* spp., and enteropathogenic strains of *Escherichia coli*, *Aeromonas hydrophila*, *Yersinia enterocolitica*, and *Staphylococcus aureus* may be present on fresh fruits and in related minimally processed refrigerated products (Breidt and Fleming, 1997). Whether these pathogens can grow and cause disease depends on product type, storage conditions, and competing microflora. However, the dominance of the spoilage association depends on the atmosphere composition and storage temperature. When the microbial load overcomes the value of 10^6 cfu (colony-forming units)/g, product quality deterioration becomes perceivable. Therefore, this critical value of the dominant spoilage microorganism load represents the acceptability limit for shelf life computation. To find the dominant microorganism load, preliminary tests at 4°C were carried out, and the growth of five microbial groups (i.e., total psychrotrophic bacteria, total mesophilic bacteria, lactic acid bacteria, yeasts, and molds) was monitored. Among the different microbial groups considered, only psychrotrophic bacteria were able to grow during storage at 4°C. Therefore, under the adopted conditions, this microbial group determined the product's shelf life.

9.2.3.2 *Testing*

Granny Smith apples were manually cored and sliced, and 100 g of slices were immediately packaged in high-barrier plastic bags under different atmospheres: vacuum, ordinary atmosphere, modified atmosphere (70% N_2, 30% CO_2), ordinary atmosphere with 100 ppm hexanal, and modified atmosphere (70% N_2, 30% CO_2) with 100 ppm hexanal. Previous studies showed that many compounds, naturally occurring in plant extracts and spices, such as phenols, aldehydes, and organic acids, may exert strong antimicrobial activity (Song et al., 1996; Gardini et al., 2001; Lanciotti et al., 2004). Table 9.1 shows the psychrotrophic bacterial loads collected during the storage of the fresh-cut apple slices packaged under different atmospheres. The cell load data (averages of at least three repetitions) were modeled according to the Gompertz equation as modified by Zwietering et al. (1990):

$$y = K + A \cdot \exp\left\{-\exp\left[\frac{\mu_{\max} \cdot e}{A}(\lambda - t) + 1\right]\right\} \tag{9.1}$$

where y is the log colony-forming units per gram at time t, K is the initial level of each microbial group (as log cfu/g), A is the maximum bacterial growth attained at

Table 9.1 Psychrotrophic Bacteria Loads Collected during Storage at 4°C of Fresh-Cut Apple Slices Packaged under Different Atmospheres

Time (days)	Psychrotrophic Counts (log cfu/g)				
	A	B	C	D	E
0	0.8	0.8	0.8	0.8	0.8
2	1.2	1.0	0.9	0.7	0.7
4	1.9	0.8	0.7	0.8	0.7
6	2.5	1.0	1.0	0.8	1.0
8	3.9	1.2	1.0	1.0	1.3
10	4.4	2.2	1.8	1.2	1.5
12	5.2	2.6	2.5	1.2	1.6
14	5.3	3.5	3.0	1.2	1.8
16	6.0	3.8	3.1	1.3	2.1
18	5.8	4.0	3.0	1.1	2.3
20	6.2	3.8	3.2	1.3	2.2

Note: A, under vacuum; B, modified atmosphere; C, modified atmosphere with hexanal; D, ordinary atmosphere with hexanal; E, ordinary atmosphere. Data are the mean of at least three repetitions. Variability coefficient ranged between 5% and 10%.

the stationary phase (as log cfu/g), μ_{max} is the maximal growth rate (log cfu/g/day), λ is the lag time (days), and t is the time (days). The same equation was used to calculate the time necessary for psychrotrophic bacteria to attain the threshold of 6.0 log cfu/g. Table 9.2 shows the Gompertz parameters obtained using the equation modified by Zwietering et al. (1990) as well as the end of the shelf life of the fresh-cut apple slices packaged under the different atmospheres. The fitting of the Gompertz equation to the experimental data is shown in Figure 9.1.

These data suggest that, with the exception of packaging under vacuum, all the atmospheres studied were able to cause a significant reduction of the maximum cell load attained at the stationary phase as well as the extension of the incubation phases. The shelf life of the fresh-cut apple slices packaged under ordinary atmosphere or modified atmosphere in the presence or in the absence of hexanal under storage at 4°C resulted in a life longer than 20 days. On the contrary, in the case of the product packaged under vacuum the psychrotrophic spoilage microorganisms reached 6.0 log cfu/g in 18 days.

9.2.3.3 Experimental Needs

- Granny Smith apples
- Sealing and packaging machine
- High-barrier plastic bags (Nylon/30-mm nylon and 120-mm polyethylene)
- Hexanal (Sigma Aldrich)
- Material (culture media, Petri dishes, tubes, pipettes) and equipment (autoclaves, incubators) necessary for microbial cultivation and analyses

Table 9.2 Gompertz Parameters Obtained Using the Equation Modified by Zwietering et al. (1990) and Shelf Life of Fresh-Cut Apple Slices Packaged under Different Atmospheres

Gompertz Parameters	A	B	C	D	E
k	0.74	0.89	0.85	0.78	0.72
A	5.56	3.15	2.30	0.45	1.76
μ_{max}	0.49	0.43	0.48	0.13	0.13
λ	1.90	7.37	8.09	6.09	4.32
Time (days) to					
6 log cfu/g	18.4	>20	>20	>20	>20

Note: A, under vacuum; B, modified atmosphere; C, modified atmosphere with hexanal; D, ordinary atmosphere with hexanal; E, ordinary atmosphere. The Gompertz parameters were calculated using the cell load averages of at least three repetitions. The estimated confidence interval for each parameter was less than 5%.

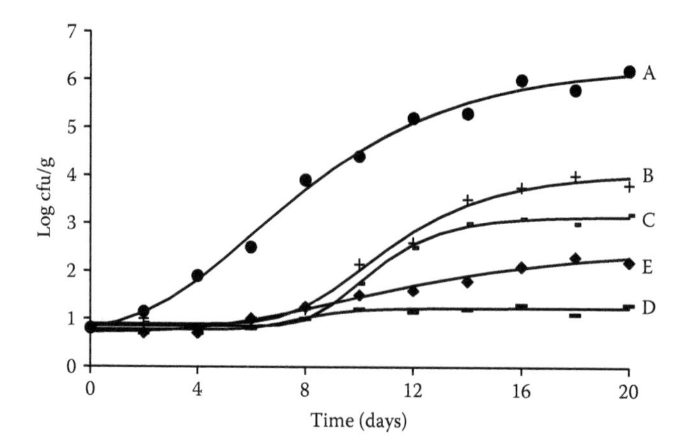

Figure 9.1 Fitting of the Gompertz equation to the experimental data. A = under vacuum; B = modified atmosphere; C = modified atmosphere with hexanal; D = ordinary atmosphere with hexanal; E = ordinary atmosphere.

9.2.3.4 Experimental Plan

Action	Time (days)									
	0	2	4	6	8	10	12	14	16	18
Apple slice preparation	■									
Microbiological analyses	■	■	■	■	■	■	■	■	■	■
Data analysis and shelf life calculation							■	■	■	■

9.3 CASE STUDY 9.2. FRUIT JUICE NEW PROCESS DESIGN AND SHELF LIFE ASSESSMENT

9.3.1 Product Presentation

Fruit juice is made with 100% freshly squeezed fruit juice or reconstituted concentrate. Since fruit juice intake has been consistently associated with reduced risk of many cancers and other chronic diseases, in recent years its consumption in developed countries has greatly increased. In fact, fruit juice is perceived by consumers as a healthy natural source of bioavailable nutrients, antioxidants, and other healthy phytochemicals. Fruit juices are traditionally stabilized through heat treatment or chemical preservatives, and the shelf life generally attributed is several months.

9.3.2 Problem

Fruit juice stabilization by thermal treatment may be responsible for loss of its health value and fresh appearance. Analogously, the use of preservatives as stabilizing agents is perceived by consumers as extraneous and not natural. As a consequence, a number of alternative, nonthermal technologies have been suggested to inactivate/control microorganisms at room or mild temperatures. Among these, high-pressure homogenization (HPH) treatments have been proposed for the stabilization of liquid food mixtures (Diels and Michiels, 2006; Pathanibul et al., 2009; Brinez et al., 2006; Patrignani et al., 2009, 2010).

This study investigated the effects of HPH treatments on the microbial decontamination of apricot juice and hence on its shelf life. Apricot juice was chosen for its high carotenoid content, whose *in vivo* antioxidant activity is well documented (Rice Evans et al., 1997). In particular, the study first focused on identifying HPH conditions able to reduce the cell load of the apricot juice significantly and subsequently assessed the shelf life of the HPH-treated product.

9.3.3 Shelf Life Study

9.3.3.1 Identification of the Early Quality Decay Event and Critical Descriptor

Yeasts are common contaminants of fruit concentrates and represent a major problem to industries that process fruits or fruit products (Hatcher et al., 2000; Sancho et al., 2000). The low pH and high sugar content of these products as well as the possibility to exclude oxygen from packages favor yeast growth; consequently, product deterioration is predominantly due to yeast activity.

Saccharomyces cerevisiae is considered to be a predominant spoilage species in concentrates, juices, and fruit beverages, and in that respect is considered to be the source of most problems associated with processed fruits (Suárez-Jacobo et al., 2010). Consequently, *S. cerevisiae* was chosen as the critical descriptor to test the effectiveness of HPH treatment in prolonging the shelf life of apricot juice. The growth of this yeast results in haze formation, CO_2 (blowing) and off-odor production, and color changes. The product changes associated with yeast growth begin when the cell loads attain levels higher than 6.0 log cfu/ml (Patrignani et al., 2010). Consequently, this threshold can be considered the acceptability limit for this kind of product.

9.3.3.2 HPH Treatment Setup

The first part of the study was focused on the identification of process conditions able to decrease the initial microbial load of the product significantly. To this purpose, 10 L of apricot juice were inoculated with the target organism and subjected to different HPH treatments. In particular, different inoculation levels were taken into consideration. The strain used was *S. cerevisiae* 635, belonging to the collection of the Department of Food Science of the University of Bologna, Italy. The inoculated fruit juice was subjected to repeated treatments at 100 MPa (from one to eight passes), in a continuous high-pressure homogenizer (PANDA, Niro Soavi, Parma, Italy) supplied with a pressure standard (PS)-type valve. The increase of temperature during the treatment was about 2.5°C/10 MPa. After each pass at 100 MPa, the fruit juice was cooled using a thermal exchanger (Niro Soavi). The maximum temperature reached by the samples did not exceed 40°C. As a control, an untreated inoculated juice was used. On each treated sample, the effects of the different HPH treatments on *S. cerevisiae* survival were evaluated.

Table 9.3 shows the effect of the repeated treatments at 100 MPa on *S. cerevisiae* 635 inactivation in the apricot juices in relation to the inoculation levels. When *S. cerevisiae* 635 was inoculated at a level of 6.1 log cfu/ml, a significant viability

Table 9.3 Effects of Repeated Treatments at 100 MPa on *Saccharomyces cerevisiae* 635 Inactivation in Relation to the Inoculum Level

Number of Passes at 100 MPa	Apricot Juice Inoculated at 3.0 log cfu/ml (±SD)	Apricot Juice Inoculated at 6.1 log cfu/ml (±SD)
Control[a]	3.0 ± 0.3	6.1 ± 0.5
1	2.4 ± 0.2	6.3 ± 0.2
2	1.9 ± 0.2	5.6 ± 0.1
3	0.4+0.5	4.9 ± 0.2
4	0.4 ± 0.4	3.9 ± 0.4
5	nd[b]	3.8 ± 0.2
6	nd	3.4 ± 05
7	nd	3.6 ± 0.3
8	nd	3.3 ± 0.3

Source: Modified from Patrignani, F., Vannini, L., Kamdem Sado, S.L., Lanciotti, R., and M.E. Guerzoni. 2009. Effect of high pressure homogenization on *Saccharomyces cerevisiae* inactivation and physico-chemical features in apricot and carrot juices. *International Journal of Food Microbiology* 136: 26–31.

[a] Untreated inoculated samples.

[b] Under the detection limit, corresponding to cell loads < 1.0 Log cfu/ml.

decrease (2.2 log cfu/ml) was obtained only with four repeated passes at 100 MPa. A further increase of the number of pushes at 100 MPa did not significantly increase the effectiveness of HPH treatment. In the samples inoculated at a level of 3.0 log cfu/ml, three passes at 100 MPa allowed reduction of cell viability under the detection limit (<1.0 log cfu/ml).

9.3.3.3 *Shelf Life Assessment in Relation to Storage Temperature*

The second part of this case study assessed the shelf life of the HPH-treated apricot juice. To this purpose, all the treated samples were analyzed over time to better understand the fate of the surviving cells. In fact, once damages are repaired, the injured cells can proliferate more rapidly than the unstressed cells. Data recorded over time clearly showed the effects of the inoculation level, storage temperature, and severity of HPH treatment on product shelf life. In fact, all the samples inoculated at the highest level, about 6.0 log cfu/ml, spoiled during storage at 25°C regardless of the passes applied (data not shown). In particular, the most severely treated samples reached a yeast level higher than 6.0 log cfu/ml within 96 hours of storage at 25°C. This result was attributable to the survival of the resistant cell fraction in the most severely treated samples. The control sample spoiled within 24 hours. By contrast, *S. cerevisiae* 635 cell load remained under the detection limit at least up to 144 hours of storage at 25°C in the samples inoculated at the lowest level (about 3.0 log cfu/ml) and subjected to four or more repeated passes at 100 MPa (Figure 9.2).

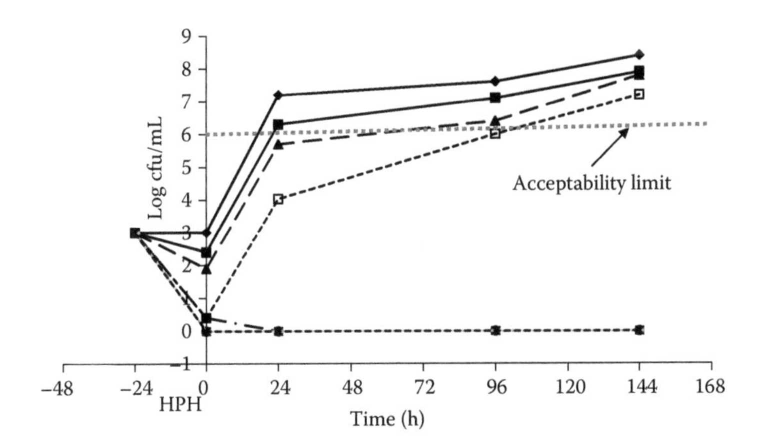

Figure 9.2 Growth of *Saccharomyces cerevisiae* 635 inoculated (3 log cfu/ml) in apricot juice at 25°C after successive passes through a high-pressure homogenizer at 100 MPa. Number of passes: control (◆), 1 (■), 2 (▲), 3 (□), 4 (●), 5 (○), 6 (◇), 7 (△), 8 (*). The inlet temperature of the samples during HPH treatment was 15°C. The data are the mean of three repetitions. Variability coefficient ranged between 5% and 7%. (Modified from Patrignani, F., Vannini, L., Kamdem Sado, S.L., Lanciotti, R., and M.E. Guerzoni. 2009. Effect of high pressure homogenization on *Saccharomyces cerevisiae* inactivation and physico-chemical features in apricot and carrot juices. *International Journal of Food Microbiology* 136: 26–31. With permission.)

Table 9.4 Effect of Repeated High Pressure Treatments at 100 MPa on Inactivation and Recovery of *Saccharomyces cerevisiae* 635 Inoculated in Apricot Juice (4 log cfu/ml), over Storage at 4°C

Number of Passes at 100 MPa	Cell Load (log cfu/ml)			
	0 h	96 h	144 h	216 h
Control[a]	3.9 ± 0.3	5.2 ± 0.4	3.9 ± 0.4	4.9 ± 0.3
1	3.6 ± 0.1	5.0 ± 0.2	3.9 ± 0.2	3.2 ± 0.2
2	3.3 ± 0.2	4.0 ± 0.2	3.8 ± 0.1	4.2 ± 0.1
3	2.9 ± 0.3	4.0 ± 0.3	3.7 ± 0.3	3.8 ± 0.5
4	1.0 ± 0.2	4.0 ± 0.1	3.3 ± 0.2	3.4 ± 0.3
5	nd[b]	4.2 ± 0.4	3.8 ± 0.4	3.7 ± 0.3
6	nd	3.5 ± 0.4	3.6 ± 0.2	3.6 ± 0.4
7	nd	3.3 ± 0.2	3.7 ± 0.1	3.3 ± 0.3
8	nd	3.7 ± 0.3	3.5 ± 0.2	2.9 ± 0.1

Source: Modified from Patrignani, F., Vannini, L., Kamdem Sado, S.L., Lanciotti, R., and M.E. Guerzoni. 2009. Effect of high pressure homogenization on *Saccharomyces cerevisiae* inactivation and physico-chemical features in apricot and carrot juices. *International Journal of Food Microbiology* 136: 26–31.

[a] Untreated inoculated sample.

[b] Under the detection limit, corresponding to cell loads < 1.0 Log cfu/ml.

Considering that the storage temperature was 25°C, this result represents an interesting goal because juice shelf life is calculated based on the time necessary to reach 6.0 log cfu/ml, also in comparison with the commercial aseptic packaged juices, which are stable but have a negligible contamination of microaerophilic fungi. The high number of data recorded over the limited time span avoided the use of a model to calculate the shelf life.

The inclusion of a further hurdle (i.e., refrigerated storage) to yeast proliferation notably increased product shelf life. In fact, a decrease in *S. cerevisiae* 635 cell viability in the apricot samples treated at 100 MPa for one to five cycles over storage was observed (Table 9.4). In the most severely treated samples, *S. cerevisiae* was under the detection limit immediately after treatment. However, after 96 hours cell loads ranging between 3.3 and 4.2 log cfu/ml were observed in the samples treated at 100 MPa for five to eight cycles. No further cell load increase was observed over the refrigerated storage, suggesting that a satisfactory extension of the juice shelf life occurred (Table 9.4). The refrigeration of the treated samples prevented cell proliferation and in some cases induced a further decrease in cell viability also in the samples with the highest inoculation levels, resulting in a further increase of shelf life.

9.3.3.4 Experimental Needs

- 10 L of apricot juice
- High-pressure homogenizer equipped with PS valve (process pressure up to 150 MPa)
- Material (culture media, Petri dishes, tubes, pipettes) and equipment (autoclaves, incubators) necessary for microbial cultivation and analyses

9.3.3.5 *Experimental Plan*

Action	Time (hours)											
	0	24	48	72	96	120	144	168	192	216	240	264
Sample inoculation and HPH treatment	■											
Microbiological analysis	■	■	■	■	■	■	■	■	■	■		
Data analysis and shelf life calculation							■	■	■	■	■	■

9.4 CASE STUDY 9.3. SHELF LIFE PREDICTION OF FROZEN PASTA MEAL

9.4.1 Product Presentation

Frozen ready-to-eat pasta with sauce represents a highly convenient ready meal. Precooked pasta can be presented separately from the sauce pellet pieces or mixed with the sauce. Generally, microwavable packaging is used to allow quick microwave reheating.

9.4.2 Problem

Frozen ready meals such as pasta with sauce are offered worldwide. However, retail temperature may vary from country to country. For instance, the storage temperature during handling adopted in the United States is -12°C, while in Europe it is equal to -18°C. These differences in the retail temperature may affect the rates of quality loss and consequently product shelf life. This case study estimated the shelf life of frozen macaroni–tomato sauce pellets at -12°C and -18°C.

9.4.3 Shelf Life Study

9.4.3.1 *Identification of the Early Quality Decay Event and Critical Descriptor*

In our study, freshly produced frozen pasta ("macaroni") with tomato sauce pellets was taken into account. To identify the proper quality indices describing the quality depletion of the frozen ready meal during storage, the main chemical and physical events first have to be considered for each food component (i.e., pasta and sauce). In particular, the quality depletion of pasta can be attributed to color changes due to superficial dehydration, whereas for tomato and oil-containing sauce, lipid

Table 9.5 Analytical Descriptors of Quality Depletion of Each Component of Frozen Pasta with Sauce

Product Component	Quality Decay	Analytical Descriptor
Pasta	Dehydration	Color
Sauce	Lipid oxidation	Peroxide value
	Carotenoid oxidation	Color

and/or carotenoid oxidation reactions are the events more likely responsible for quality loss. Table 9.5 shows the analytical descriptors that can be used to monitor quality depletion of this frozen ready meal.

To identify the earliest quality decay event, the macaroni–tomato sauce pellets were submitted to an accelerated stability test at -7°C. In particular, two packages (made of oriented polypropylene), containing 400 g of frozen product, were stored at -7°C in a thermostatic cell. One additional package was taken as a reference sample at time zero. At interval times, one package was removed from the cell, and the analytical determinations reported in Table 9.5 were carried out. In particular, the peroxide value was assessed on the lipid fraction previously extracted from the sauce according to the European Official Methods of Analysis (1991); color analysis was performed using a tristimulus colorimeter (Chromamater-2 Reflectance, Minolta, Japan). Measurements were carried out at five points on five different thawed pasta or sauce pieces taken from the same package. In the case of pasta, measurements were performed on the previously ground sample. Color was expressed as L^*, a^*, b^* Hunter scale parameters and relevant ratios.

Figure 9.3 shows the changes in peroxide value and color of the tomato sauce pellets. The peroxide value did not change during storage at -7°C. This is not a surprising result. In fact, it has already been demonstrated that the naturally occurring carotenoids might protect the lipid fraction from oxidative reaction by virtue of their strong antioxidant activity (Nicoli et al., 1999). However, the protective action exerted by the carotenoids goes with a loss in redness. Therefore, the color bleaching represents a good indicator of quality loss of the tomato sauce pellets.

Figure 9.4 shows the color decrease, expressed by the b^* parameter (yellowness) of pasta during storage at -7°C.

Results indicated that macaroni dehydration also may be considered an early critical event responsible for quality loss of a frozen pasta meal.

In conclusion, the results of this stability test highlighted that the quality loss of the considered frozen pasta meal were due to two different phenomena (i.e., pasta dehydration and carotenoid bleaching), which in turn can be described by easily measurable indexes. To define the quality descriptor mainly affecting consumer acceptability, a small-scale consumer panel test was performed. Thirty panelists were presented with 50 g of reconstituted samples of macaroni–tomato sauce pellets that had been stored at -7°C for 4 months and were asked to evaluate the meal for the pasta and tomato sauce colors. The results showed that tomato bleaching was

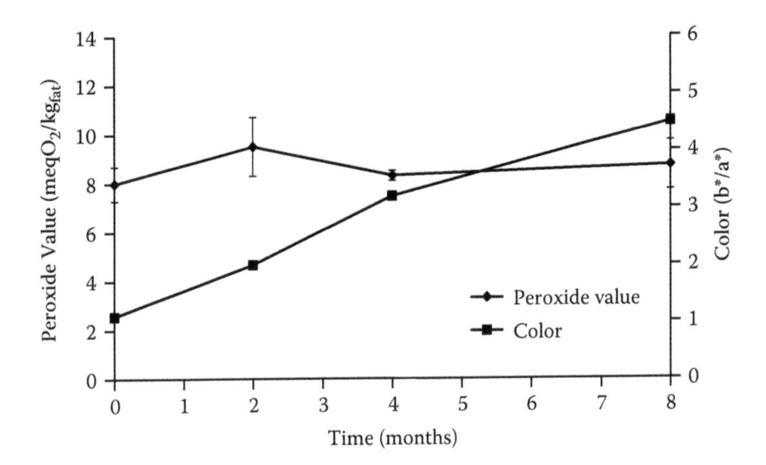

Figure 9.3 Peroxide and color values of the tomato sauce pellets during storage at -7°C.

considered more critical for the consumer than the pasta color change. In fact, the meal reconstitution before consumption is responsible for a "masking" effect of yellow fading of pasta. In light of these data, it was decided that the index describing the quality loss of the macaroni-tomato sauce during storage was the tomato color.

The tomato color (b^*/a^*) changes at -7°C as a function of time t (Figure 9.3) were found to follow pseudo-zero-order kinetics (Equation 9.2):

$$b^*/a^* = 0.429 \times t + 1.2 \tag{9.2}$$

The coefficient of determination R^2 was 0.983.

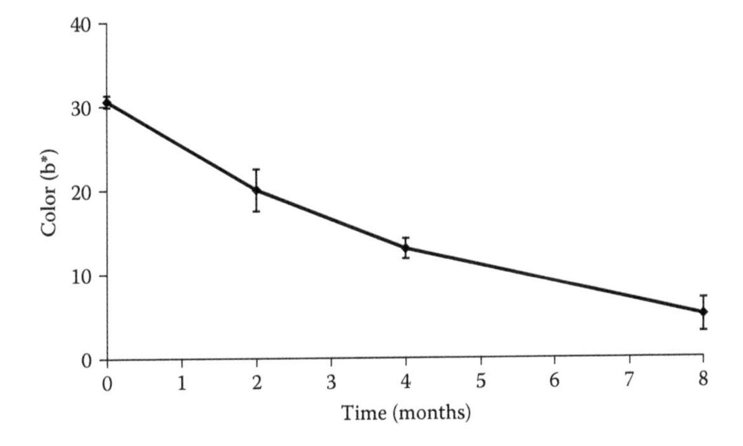

Figure 9.4 Color of pasta during storage at -7°C.

9.4.3.2 *Identification of the Acceptability Limit*

In this study, the acceptability limit was preliminarily assessed by sensory analysis. As the tomato color (*b*/a**) was the critical indicator, the obtained result was subsequently transformed into a color limit (see Chapters 3 and 4).

Sixty panelists were screened according to the criterion that they usually consumed pasta. Aliquots of 50 g of tomato sauce, previously stored at -7°C for increasing lengths of time and separated from the pasta, were placed in plastic white dishes just after reheating and presented to the panelists. It was assumed that the tomato sauce bleaching limit was not affected by the storage temperature. For each sample, panelists were asked to observe the color of the tomato sauce and answer the question: "Would you normally consume this product? Yes or no?" A sensory session was performed at each storage time.

The sensory data were analyzed by applying survival analysis (Hough et al., 2003; Chapter 7). The CensorReg procedure from S-PLUS (Insightful Corporation, Seattle, WA, USA, ver. 7) was used according to the work of Hough et al. (2003) and Hough (2010). Examples of acceptance/rejection data as well as of censoring definition for four typical panelists are shown in Table 9.6. In our experimental conditions, 23 panelists provided data that were left censored, 25 provided right censored, and 10 provided interval censored. The results of two panelists were not considered as they rejected the fresh sample (see the result of panelist 4 in Table 9.6).

The censored data were satisfactorily modeled by Weibull distribution, and the storage time corresponding to 50% of consumers rejecting the product was estimated by maximizing the likelihood function. As pointed out in this book, a 50% sensory rejection is a well-accepted limit in shelf life studies (Cardelli and Labuza, 2001). The maximum likelihood estimates of the parameters of the Weibull distribution were as follows (in parentheses are the corresponding standard errors):

$$\mu = 1.877\ (0.0973)$$

$$\log \sigma = -0.501\ (0.171)$$

Table 9.6 Example of Acceptance/Rejection Data for Panelists Who Tested Tomato Sauce Pellets with Different Storage Times at -7°C

	Time (months)					
Panelist	**0**	**2**	**4**	**6**	**8**	**Censoring**
1	Yes	No	No	No	No	Left: ≤2
2	Yes	Yes	Yes	Yes	Yes	Right: >8
3	Yes	Yes	No	No	No	Interval: >2 and <4
4	No	Yes	No	Yes	Yes	Not considered

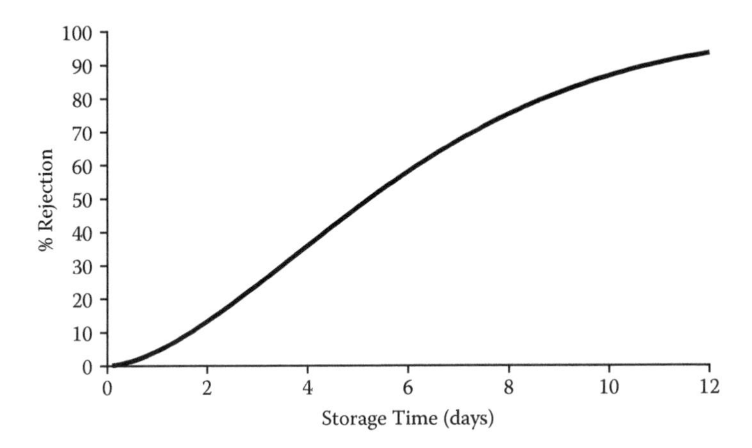

Figure 9.5 Weibull distribution relevant to tomato sauce pellets stored at -7°C.

Figure 9.5 shows the fitted Weibull distribution relevant to the frozen meal.

The end of the sensory shelf life of the frozen meal, defined as storage time corresponding to 50% of consumers rejecting the sample, was 5.2 months (95% confidence interval [CI] 4.4–6.2). This result is already a shelf life value referred to the product stored at -7°C. As reported in Chapter 7, survival analysis allows direct shelf life assessment once a given percentage of consumer rejection has been selected as the acceptability limit. However, in this case, survival analysis was performed in accelerated conditions only to assess the sensory acceptability limit, which had to be converted into a color limit. Using Equation 9.2, it was possible to determine the b^*/a^* value at 5.2 months, which represents the time value corresponding to 50% of consumers rejecting the product. Such a value was equal to 3.4 (95% CI 2.9–4.0). To summarize, the acceptability limit, expressed as b^*/a^*, was 3.4. This analytical limit corresponded to 50% of consumers rejecting the product.

9.4.3.3 *Testing by Instrumental Analysis*

To estimate the shelf life of frozen macaroni–tomato sauce pellets at actual storage temperatures, color changes versus time were monitored on samples stored at -12°C and -18°C. As mentioned, the color analysis was performed on previously thawed tomato sauce samples. Figure 9.6 shows the b^*/a^* values of the tomato sauce during storage at -12°C and -18°C. The changes of tomato color (b^*/a^*) at -12°C and -18°C as a function of time t are described by the following equations:

$$b^*/a^*_{(-12^\circ C)} = 0.276 \times t + 1.31 \tag{9.3}$$

$$b^*/a^*_{(-18^\circ C)} = 0.202 \times t + 0.92 \tag{9.4}$$

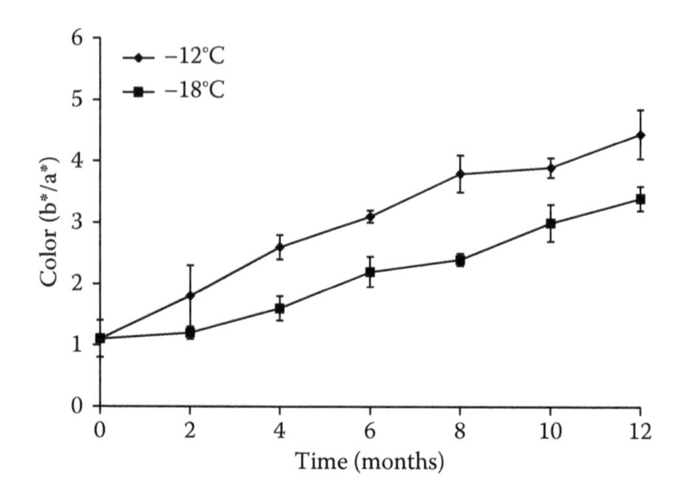

Figure 9.6 Color changes of tomato sauce during storage at -12°C and -18°C.

The coefficients of determination R^2 were 0.973 and 0.980, respectively.

From these equations it was possible to calculate the shelf life of the frozen meals at -12°C and -18°C by attributing to b^*/a^* the value 3.4. The times corresponding to the end of the shelf life were 7.2 and 11.8 months, respectively, at -12°C and -18°C.

9.4.3.4 Further Calculations

In this case study, the color changes of the tomato sauce during storage at -7°C, -12°C, and -18°C were calculated. As previously reported, they followed a pseudo-zero-order kinetics (Table 9.7). These data allowed the estimation of the temperature dependence of color bleaching rates by applying the Arrhenius equation (see Chapter 5).

Figure 9.7 shows the Arrhenius plot of the color change rate constants as a function of absolute temperature.

The best-fitting equation ($R^2 = 0.974$) resulted in the following:

$$\ln k = 16.4 - 4602 \times \frac{1}{T} \tag{9.5}$$

This equation can be used to predict the temperature dependence of color change in the temperature range from -7°C to -18°C. Extrapolation of kinetic rate constant values at temperatures outside the temperature domain considered should be carefully considered because of possible deviations from the Arrhenius equation (see Chapter 5).

Table 9.7 Pseudo-Zero-Order Rate Constants of Color Changes (*k*), Intercept, and Corresponding Determination Coefficients and *p* Values of Tomato Sauce Pellets Stored at Different Temperatures

T (°C)	*k* (*b**/*a** month^{-1})	Intercept	R^2	*p*
-7	0.429	1.20	0.983	0.009
-12	0.276	1.31	0.973	4.19E-05
-18	0.202	0.92	0.980	1.98E-05

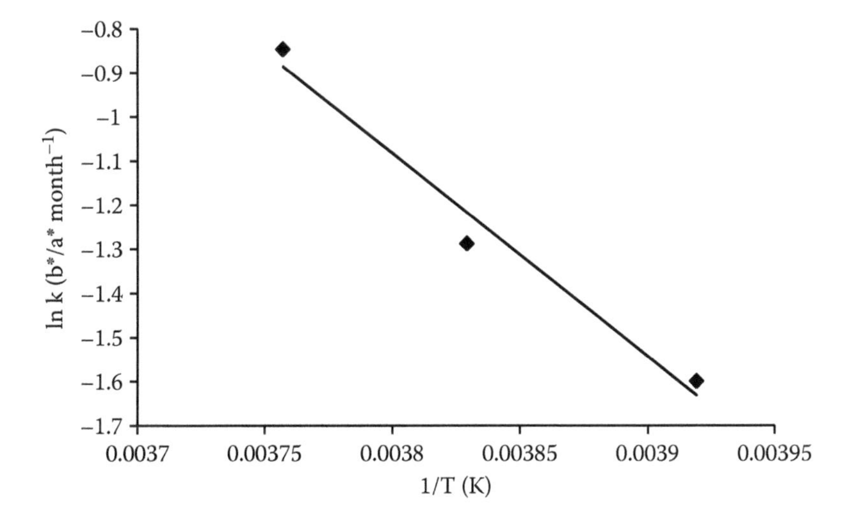

Figure 9.7 Apparent zero-order rate constants of color change of tomato sauce as a function of temperature.

9.4.3.5 Experimental Needs

- 60 packages containing about 400 g of sample each
- Three thermostatic cells working at -7°C, -12°C, and -18°C
- One freezer at -80°C
- Tristimulus colorimeter
- Domestic homogenizer
- Solvent evaporator
- Organic solvents for fat extraction
- Analytical/technical balances
- Glassware and other lab consumables
- 300 white plastic dishes for sensory analysis
- 60 panelists
- A sensory lab with individual booths and artificial daylight-type illumination, temperature control, and air circulation
- Personal computer (PC) and statistical packages

9.4.3.6 *Experimental Plan*

Action Sample Withdrawal for	Storage Temperature (°C)	Storage Time (months)						
		0	2	4	6	8	10	12
Identification of analytical descriptors	-7							
Small-scale panel test	-7							
Sensory testing	-7							
Instrumental testing	-12							
	-18							

9.5 CASE STUDY 9.4. SHELF LIFE ESTIMATION OF PASTEURIZED ORANGE JUICE

9.5.1 Product Presentation

Since their appearance on the market, heat-pasteurized fruit juices have received consensus among consumers because of their high convenience. Thermally processed orange juice is the most popular fruit juice, accounting for more than 50% of commercial juices consumed worldwide (Varnam and Sutherland, 1999). Approximately 90% of orange juice is made from frozen concentrate. Orange juices are generally sold in multilayer PET (polyethylene terephthalate), which has low permeability to oxygen. As known, pasteurized fruit juices are ambient stable products.

9.5.2 Problem

Consumers' expectation from consuming fruit juices is high vitamin intake. It is known that thermal processing significantly affects the initial vitamin (mainly ascorbic acid) content, which further decreases during storage depending on storage conditions, such as temperature, oxygen content, and light. Also, during storage the product may undergo reactions (e.g., ascorbic acid degradation reactions) that are responsible for the development of undesired changes of sensory properties.

9.5.3 Shelf Life Study

9.5.3.1 *Identification of the Early Quality Decay Event and Critical Descriptor and Relevant Acceptability Limit*

As a sensitive compound, ascorbic acid is generally used as a quality indicator. In fact, it provides an indication of the loss of other vitamins and, as previously mentioned, is one precursor of degradation reactions that may lead to undesired color and flavor development.

Even if not compulsory, the acceptability limit for ascorbic acid depletion can be derived from relevant E.U. and U.S. guidelines. According to the European Scientific Committee for Food (SCF, 1993), a population reference intake of 45 mg of ascorbic acid per day for adults, with an increase to 55 mg/day in pregnancy and to 70 mg/day during lactation, is recommended. In the United States, the recommended daily allowances (USRDAs) for ascorbic acid are 75 mg/day for females and 90 mg/day for males (United States Department of Agriculture, 2010).

In this case study, the shelf life of thermally processed orange juice was estimated in accordance with E.U. and U.S. guidelines, assuming a daily consumption of at least 300 ml of orange juice. This means that the ascorbic acid content of orange juice should be higher than 15 and 30 mg/100 ml at the expiration date to satisfy the E.U. and U.S. requirements for adult population, respectively.

9.5.3.2 *Shelf Life Testing*

To estimate the shelf life of pasteurized orange juice, 10 multilayer PET cartons containing 1,000 ml of freshly produced product were stored at 25°C in a thermostatic cell up to 1 month. At interval times, two packages were removed from the cell, and the orange juices were analyzed for their ascorbic acid concentration. Ascorbic acid content was determined following the high-performance liquid chromatographic (HPLC) analytical procedure outlined by Lee and Coates (1999). Sample aliquots, previously centrifuged in the presence of 2.5% metaphosphoric acid and subsequently filtered through 0.45-mm polytetrafluoroethylene (PTFE) syringe filters, were injected in a C18 column using 25 m*M* KH_2PO_4 as the mobile phase. The elute was monitored by an ultraviolet (UV) detector at 245 nm.

Figure 9.8 shows the changes of ascorbic acid concentration of the pasteurized orange juice during storage at 25°C. The ascorbic acid concentration changes were found to follow a pseudo-first-order kinetics, in accordance with literature reports (Johnson et al., 1995). The pseudo-first-order rate constant, the intercept, and the goodness-of-fit measures are shown in Table 9.8.

According to Chapter 5, the shelf life (*SL*) of the pasteurized orange juice can be easily calculated from the pseudo-first-order rate constant:

$$SL = \frac{\ln AA - 3.82}{0.021} \tag{9.6}$$

where *AA* is the ascorbic acid concentration corresponding to the acceptability limit.

Based on the E.U. and U.S. recommended daily allowances for adults and assuming a daily consumption of 300 ml, the shelf life of the pasteurized orange juice can be estimated by attributing to the ascorbic acid concentration the values of 15 or 30 mg/100 ml, respectively. The shelf life values were found to be 53 (95% CI 47–58) and 20 (95% CI 19–21) days accordingly.

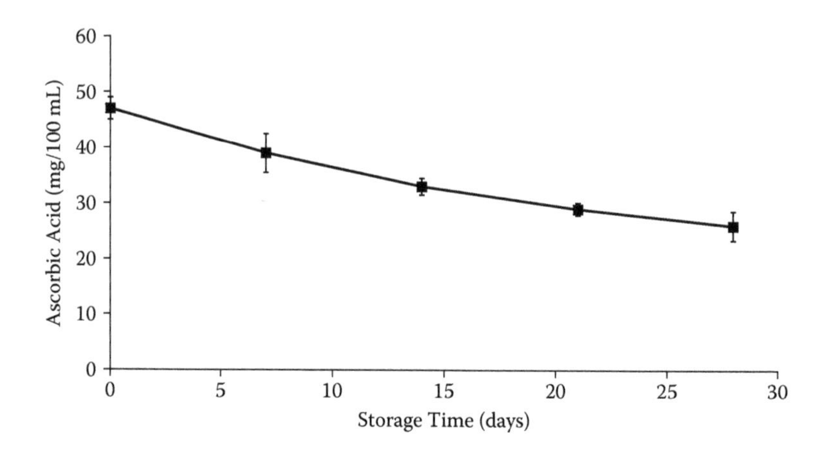

Figure 9.8 Changes in ascorbic acid concentration of pasteurized orange juice during storage at 25°C.

Table 9.8 Pseudo-First-Order Rate Constant of Ascorbic Acid Concentration Changes (*k*), Intercept, and Corresponding Determination Coefficient and *p* Value of Pasteurized Orange Juice during Storage at 25°C

Parameters	Estimate	95% CI
k (days^{-1})	-0.021	-0.025 to -0.017
Intercept	3.82	3.75 to 3.90
R^2	0.988	
p value	0.0006	

9.5.3.3 *Experimental Needs*

- 10 packages containing 1 L of sample each
- One thermostatic cell working at 25°C
- Analytical/technical balances
- Glassware and other lab consumables
- HPLC equipped with a C18 column and a UV detector
- Other lab facilities

9.5.3.4 *Experimental Plan*

Action Sample Withdrawal for	Storage Time (days)				
	0	7	14	21	28
Ascorbic acid analysis					

9.6 CASE STUDY 9.5. SHELF LIFE PREDICTION OF GORGONZOLA CHEESE BREADSTICKS

9.6.1 Product Presentation

Breadsticks are dry sticks of bread that originated in Italy (Torino and surrounding areas), probably in the 14th century, and are now popular in Europe, the Americas, Australia, and some parts of Asia. They might be consumed alone or in combination with salty (cheese, ham, vegetables) or sweet (honey, jam) foods. Breadsticks are proposed for consumption as appetizers, bread replacers, or between-meal snacks. In recent years, different breadstick formulations have appeared in the market with recipes that contain various extra ingredients, such as cheese, tomato, garlic, sesame seeds, and so on. Breadsticks show long-term shelf life. In fact, the shelf life generally attributed to traditional breadsticks is around 6 to 8 months. However, depending on their nature, the presence of extra ingredients can greatly affect product shelf life.

9.6.2 Problem

The incorporation of fat-containing extra ingredients with the breadstick formula may be responsible for accelerated quality loss. Therefore, a primary need for breadstick producers is to estimate the shelf life of the new formula. The prediction of long-term shelf life can be obtained by accelerated shelf life tests (ASLTs), that is, performing experiments under environmental conditions (generally different temperatures) able to speed up quality depletion (see Chapter 5). To predict the shelf life successfully by applying ASLTs, the working temperature range should be carefully chosen (Calligaris et al., 2007; Manzocco et al., 2010; Chapter 5). This case study assessed the shelf life of gorgonzola cheese breadsticks at a retail temperature of 25°C under accelerated storage conditions.

9.6.3 Shelf Life Study

9.6.3.1 Identification of the Early Quality Decay Event and Critical Descriptor

It is well known that the quality depletion of breadsticks is mainly attributable to lipid oxidation reactions, which lead to the formation of off-flavors accounting for consumers rejecting the product. As the sensory tests are expensive and difficult to perform at the industrial level, a cheaper and simpler analytical measurement should be found to account for quality loss. The peroxide value has been reported to be a suitable chemical index to monitor quality depletion of breadsticks (Calligaris et al., 2008). Also, peroxide value changes were found to correlate well to sensory consumer acceptance.

As lipid oxidation proceeds fairly slowly at room temperature, an ASLT has been employed. Here, we present the results obtained following this approach. Thus, a mathematical model able to predict the product shelf life using the chemical index (peroxide value) in a wide temperature range was developed.

Freshly prepared gorgonzola cheese breadsticks containing 13% fat were used. In particular, 24 packages (made of oriented polypropylene), containing 200 g of breadsticks each, were divided in four aliquots and stored at 20°C, 30°C, 37°C, and 45°C in thermostatic cells. One more package was used as a reference sample at time zero. At interval times, one package was removed from each cell, and the peroxide value was measured. Nine additional packages were also stored at 45°C and used to perform the sensory analysis. Previous studies showed that the peroxide value limit was not affected by storage temperature (Calligaris et al., 2007, 2008). After removal from the thermostatic cell, the samples were kept at -18°C to perform the sensory analysis in one session. Although storage at lower temperatures would be preferable to inhibit oxidation reactions, it was decided to keep samples at -18°C because freezers working at this temperature are diffusely available in most laboratories. Moreover, it was demonstrated that keeping samples at -18°C did not affect significantly oxidation level within the time the shelf life test was carried out.

9.6.3.2 *Identification of the Acceptability Limit*

In this study, the acceptability limit was preliminarily assessed by sensory analysis. The sensory acceptability limit was further converted into an analytical acceptability limit, here represented by the peroxide value (see Chapters 3 and 4).

Eighty panelists were screened according to the criterion that they usually consumed breadsticks. Aliquots of 10 g of breadsticks, previously stored at 45°C and subsequently kept at -18°C, were introduced in 50-ml plastic containers and sealed with a pressure cap. Samples were equilibrated at room temperature before the sensory test. For each sample, panelists were asked to sniff the breadsticks and answer the question: "Would you normally consume this product? Yes or no?" During the sensory session, each panelist sniffed six samples with different storage times presented in random order. Time between each sample sniffing was approximately 1 minute.

The sensory data were analyzed by applying survival analysis (Hough et al., 2003; Chapter 7). Such a method allows simultaneous assessment of acceptability limit and shelf life (see Chapter 3). The CensorReg procedure from S-PLUS (Insightful Corporation) was used according to the work of Hough et al. (2003) and Hough (2010). All the observations were censored because for each observation time the product was either acceptable (i.e., spoilage has not yet reached the acceptability limit) or not (depletion had already occurred). Examples of acceptance/rejection data as well as of censoring definition for four typical panelists are shown in Table 9.9. In our experimental conditions, data for 44 panelists were left censored, and for 30 they were interval censored; right censoring was not present. The results of 6 panelists were not considered as they rejected the fresh sample (see the result of panelist 4 in Table 9.9).

Table 9.9 Example of Acceptance/Rejection Data for Four Panelists Who Tested the Gorgonzola Cheese Breadsticks Having Different Storage Times at 45°C

	Time (days)									
Panelist	**0**	**15**	**30**	**45**	**60**	**75**	**90**	**105**	**120**	**Censoring**
1	Yes	No	No	No	No	No	No	No	No	Left: ≤15
2	Yes	Yes	Yes	Yes	Yes	Yes	Yes	Yes	Yes	Right: >120
3	Yes	Yes	No	Yes	No	No	No	No	No	Interval: >15 and ≤60
4	No	No	Yes	Yes	Yes	No	No	No	No	Not considered

The Weibull distribution was chosen to model the censored data, and the storage times corresponding to 10% (very low risk), 25% (low risk), 50% (medium risk), and 75% (high risk) of consumers rejecting the product were estimated by maximizing the likelihood function (see Chapters 3 and 7). The maximum likelihood estimates of the parameters of the Weibull distribution were as follows (in parentheses are the corresponding standard errors):

$$\mu = 3.150\ (0.142)$$

$$\log \sigma = -0.0887\ (0.137)$$

Figure 9.9 shows the fitted Weibull distribution relevant to the breadsticks.

Table 9.10 shows the storage time limits (i.e., the shelf life) of the breadsticks stored at 45°C corresponding to 10%, 25%, 50%, and 75% of consumers rejecting the sample. The end of the sensory shelf life of the breadsticks was 3, 7, 17, and 31 days, respectively. These times were lower than those found for traditional breadsticks (Calligaris et al., 2008), probably because of their higher fat content.

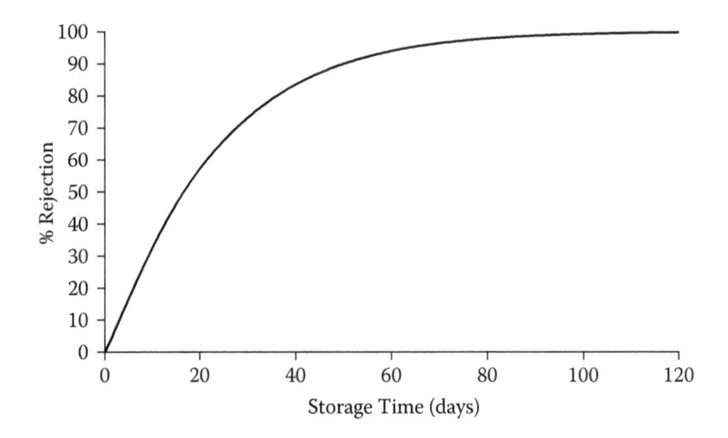

Figure 9.9 Weibull distribution fitted to gorgonzola cheese breadstick sensory data.

Table 9.10 Shelf Life of Gorgonzola Cheese Breadsticks, Defined as Storage Time at 45°C, Corresponding to 10%, 25%, 50%, and 75% of Consumers Rejecting the Sample

Consumers Rejecting the Sample (%)	Shelf Life (days)		
	95% Lower Limit	Estimated Quantile	95% Upper Limit
10	1.61	2.98	5.49
25	4.89	7.46	11.39
50	12.63	16.68	22.03
75	25.52	31.45	38.77

Despite its powerful applications, survival analysis requires a wide number of panelists and samples to be tested. For this reason, it was applied only to samples stored in accelerated conditions (45°C) to identify the acceptability limit and the corresponding time limits (i.e., shelf life). Further experiments were performed to assess the product shelf life in a wider temperature range, using peroxide value as critical indicator. To this aim, the conversion of the sensory acceptability limit into peroxide acceptability limit was done.

9.6.3.3 *Testing by Instrumental Analysis*

The peroxide value analysis was carried out on breadstick fat, previously obtained by solid–liquid extraction (Calligaris et al., 2008). The peroxide value determination was carried out according to the European Official Methods of Analysis (1991).

Figure 9.10 shows the changes in peroxide value of the breadsticks stored at different temperatures. As expected, the peroxide value increased during storage, and the increase was faster as the temperature increased. No lag phase was observed.

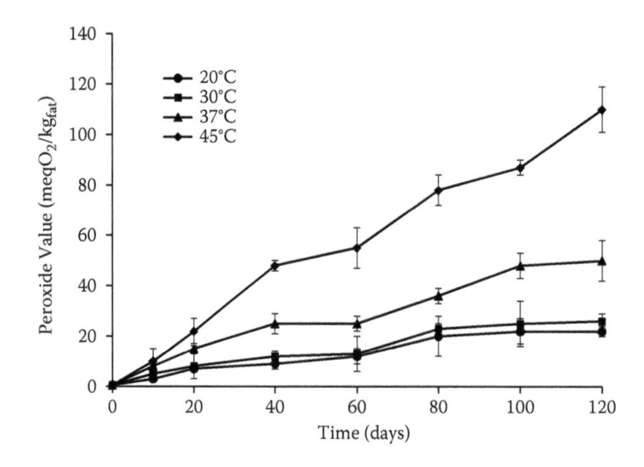

Figure 9.10 Changes in peroxide value of gorgonzola cheese breadsticks during storage at different temperatures.

Table 9.11 Pseudo-Zero-Order Rate Constants of Peroxide Formation (*k*), Intercept, and Corresponding Determination Coefficients and *p* Values of Gorgonzola Cheese Breadsticks Stored at Different Temperatures

T (°C)	*k* (meq O_2/kg_{fat} day)	Intercept	R^2	*p*
20	0.191	1.67	0.955	2.85E-5
30	0.215	2.49	0.955	2.91E-5
37	0.404	4.23	0.967	1.09E-5
45	0.884	3.79	0.987	7.05E-7

The peroxide value changes were found to follow a pseudo-zero-order kinetics, and the pseudo-zero-order rate constants are shown in Table 9.11.

From the pseudo-zero-order rate constants, the peroxide values corresponding to the acceptance limits obtained by the sensory analysis (Table 9.10) can be easily calculated (see Chapter 5):

$$\text{Peroxide value} = kt + q \tag{9.7}$$

where k is the pseudo-zero-order rate constant of peroxide formation, q is the intercept that is the peroxide value at time zero, and t is the time value corresponding to the shelf life.

9.6.3.4 Relationship between Sensory Acceptability and Peroxide Formation

To convert the sensory acceptability limit, previously estimated by survival analysis, into the acceptability limit, expressed as peroxide value, Equation 9.7 was used. Time limit values reported in Table 9.10 corresponding to different percentages of consumer rejections were used to calculate relevant peroxide value limits. Results are shown in Table 9.12.

As pointed out, since the peroxide value limit at any percentage of sensory rejection is not temperature dependent, these values represent the peroxide formation limits to estimate the shelf life of the breadsticks. Reporting the peroxide value

Table 9.12 Peroxide Values Corresponding to 10%, 25%, 50%, and 75% Sensory Rejection

Consumers Rejecting the Sample (%)	Time (days)	Peroxide Value (meq O_2/kg_{fat})		
		95% Lower Limit	Estimated Quantile	95% Upper Limit
10	2.98	0.00	6.43	13.06
25	7.46	4.10	10.39	16.67
50	16.68	12.92	18.54	24.16
75	31.45	26.86	31.60	36.34

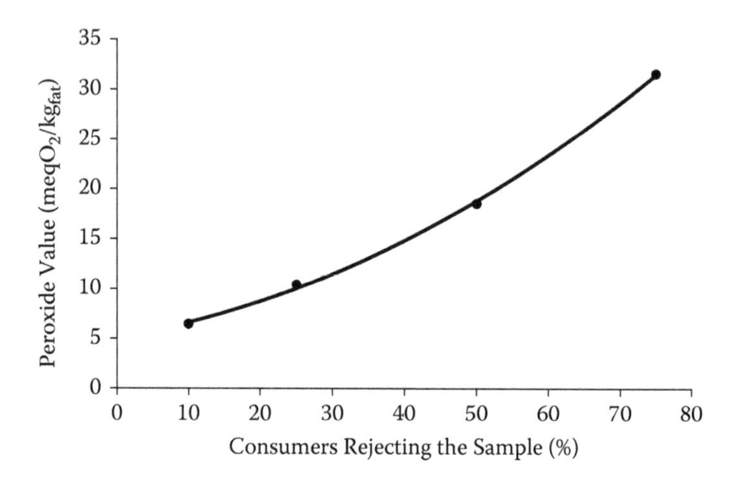

Figure 9.11 Peroxide value limits as a function of the percentage of consumers rejecting the sample.

limits PV_{lim} as a function of the percentage of consumers rejecting the sample $\%R$ (Figure 9.11), locally a quadratic relationship was found ($R^2 = 0.999$; $p < 0.05$):

$$PV_{\text{lim}} = 0.0031(\%R)^2 + 0.12(\%R) + 5.11 \tag{9.8}$$

Depending on the risk level the industry is willing to run, different peroxide values indicating the end of the product shelf life can be calculated.

9.6.3.5 *Development of Predictive Mathematical Model*

This section deals with the development of a mathematical model to predict the product shelf life using the chemical index (peroxide value) instead of sensory acceptability.

Because peroxide formation was found to follow a pseudo-zero-order kinetics, the shelf life of the breadsticks SL, expressed in days, can be predicted using the following equation:

$$SL = \frac{PV_{\text{lim}} - PV_{T0}}{k_T} \tag{9.9}$$

where PV_{lim} is the peroxide value corresponding to the limit of breadstick sensory acceptability, PV_{T0} is the peroxide value at zero storage time, and k_T represents the pseudo-zero-order rate constant at the selected storage temperature T.

In our study, we wanted to predict the shelf life of gorgonzola cheese breadsticks at a retail temperature of 25°C. The k value at 25°C can be extrapolated from the Arrhenius plot (Figure 9.12), which was obtained according to the Arrhenius equation:

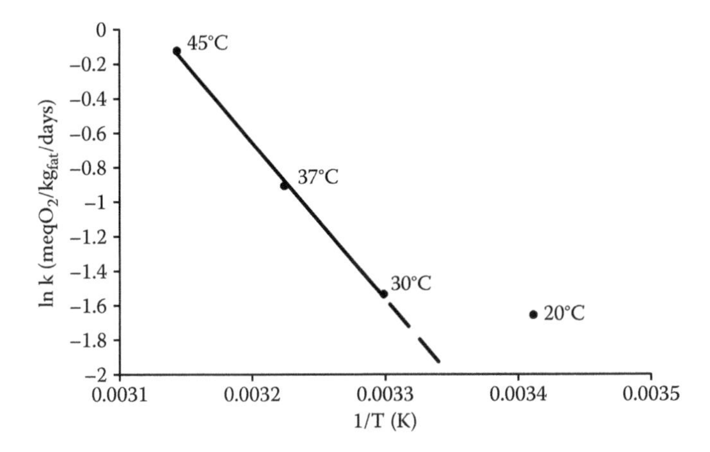

Figure 9.12 Apparent zero-order rate constants of peroxide formation of gorgonzola cheese breadsticks as a function of temperature.

$$\ln k = \ln k_0 - \frac{E_a}{RT} \tag{9.10}$$

where k is the reaction rate constant, k_0 is the pre-exponential factor of the frequency factor, R is the molar gas constant (8.31 J/K mol), T is the absolute temperature (K), and E_a is the activation energy (J/mol).

A nonlinearity in the Arrhenius plot can be observed. In fact, while the plot behavior is linear in the first tract, the experimental rate constant at 20°C was higher than expected under the Arrhenius equation. This deviation from the Arrhenius model has been attributed to an increase in the oxidation rate at lower temperatures due to a change in the reactant concentration caused by a transition of lipids from liquid to a crystallized phase. The increase of reactant concentration in the liquid phase is responsible for speeding up the reaction rate (Calligaris et al., 2007). In other words, these results indicate that the rate of oxidative reactions depends not only on temperature but also on concentration of reactants involved in the oxidation reaction.

Therefore, to evaluate the shelf life of the gorgonzola cheese breadsticks at 25°C, three different alternative approaches can be followed:

a. Evaluate the shelf life under the actual storage temperature (i.e., 25°C) instead of predicting it by ASLT. This approach, although easy to perform, is time consuming.
b. Evaluate the shelf life using a modified Arrhenius equation accounting for the changes in the fat physical state. To apply this approach, one should (1) have deep knowledge of the phenomena occurring in the lipid portion of the food matrix as a consequence of temperature changes; (2) be able to describe such phenomena mathematically to find out a corrective factor accounting for the influence of variables, other than temperature, on the oxidation rate; (3) include the corrective factor in the Arrhenius equation. An application of this approach was described by Calligaris et al. (2007) for the shelf life prediction of biscuits.

c. Extrapolate an empiric equation from the data obtained by the ASLT. This is an easy approach that, however, does not take into account the mechanisms underlying the phenomena. In addition, to give reliable results, this approach presupposes (1) a relatively high number of data, that is that the ASLT should be performed at a relatively high number of different temperatures (preferably more than four); (2) good knowledge of the phenomena to be able to choose an appropriate equation among those characterized by a good fit. In fact, the best-fitting equation would not necessarily meet the requirements needed to describe the dependence of the temperature sensitivity of the reaction rates appropriately.

Here, we show how to predict the shelf life of breadsticks using the obtained empiric equation. To this purpose, the Tablecurve 2D (Jandel Scientific, San Rafael, CA) program was used. Figure 9.13 shows the curve best fitting (dashed line) the experimental rate constants obtained from the ASLT study versus temperature as well as the one that, among the best-fitting curves, is appropriate to describe the behavior of rate constants (continuous line).

The latter was chosen although it was less satisfactory based on purely statistical considerations. In fact, as can be observed, the best-fitting curve presents a minimum around 25°C, which is not appropriate because the relationship between rate constants and temperatures should be monotonic. The selected equation ($R^2 = 0.998$) fitting the rate constants was the following power law:

$$k = a + bT^c \tag{9.11}$$

where $a = 0.173$, $b = 9.1\ 10^{-11}$, $c = 5.98$, and T is the temperature (°C).

Using this equation, it was possible to estimate the k value (k = meq O_2/kg$_{fat}$/days) at 25°C. Such a k value was equal to 0.194 meq O_2/kg$_{fat}$/day, which is an intermediate value between those found corresponding to 20°C and 30°C.

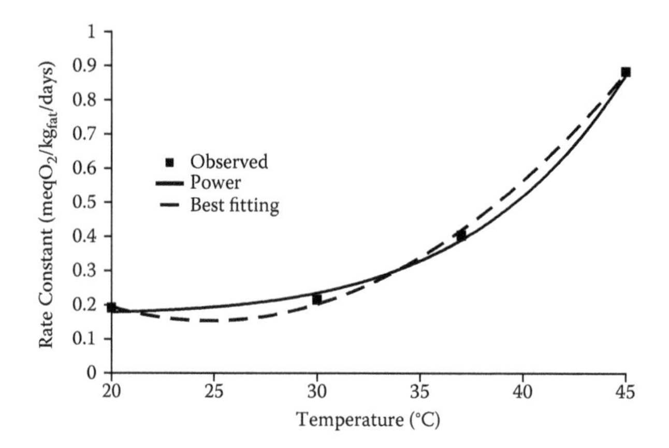

Figure 9.13 Curves best fitting and best describing the experimental rate constants as a function of temperature.

Equation 9.11 can be applied to predict the shelf life of the gorgonzola cheese breadsticks at 25°C as follows:

$$SL = \frac{(0.37(\%R) + 1.7) - PV_i}{0.194} \tag{9.12}$$

For instance, choosing the peroxide value corresponding to 50% sensory rejection, which is a well-accepted limit in shelf life studies (Cardelli and Labuza, 2001), and with PV_i equal to 2 meq O_2/kg_{fat}, the shelf life of the gorgonzola cheese breadsticks was 87 days.

It must be pointed out that if a shelf life estimation in the temperature range between 30°C and 45°C is required, the Arrhenius equation previously estimated can be used.

9.6.3.6 *Experimental Needs*

- 34 packages containing about 200 g of sample each
- 480 plastic containers (50 ml) with cap for sensory analysis
- Four thermostatic cells working at 20°C, 30°C, 37°C, and 45°C
- One freezer at -18°C
- Solvent evaporator
- Organic solvents for fat extraction
- Analytical/technical balances
- Glassware and other lab consumables
- 80 panelists
- A sensory lab with individual booths and artificial daylight-type illumination, temperature control, and air circulation
- PC and statistical packages

9.6.3.7 *Experimental Plan*

See plan on the next page.

9.7 CASE STUDY 9.6. ASSESSMENT OF SECONDARY SHELF LIFE OF GROUND ROASTED COFFEE

9.7.1 Product Presentation

Roasted whole and ground coffee are among the most present coffee products in the market. Coffee is generally presented packaged under vacuum in hermetically sealed aluminum packages and distributed at ambient temperature. Although generally recognized as a shelf-stable product, roasted coffee undergoes important chemical and physical changes that affect the quality and acceptability of the coffee brew (Nicoli

Experimental Plan

Action	Storage Temperature (°C)	Storage Time (days)																								
		0	5	10	15	20	25	30	35	40	45	50	55	60	65	70	75	80	85	90	95	100	105	110	115	120
Sample withdrawal for peroxide analysis	20																									
	30																									
	37																									
	45																									
Sample withdrawal for sensory analysis	45																									

et al., 2009). Therefore, the shelf life generally attributed to the roasted coffee is about 6–10 months, depending on the composition of the packaging atmosphere.

9.7.2 Problem

At a catering or home level, coffee is generally not used at one time. Instead, each time it is needed, a small portion of coffee is withdrawn from the package; the remaining is stored in the pack. Therefore, during usage coffee may undergo accelerated staling (i.e., loss of sensory properties) due to the changed storage conditions (e.g., higher O_2 availability, a_w, or temperature changes). This case study shows how to predict the coffee *secondary shelf life,* which is the length of time after package opening during which coffee does maintain acceptable hygienic, nutritional, and sensory properties. It is noteworthy that the hygienic issue is not relevant in the case of coffee due to very low water activity.

9.7.3 Secondary Shelf Life Study

9.7.3.1 *Identification of the Early Quality Decay Event and Critical Descriptor*

It is well documented that the main critical event responsible for quality loss of roasted coffee is likely represented by oxidative reactions that may lead to a change in flavor profile (Nicoli et al., 2004, 2009, Manzocco et al., 2011). A proper descriptor of coffee quality depletion is sensory acceptability. As the sensory tests are time and money consuming, coffee quality loss during usage could be followed by the cheaper and faster instrumental analysis of the loss of volatile compounds. In fact, previous studies showed that the release of volatile compounds well describes coffee staling, whereas oxidation reactions do not (Anese et al., 2006).

To predict the secondary shelf life of roasted coffee, home storage conditions were simulated. Commercial ground roasted coffee (250 g) packaged under vacuum in hermetically sealed aluminum packages was considered. The coffee samples were stored at 25°C for 40 days in packages that were periodically opened for a short time and subsequently closed again. To facilitate the open/close procedure, coffee (250 g) was previously transferred into 500-ml screw-capped glass jars.

9.7.3.2 *Testing by Sensory Analysis*

Sensory analysis was carried out by applying survival analysis (Hough et al., 2003; Hough, 2010; Chapter 7). Such analysis allows simultaneous assessment of the acceptability limit and shelf life (see Chapter 3). Thirty panelists were screened according to the criterion that they consumed at least one cup of coffee every day. They were presented with six coffee samples (0, 5, 12, 20, 30, 40 days of storage at 25°C) in random order. Aliquots of 3 g of ground roasted

Table 9.13 Example of Acceptance/Rejection Data for Four Panelists Who Tested the Coffee Samples Having Different Storage Times at 25°C

	Time (days)						
Panelist	**0**	**5**	**12**	**20**	**30**	**40**	**Censoring**
1	Yes	No	No	No	No	No	Left: ≤5
2	Yes	Yes	Yes	Yes	Yes	Yes	Right: >40
3	Yes	Yes	No	Yes	No	No	Interval: >5 and ≤30
4	No	No	Yes	Yes	Yes	No	Not considered

coffee were introduced in 30-ml plastic containers and sealed with a pressure cap. Before the sensory test, samples were equilibrated at room temperature for 1 hour. Panelists were asked to sniff the samples and answer the question: "Would you normally use this coffee powder to prepare the beverage product? Yes or no?" Time between each sample testing was approximately 1 minute. The CensorReg procedure from S-PLUS (Insightful Corporation) was used according to the work of Hough et al. (2003). Examples of acceptance/rejection data as well as of censoring definition for four typical panelists are shown in Table 9.13. In our experimental conditions, 2 panelists provided data that were left censored, 12 provided right-censored data, and 13 gave interval-censored data. The results for three panelists were not considered as they rejected the fresh sample (see the result of panelist 4 in Table 9.13).

The log-normal distribution was chosen to fit the censored data and storage times corresponding to 10%, 25%, 50%, and 75% of consumers rejecting the product were estimated by maximizing the likelihood function (see Chapter 7). The estimates of the parameters of the log-normal distribution were as follows (in parentheses are the corresponding standard errors):

$$\mu = 3.199\ (0.231)$$

$$\log \sigma = -0.0151\ (0.210)$$

Figure 9.14 shows the log-normal distribution relevant to the roasted coffee.

Table 9.14 shows the secondary shelf life of the roasted coffee defined as storage time corresponding to 10%, 25%, 50%, and 75% of consumers rejecting the sample. The end of the sensory shelf life of the roasted coffee was 7, 13, 25, and 48 days, respectively.

This approach is sufficient by itself for determining the pantry shelf life of coffee by choosing a proper consumer rejection percentage. However, considering the case in which the secondary shelf life should be estimated for different ground and roasted coffees packaged under different conditions, the search for an analytical index to be related to sensory acceptability data is crucial to speed up the shelf life study and reduce experimental costs.

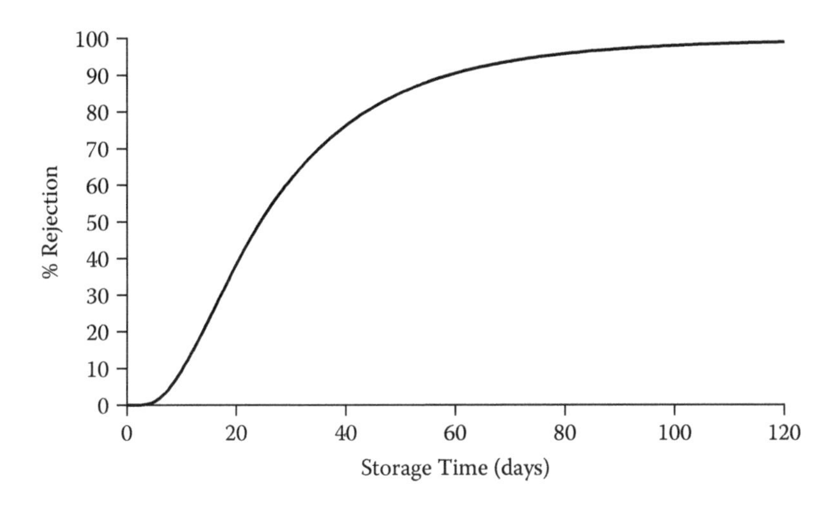

Figure 9.14 Log-normal distribution fitted to roasted coffee sensory data.

Table 9.14 Secondary Shelf Life of the Roasted Coffee Defined as Storage Time at 25°C Corresponding to 10%, 25%, 50%, and 70% of Consumers Rejecting the Sample

Consumers Rejecting the Sample (%)	Secondary Shelf Life (days)		
	95% Lower Limit	Estimated Quantile	95% Upper Limit
10	4.20	6.93	11.46
25	8.52	12.61	18.67
50	16.76	24.50	35.83
75	29.17	47.62	77.73

9.7.3.3 *Testing by Instrumental Analysis*

Gas chromatographic (GC) analysis of the volatile compounds in the headspace of the roasted coffee was performed. Aliquots of about 1.5 g of roasted coffee were taken from the screw-capped glass jars at increasing lengths of storage time at 25°C and immediately introduced in 10-ml vials, which were hermetically closed with butyl septa and metallic caps. After equilibration at 40°C, 0.5 ml of the sample headspace were injected in a gas chromatograph equipped with a flame ionization detector (López-Galilea et al., 2006).

Figure 9.15 shows the changes in the peak area of total volatile compounds in the headspace of the roasted coffee during pantry life.

As expected, the headspace volatile peak area progressively decreased with increasing time. Such a change followed a pseudo-zero-order kinetics (Equation 9.13).

$$A = -2.3 \times t + 96.8 \tag{9.13}$$

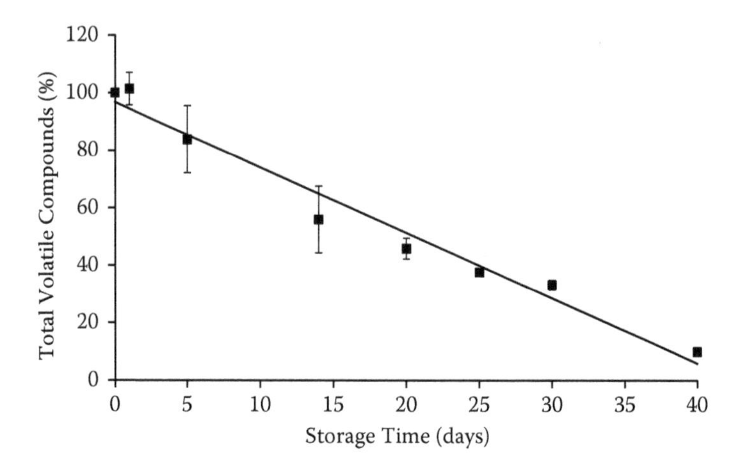

Figure 9.15 Total volatile peak area of roasted coffee as a function of storage time at 25°C.

where A is the residual percentage of the total volatile compounds peak area, and t is the storage time. The coefficient of determination R^2 was 0.972; diagnostics based on residuals gave no evidence of violations of the assumptions for linear regression.

9.7.3.4 Conversion of the Sensory Acceptability Limit into the Analytical One

Using this equation, it was possible to determine the area of the total volatiles in the headspace corresponding to the end of sensory shelf life. As 50% of consumers rejecting the product is a well-accepted limit in shelf life studies (Cardelli and Labuza, 2001), the percentage of residual volatile compounds corresponding to 24.5 days of storage was calculated (see Table 9.14). A total volatile area of 40.5% (95% CI 35.2–47.1) was found with respect to that of the coffee at the moment of opening the package.

9.7.3.5 Secondary Shelf Life Assessment by Instrumental Analysis

Because the volatile compounds were found to follow pseudo-zero-order kinetics, the secondary shelf life of coffee SL, expressed in days, can be predicted for different temperatures using the following equation:

$$SL = \frac{A_f - A_i}{k} \tag{9.14}$$

where A_f is the percentage area of volatile compounds corresponding to the limit of coffee sensory acceptability (i.e., 40.5% for 50% consumer rejection), A_i is the percentage area of volatile compounds at zero storage time (i.e., at the moment of package opening), and k represents the rate of change of the area of the volatile compounds in the headspace at the temperature of interest.

9.7.3.6 *Experimental Needs*

- 600 g ground roasted coffee
- 180 plastic containers (30 ml) with cap for sensory analysis
- Two 500-ml screw-capped glass bottles
- 16 vials with septa and metallic caps for instrumental analysis
- A thermostatic cell working at 25°C
- A gas chromatograph equipped with a flame ionization detector and a capillary column
- Analytical/technical balances
- Other lab facilities
- 30 panelists
- A sensory lab with individual booths and artificial daylight-type illumination, temperature control, and air circulation
- PC and statistical packages

9.7.3.7 *Experimental Plan*

See plan on the next page.

9.8 CASE STUDY 9.7. SHELF LIFE PREDICTION OF FROZEN SHRIMP

9.8.1 Product Presentation

Frozen raw or cooked shrimp are valuable as a commercial product due to their competitive price and extended shelf life. Frozen shrimp prepared from a sound product, after any suitable preparation, are subjected to a quick freezing process and wrapped in vapor-impermeable packaging to avoid dehydration. The frozen shrimp are maintained at a nominal temperature of -18°C during transportation, storage, and distribution. The shelf life generally attributed to this product is around 9 months.

9.8.2 Problem

Significant temperature fluctuations generally occur during retail storage of frozen shrimp; this can be responsible for accelerated product quality decay. To estimate the effect of nonisothermal handling of the product, producers have developed empirical formulas. For instance, it is widely accepted that quality changes occurring after 1 month of storage at -18°C correspond to those that happen in 1 week of storage at -7°C. This case study considered the development of a predictive model to verify such a relationship.

Experimental Plan

Action	Storage Time (days) at 25°C																																								
	0	1	2	3	4	5	6	7	8	9	10	11	12	13	14	15	16	17	18	19	20	21	22	23	23	24	25	26	27	28	29	30	31	32	33	34	35	36	37	38	39	40
Opening/closure																																										
Sample withdrawal for GC analysis																																										
Sample withdrawal for sensory analysis																																										

9.8.3 Shelf Life Study

9.8.3.1 *Identification of the Early Quality Decay Event and Critical Descriptor*

The most important quality changes responsible for the quality decay of frozen shrimp are protein denaturation, lipid oxidation, ice sublimation and recrystallization, and color fading (Ghosh and Nekar, 1991; Reddy et al., 1981; Bhobe and Pai, 1986; Londahl, 1997). In particular, the last consists of progressive yellowing of the product under frozen storage due to carotenoid oxidation, which can be well described by instrumental measurements of the chromatic parameter b^* (Tsironi et al., 2009). Here, we present the results obtained by quantifying the color changes of the frozen shrimp by measuring the b^* value. The aim was to find the ratio between the shelf life at a given temperature (i.e., -7°C) and the shelf life at a reference temperature (i.e., -18°C). Since the desired output is a ratio, the search for an acceptability limit is not needed.

9.8.3.2 *Testing by Instrumental Analysis*

Thirty-four packages (cardboard lined with high-density polyethylene [HDPE] film) each containing 125 g of freshly produced frozen shrimp were divided in four aliquots and stored at -7°C, -12°C, -18°C, and -40°C in thermostatic cells. One more package was taken as a reference sample at time zero. At interval times, one package was removed from each cell, and the b^* value was measured (Tsironi et al., 2009). In particular, the analysis was carried out by using a tristimulus colorimeter (Chromameter-2 Reflectance) with an 8-mm measuring area. Measurements were carried out at three surface points on five different shrimp taken from the same package. Samples were analyzed immediately after thawing.

Figure 9.16 shows the changes in the b^* value of thawed frozen shrimp as a function of storage time at -7°C, -12°C, -18°C, and -40°C. The color change (from red to yellow) of the shrimp was observed during storage at all the temperatures considered, although it decreased with the decrease of storage temperature. No lag phase was observed.

The color changes were found to follow pseudo-zero-order kinetics, and the pseudo-zero-order rate constants are shown in Table 9.15.

To describe the temperature dependence of color change rate, the Arrhenius equation can be used:

$$\ln k = \ln k_0 - \frac{E_a}{RT} \tag{9.15}$$

where k is the reaction rate constant, k_0 is the pre-exponential factor of the frequency factor, R is the molar gas constant (8.31 J/K mol), T is the absolute temperature (K), and E_a is the activation energy (J/mol).

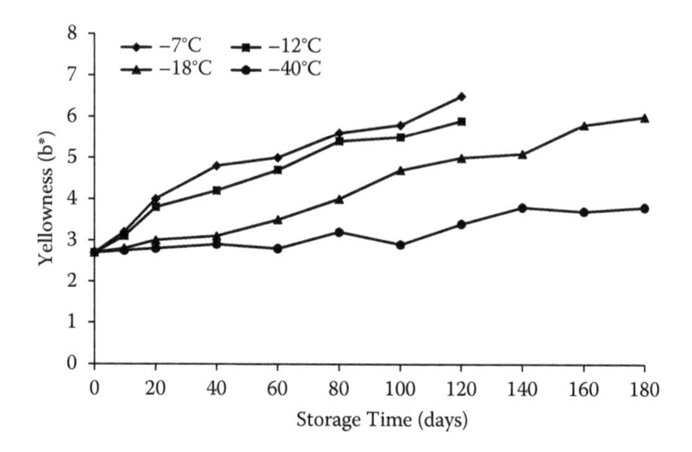

Figure 9.16 Changes in the b^* value of thawed frozen shrimp as a function of storage time at -7°C, -12°C, -18°C, and -40°C.

Table 9.15 Pseudo-Zero-Order Rate Constants of Frozen Shrimp Yellowness as a Function of Storage Temperature

T (°C)	k (b^*/days)	Intercept	R^2	p
-7	0.0291	3.130	0.944	5.68E-5
-12	0.0261	3.009	0.951	2.31E-5
-18	0.0194	2.549	0.981	2.53E-9
-40	0.0066	2.616	0.844	3.98E-7

Figure 9.17 shows the Arrhenius plot of the yellowness rate constants as a function of temperature.

The best-fitting equation resulted in the following:

$$\ln k = 7.33 - 2880 \times \frac{1}{T} \tag{9.16}$$

Due to the goodness of fit ($R^2 = 0.995$; p value = 0.0016), this model can be used to predict the temperature dependence of shrimp yellowness in the temperature range from -7°C to -40°C. The activation energy E_a was equal to 23.9 kJ/mol.

Since the color changes were found to follow pseudo-zero-order kinetics, frozen shrimp shelf life can be computed as follows (see Chapter 5):

$$SL = \frac{b^*_{\text{lim}} - b^*_{T0}}{k_T} \tag{9.17}$$

where b^*_{T0} is the initial b^* value, b^*_{lim} is the b^* value corresponding to the acceptability limit, and k_T is the rate constant at temperature T.

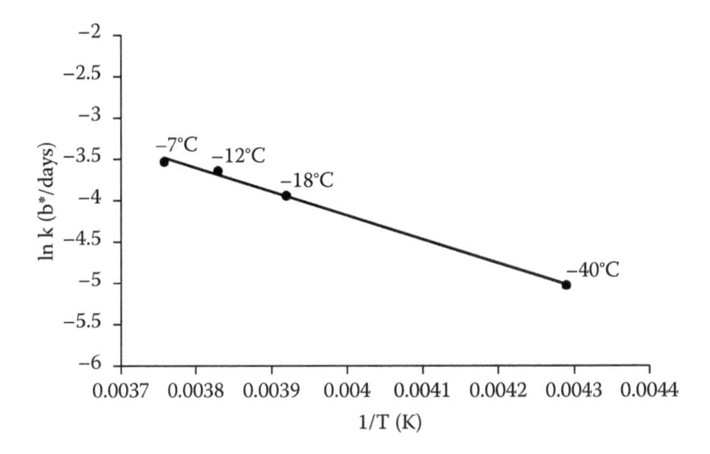

Figure 9.17 Arrhenius plot of frozen shrimp yellowness.

The shelf life factor, defined as the ratio between the shelf life value at a given temperature T and the shelf life at a reference temperature T_{ref}, can be calculated:

$$\text{Shelf life factor} = \frac{SL_T}{SL_{Tref}} = \frac{k_{ref}}{k_T} \tag{9.18}$$

Considering T_{ref} = -18°C, the shelf life factors were calculated for -7°C, -12°C, and -40°C. Results are shown in Table 9.16. According to these results, the frozen shrimp stored at -7°C had a shelf life that was 0.66 times that measured at the reference temperature (-18°C).

Since the Arrhenius plot was developed, the shelf life factors referred to any other subzero temperature of interest can be easily calculated.

9.8.3.3 *Experimental Needs*

- 35 packages containing about 125 g of sample each
- Four thermostatic cells working at -7°C, -12°C, -18°C, and -40°C
- Tristimulus colorimeter
- PC and statistical packages

Table 9.16 Shelf Life Factors Calculated for Frozen Shrimp Based on *b Value Changes and Considering T_{ref} = -18°C**

	Temperature (°C)			
	-40	**-18**	**-12**	**-7**
Shelf life factor	2.96	1	0.74	0.66

9.8.3.4 *Experimental Plan*

Action	Storage Temperature (°C)	Storage Time (days)									
		0	20	40	60	80	100	120	140	160	180
Sample withdrawal for color analysis	-7										
	-12										
	-18										
	-40										

9.9 CASE STUDY 9.8. SHELF LIFE ASSESSMENT OF FRUIT-BASED NONCARBONATED SOFT DRINKS

9.9.1 Product Presentation

Fruit-based noncarbonated soft drinks are generally made from a concentrated source of fruit and do not have added carbon dioxide. In recent years, marketing of noncarbonated beverages has grown extensively, also fueled by the perception that these products are healthier than the carbonated ones. As a consequence, soft drink producers are diversifying their offerings to include noncarbonated beverages. Because of the lack of protection against spoilage offered by carbonation, noncarbonated beverages are generally pasteurized either in bulk or by continuous flash pasteurization prior to filling or in the bottle. Alternatively, their stabilization can be achieved through the use of preservatives, usually belonging to the weak acid group, such as sorbate, benzoate, lactate, propionate, acetate, and so on. Consequently, the shelf life assigned to these products is rather long (several months).

9.9.2 Problem

When the producer's task is to improve appearance and preserve taste and healthy properties, noncarbonated soft drink stabilization cannot be achieved by the application of thermal treatment, but through the addition of weak acids or other preservatives. However, currently producers are searching for alternatives to such preservatives because they are perceived by consumers as extraneous and not natural. Aroma compounds and essential oils, whose antimicrobial potential is well known, can be an interesting alternative. The main limitations to the industrial use of the essential oils as preservatives are their sensory impact and variable composition (and in turn their antimicrobial activity) (Burt, 2004; Lanciotti et al., 2004). The combination of essential oils or their components with a mild heat treatment could allow overcoming such limitations.

In this case study, a shelf life modeling approach different from those described in Chapter 6 is shown. In fact, when the goal is to assess the individual or combined effect of different processing variables, which in turn may affect the overall composition properties of the product and hence the quality decay rates, the proposed approach could be useful, as it allows reduction of the number of experiments and thus saves time and money. Specifically, in this case study the individual and combined effects of different process parameters (i.e., initial microbial concentration, heating time, and preservative concentration) were investigated by an experimental design on the growth of a target microorganism. The purpose was to identify the best processing conditions able to guarantee microbial shelf life (3 months) as requested by the producers.

9.9.3 Shelf Life Study

9.9.3.1 Identification of Early Quality Decay Event and Critical Descriptor

In our study, apple- and orange-based noncarbonated drinks were taken into account. Even if these products cannot support the growth of pathogenic or toxin-producer species, the multiplication of spoilage microflora can cause the formation of microbial clouds, sediments, alterations of flavor, and blowing, which may be responsible for serious economic losses (International Commission on Microbiological Specifications for Food, 2005). The spoilage, caused mainly by yeasts and molds, becomes perceivable when cell loads higher than 6.0 log cfu/ml are attained. For these reasons, yeasts and molds were considered the critical descriptors of product quality loss over storage time, and the relevant acceptability limit was identified in 6.0 log cfu/ml.

9.9.3.2 Testing

The probability of having unspoiled products in relation to the modulation of three specific process variables was investigated (Belletti et al., 2007). To this purpose, three variables were considered: (a) the inoculum level of a yeast strain (*Saccharomyces cerevisiae* SPA) isolated during an industrial spoilage accident; (b) the duration of the thermal treatment; and (c) the concentration of three different aroma compounds [i.e., (E)-2-hexenal, citron essential oil, and citral].

The trials were carried out on beverages prepared by diluting with distilled water industrial orange and apple juice concentrates, which were subsequently bottled in 500-ml PET bottles. One batch (200 bottles) of the apple-based beverage and a second one (400 bottles) of the orange-based beverage were obtained. According to the experimental plan (Table 9.17), the beverages were inoculated with *S. cerevisiae* SPA, whose concentration ranged from 10^1 to 10^5 cfu/bottle (the usual yeast contamination for such products can reach 1–2 cfu/ml). Then, increasing amounts of the aroma compounds were added to the beverages (Table 9.17), namely, (E)-2-hexenal to the apple-based beverages, citron essential oil to half of the orange-based beverages, and citral to the remaining bottles. Finally, the bottles were heated at 55°C (such a

Table 9.17 Experimental Designs Adopted for the Evaluation of Stability of the Three Types of Beverages

	Experimental Design					Results (percentage of spoiled bottles)		
	Aroma Compound (ppm)							
Run	(*E*)-2-Hexenal	Citral	Citron Essential Oil	Thermal Treatment (min)	Inoculum (log cfu/bottle)	(*E*)-2-Hexenal	Citral	Citron Essential Oil
1	0	0	0	0	1	100	100	100
2	0	0	0	20	3	100	100	100
3	0	0	0	20	5	100	100	100
4	0	0	0	10	3	100	100	100
5	10	30	125	5	2	100	100	20
6	10	30	125	15	2	100	100	0
7	10	30	125	5	4	100	100	80
8	10	30	125	15	4	100	100	0
9	20	60	250	10	3	100	100	0
10	20	60	250	10	3	100	80	0
11	20	60	250	10	3	100	100	0
12	20	60	250	0	3	100	100	100
13	20	60	250	20	3	80	20	0
14	20	60	250	10	1	80	0	0
15	20	60	250	10	5	100	100	0
16	30	90	375	5	2	0	100	0
17	40	90	375	15	2	0	0	0
18	40	90	375	5	4	40	100	0
19	40	90	375	15	4	0	80	0
20	50	120	500	10	3	0	20	0

Source: Belletti, N., Sado Kamdem, S., Patrignani, F., Lanciotti, R., Covelli, A., and F. Gardini. 2007. Antimicrobial activity of aroma compounds against *Saccharomyces cerevisiae* and improvement of microbiological stability of soft drinks as assessed by logistic regression. *Applied and Environmental Microbiology* 73: 5580–5586. With permission.

Note: While the length of thermal treatment and the inoculum were the same in each run of the design independently of the beverage, the concentration of aroma compounds varied according to their antimicrobial activity and sensory impact. Results (expressed as percentages of spoiled bottles) observed in each run of the experimental design in relation to the type of beverage (i.e., to the aroma compound added).

temperature is markedly lower than those usually applied) for up to 20 minutes. Ten repetitions for each condition were carried out.

After 3 months of storage at room temperature (28 ± 3°C), the samples were analyzed to assess the yeast growth or nongrowth. The results were treated in order to determine the probability of microbial stability of the beverage in relation to the conditions applied. Each observation (i.e., the result obtained for each bottle) was transformed into positive and negative growth responses over time. In particular, the value 0 was assigned to the bottles in which yeast growth was not observed, while 1 meant that growth occurred. A logistic regression analysis was conducted on the raw data to assess the probability of growth after the storage period as a function of aroma compound concentration, length of thermal treatment, and inoculum level. The significance of the selected variables was evaluated first by the relation between each variable alone and the probability of absence of growth examined by a likelihood ratio test; the reduction in deviance (-2 × log likelihood) when entering the variable into a model with no other variables was tested against a χ^2 value. In addition, the significance of each variable was tested by removing it from a complete model with all variables included. Interactive and quadratic effects of significant variables were tested in the same way. The evaluation of the goodness of fit of the model was also performed by assessing the percentage of correct predictions compared to the observed experimental results. According to this procedure, logit(P) models describing the probability of spoilage were obtained for each type of beverage and, consequently, for each type of aroma compound studied (Table 9.18).

(E)-2-Hexenal. The logit model describing the microbial spoilage probability of the beverages supplemented with (*E*)-2-hexenal allowed the correct classification of 192 observations out of 200 (96.0% of correct classifications). Figure 9.18 reports the predicted probability for having unspoiled bottles after 3 months with an initial yeast inoculum of 10^3 cfu/bottle (corresponding to 2 cfu/ml).

Table 9.18 Logistic Models Predicting the Spoilage of the Beverages Added with Three Aroma Compounds

Aroma Compound	Logit Equation	χ^2	Deviance
(*E*)-2-hexenal	57.257 - 2.500[(*E*)-2-hexenal] - 1.162[time] + 5.809[inoculum]	191.46 ($p <$ 0.0001)	33.55
Citral	-5.411 + 0.053[citral] + 0.321[time] + 4.094[inoculum] - 0.0147[citral][time]	146.70 ($p <$ 0.0001)	53.41
Citron essential oil	1.151 - 0.010[citron] - 0.038[time] + 1.602[inoculum] - 0.0065[citron][time]	221.53 ($p <$ 0.0001)	22.82

Source: Belletti, N., Sado Kamdem, S., Patrignani, F., Lanciotti, R., Covelli, A., and F. Gardini. 2007. Antimicrobial activity of aroma compounds against *Saccharomyces cerevisiae* and improvement of microbiological stability of soft drinks as assessed by logistic regression. *Applied and Environmental Microbiology* 73: 5580–5586. With permission.

Note: For each equation, the relative χ^2 and deviance (-2 times log likelihood) are also reported.

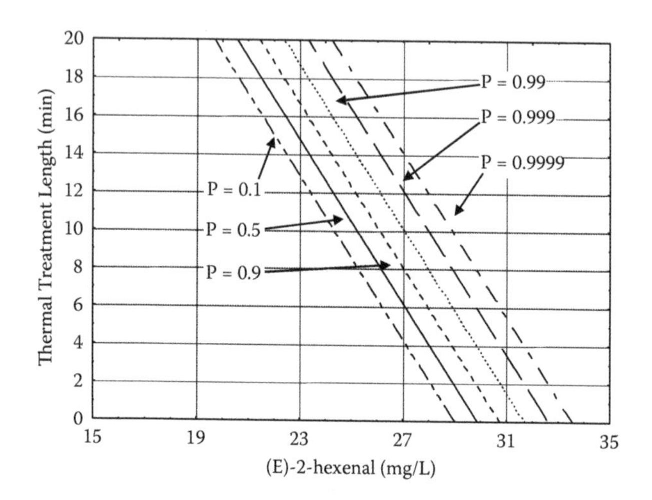

Figure 9.18 Combination of (E)-2-hexenal concentration and length of thermal treatment (at 55°C) corresponding to different probabilities of absence of yeast growth with an initial inoculum of 2 cells/ml. (Modified from Belletti, N., Sado Kamdem, S., Patrignani, F., Lanciotti, R., Covelli, A., and F. Gardini. 2007. Antimicrobial activity of aroma compounds against *Saccharomyces cerevisiae* and improvement of microbiological stability of soft drinks as assessed by logistic regression. *Applied and Environmental Microbiology* 73: 5580–5586.)

(*E*)-2-hexenal alone (in the absence of thermal treatment) was able to ensure high spoilage inhibition (>99.99%) when added at concentrations higher than 34 ppm. The same result can be achieved with lower (E)-2-hexenal concentrations combined with thermal treatment. For instance, by combining 24 ppm with 20 minutes of heating at 55°C, it was possible to inhibit more than 99.99% of yeast growth. Interestingly, once exceeding a critical threshold after which (*E*)-2-hexenal exerted a marked antimicrobial activity, very small increases of this molecule highly increased the probability of avoiding beverage spoilage.

Citral. The elaboration of the data reported in Table 9.17 with the logit model included the interaction between citral concentration and duration of the thermal treatment (Table 9.18). In fact, the inclusion of this term determined a significant improvement of the value of χ^2 and -2 × log likelihood. Eight observations of 200 were misclassified by the model (96.0% of correct assignments). Figure 9.19 shows the predicted probability to have unspoiled bottles after 3 months with an initial yeast inoculum of 10^3 cfu/bottle (corresponding to 2 cfu/ml) in relation to the length of the thermal treatment and citral concentration.

A satisfactory probability (>99%) of reaching the desired shelf life can be achieved only with the highest concentration of citral considered in this trial and only when heating at 55°C was prolonged over 15 minutes.

Citron essential oil. As for citral, in addition to the linear terms of the variables, the insertion in the model of the interaction between heating time and essential oil concentration improved it significantly. This model allowed correct classification of 98% of the observations. Figure 9.20 shows the predicted probability to ensure the absence of spoilage after 3 months with an initial yeast inoculum of 2 cfu/ml in relation to the duration of heating and citron essential oil concentration.

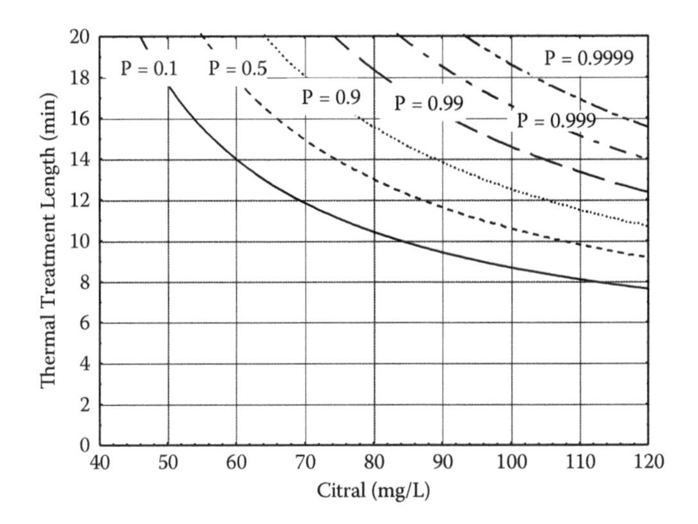

Figure 9.19 Combination of citral concentration and length of thermal treatment (at 55°C) corresponding to different probabilities of absence of yeast growth with an initial inoculum of 2 cells/ml (Modified from Belletti, N., Sado Kamdem, S., Patrignani, F., Lanciotti, R., Covelli, A., and F. Gardini. 2007. Antimicrobial activity of aroma compounds against *Saccharomyces cerevisiae* and improvement of microbiological stability of soft drinks as assessed by logistic regression. *Applied and Environmental Microbiology* 73: 5580–5586.)

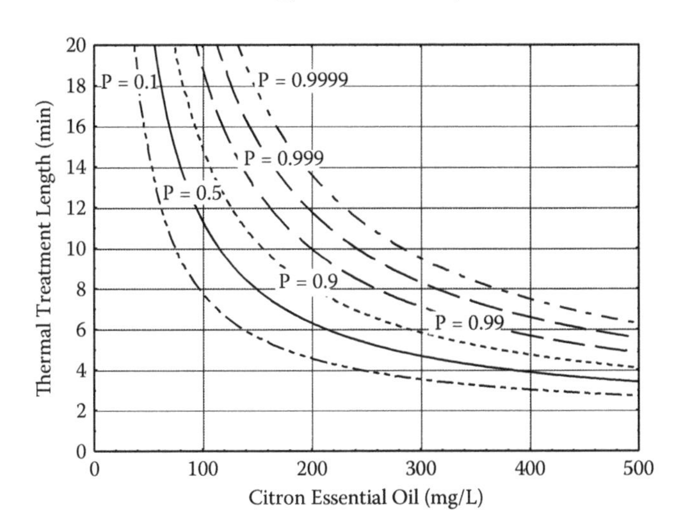

Figure 9.20 Combination of citron essential oil concentration and length of thermal treatment (at 55°C) corresponding to different probabilities of absence of yeast growth with an initial inoculum of 2 cells/ml. (Modified from Belletti, N., Sado Kamdem, S., Patrignani, F., Lanciotti, R., Covelli, A., and F. Gardini. 2007. Antimicrobial activity of aroma compounds against *Saccharomyces cerevisiae* and improvement of microbiological stability of soft drinks as assessed by logistic regression. *Applied and Environmental Microbiology* 73: 5580–5586.)

Neither the thermal treatment nor the essential oil, if used alone, was able to prevent yeast growth. On the contrary, a strong interactive effect was observed by combining the two treatments. For instance, the desired shelf life (3 months) can be achieved with high probability (>99.99%) using high concentrations of citron essential oil (equal to or higher than 500 ppm, which is compatible with the concentration at which this oil is used as a flavoring agent) in the presence of very short thermal treatments (less than 7 minutes), while prolonged heating (20 minutes) at 55°C reduced the essential oil concentration needed to reach the same result at about 120 ppm.

The procedure here reported does not allow prediction of the shelf life of a food under specific conditions. However, it is useful for verification if the different conditions satisfy the producer‘s shelf life expectations. The collection of binary data (spoiled/unspoiled) can be simple, but the reliability of the model is strongly influenced by the number of observations, as it happens in all the probabilistic models. Moreover, data elaboration requires the supervision of an expert microbiologist particularly skilled in the field of predictive microbiology.

9.9.3.3 Experimental Needs

- 600 bottles of noncarbonated beverages (200 for each aroma compound)
- Material (culture media, Petri dishes, tubes, pipettes) and equipment (autoclaves, incubators) necessary for microbial cultivation and analyses
- Statistical packages able to run the logit equation

9.9.3.4 Experimental Plan

	Time (days)		
Action	**0**	**90**	**>90**
Sample inoculation	■		
Storage		■	
Growth/nongrowth assessment and modeling			■

9.10 REFERENCES

Anese, M., Manzocco, L., and M.C. Nicoli. 2006. Modeling the secondary shelf life of ground roasted coffee. *Journal of Agriculture and Food Chemistry* 54: 5571–5576.

Belletti, N., Sado Kamdem, S., Patrignani, F., Lanciotti, R., Covelli, A., and F. Gardini. 2007. Antimicrobial activity of aroma compounds against *Saccharomyces cerevisiae* and improvement of microbiological stability of soft drinks as assessed by logistic regression. *Applied and Environmental Microbiology* 73: 5580–5586.

Beuchat, L.R. 2002. Ecological factors influencing survival and growth of human pathogens on raw fruits and vegetables. *Microbes and Infection* 4: 413–423.

Bhobe, A.M., and J.S. Pai. 1986. Study of the properties of frozen shrimps. *Journal of Food Science and Technology* 23: 143–147.

Breidt, F., and H.P. Fleming. 1997. Using lactic acid bacteria to improve the safety of minimally processed fruits and vegetables. *Food Technology* 51: 44–51.

Brinez, W.J., Roig-Sagues, A.X., Hernandez-Herrero, M.M., and B. Guamis-Lopez. 2006. Inactivation of *Listeria innocua* inoculated into milk and orange juice using ultra-high-pressure homogenization. *Journal of Food Protection* 69: 86–92.

Burt, S. 2004. Essential oils: their antibacterial properties and potential applications in foods—a review. *International Journal of Food Microbiology* 94: 223–253.

Buta, J.G., Moline, H.E., Spaulding, D.W., and C.Y. Wang. 1999. Extending shelf-life of fresh-cut apples using natural products and their derivatives. *Journal of Agriculture and Food Chemistry* 47: 1–6.

Calligaris, S., Da Pieve, S., Kravina, G., Manzocco, L., and M.C. Nicoli. 2008. Shelf life prediction of breadsticks using oxidation indices: a validation study. *Journal of Food Science* 73: E51–E56.

Calligaris, S., Manzocco, L., Kravina, G., and M.C. Nicoli. 2007. Shelf life modeling of bakery products by using oxidation indices. *Journal of Agricultural and Food Chemistry* 55: 2004–2009.

Cardelli, C., and T.P. Labuza. 2001. Application of Weibull hazard analysis to the determination of the shelf life of roasted and ground coffee. *Lebensmittel Wissenschaft und Technologie* 34: 273–278.

Diels, A.M.J., and C.W. Michiels. 2006. High-pressure homogenization treatment as a nonthermal technique for the inactivation of microorganisms. *Critical Reviews in Microbiology* 32: 201–216.

European Official Methods of Analysis, Regulation 2568/91. 1991. *Official Journal of European Community* L. 248

Gardini, F., Lanciotti, R., and M.E. Guerzoni. 2001. Effect of (E)-2-hexenal on the growth of *Aspergillus flavus* in relation to its concentration, temperature and water activity. *Letters of Applied Microbiology* 33: 50–55.

Ghosh, S., and D.P. Nekar. 1991. Preventing discoloration in small dried shrimps. *Fleischwirtschaft* 71: 834–835.

Hatcher, W.S., Jr., Parish, M.E., Weih, J.L., Splittstoesser, D.F., and B.B. Woodward. 2000. Fruit beverages. In *Methods for the microbiological examination of foods*, F.P. Downes and K. Ito, Eds., 565–568. Washington, DC: American Public Health Association.

Hough, G. 2010. *Sensory shelf life estimation of food products*. Boca Raton, FL: Taylor & Francis.

Hough, G., Lamgohr, K., Gomez, G., and A. Curia. 2003. Survival analysis applied to sensory shelf life of foods. *Journal of Food Science* 68: 359–362.

International Commission on Microbiological Specifications for Food (ICMSF). 2005. *Microorganisms in foods 6. Microbial ecology of food commodities. Cap 13—soft drinks, fruit juices, concentrates, and fruit preserves*. 544–573. New York: Kluwer Academic/Plenum.

Johnson, J.R., Braddock, R.J., and C.S. Chen. 1995. Kinetics of ascorbic acid loss and nonenzymatic browning in orange juice serum: experimental rate constants. *Journal of Food Science* 60: 502–505.

Lanciotti, R., Corbo, M.R., Gardini, F., Sinigaglia, M., and M.E. Guerzoni. 1999. Effect of hexanal on the shelf-life of fresh apple slices. *Journal of Agricultural and Food Chemistry* 47: 4769–4776.

Lanciotti, R., Gianotti, A., Patrignani, F., Belletti, N., Guerzoni, M.E., and F. Gardini. 2004. Use of natural aroma compounds to improve shelf life and safety of minimally processed fruits. *Trends in Food Science and Technology* 15: 201–208.

Lee, H.S., and G.A. Coates. 1999. Vitamin C in frozen, fresh squeezed, unpasteurized, polyethylene-bottled orange juice: a storage study. *Food Chemistry* 65: 165–168.

Londahl, G. 1997. Technological aspects of freezing and glazing shrimps. *Infofish International* 3: 49–56.

López-Galilea, L., Fournier, N., Cid, C., and E. Guichard. 2006. Changes in headspace volatile concentrations of coffee brews caused by the roasting process and the brewing procedure. *Journal of Agriculture and Food Chemistry* 54: 8560–8566.

Manzocco, L., Calligaris, S., and M.C. Nicoli. 2010. Methods of shelf life determination and prediction. In *Oxidation in foods and beverages and antioxidant applications*, E. Decker, R. Elias, and D.J. McClements, Eds., 196–222. Oxford, UK: Woodhead.

Manzocco, L. Calligaris, S., and M.C. Nicoli. 2011. Coffee. In *Food and beverage shelf-life and stability*, D. Kilcast and P. Subramaniam Eds., 615–640. Oxford, UK: Woodhead.

Nicoli, M.C., Anese M., and L. Manzocco. 1999. Oil stability and antioxidant properties of an oil tomato food system as affected by processing. *Advances in Food Science* 21: 10–14.

Nicoli, M.C., Calligaris, S., and L. Manzocco. 2009. Shelf life testing of coffee and related products: uncertainties, pitfalls, and perspectives. *Food Engineering Review* 1: 159–168.

Nicoli, M.C., Toniolo, R., and M. Anese. 2004. Relationship between redox potential and chain-breaking activity of model systems and foods. *Food Chemistry* 88: 79–83.

Pathanibul, P., Taylor, T.M., Davidson, P.M., and F. Harte. 2009. Inactivation of *Escherichia coli* and *Listeria innocua* in apple and carrot juices using high pressure homogenization and nisin. *International Journal of Food Microbiology* 129: 316–320.

Patrignani, F., Vannini, L., Kamdem Sado, S.L., Lanciotti, R., and M.E. Guerzoni. 2009. Effect of high pressure homogenization on *Saccharomyces cerevisiae* inactivation and physico-chemical features in apricot and carrot juices. *International Journal of Food Microbiology* 136: 26–31.

Patrignani, F., Vannini, L., Kamdem Sado, S.L., Lanciotti, R., and M.E. Guerzoni. 2010. Potentialities of high pressure homogenization to inactivate *Zygosaccharomyces bailii* in fruit juices. *Journal of Food Science* 75: M116–M120.

Reddy, S.K., Nip, W.K., and C.S. Tang. 1981. Changes in fatty acids and sensory quality of fresh water shrimp (*Macrobrachium rosenbergii*) stored under frozen conditions. *Journal of Food Science* 46: 353–356.

Rice Evans, C.A., Sampson, J., Bramley, P.M., and D.E. Holloway. 1997. Why do we expect carotenoids to be antioxidants in vivo? *Free Radical Research* 26: 381–398.

Sancho, T., Giménez-Jurado, G., Malfeito-Ferreira, M., and V. Loureiro. 2000. Zymological indicators: a new concept applied to the detection of potential spoilage yeast species associated with fruit pulps and concentrates. *Food Microbiology* 17: 613–624.

Scientific Committee for Food (SCF). 1993. Nutrient and energy intakes for the European Community. Reports of the Scientific Committee for Food (31st series). European Commission, Luxembourg.

Song, J., Leepipattanawit, R., and R. Beaudry. 1996. Hexanal vapour is a natural, metabolizable fungicide: inhibition of fungal activity and enhancement of aroma biosynthesis in apple slice. *Journal of American Society for Horicultural Science* 121: 937–942.

Suárez-Jacobo, A., Gervilla, R., Guamis, B., Roig-Sagués, A.X., and J. Saldo. 2010. Effect of UHPH on indigenous microbiota of apple juice: a preliminary study of microbial shelf-life. *International Journal of Food Microbiology* 136: 261–267.

Tsironi, T., Dermesolouoglou, E., Giannakourou, M., and P. Taoukis. 2009. Shelf life modeling of frozen shrimps at variable temperature conditions. *LWT—Food Science and Technology* 42: 664–671.

United States Department of Agriculture (USDA). 2010. http://iom.edu/Activities/Nutrition/SummaryDRIs/~/media/Files/Activity%20Files/Nutrition/DRIs/DRI_Vitamins.ashx

Varnam, A.H., and J.P. Sutherland. 1999. *Beverage: technology, chemistry and microbiology*. Gaithersburg, MD: Aspen.

Zwietering, M.H., Jongenberger, I., Roumbouts, F.M., and K. van't Riet. 1990. Modelling of bacterial growth curve. *Applied Environmental Microbiology* 56: 1875–1881.

Index

F

G

H

I

K

L

M

N

O

P

Q

R

S

For Product Safety Concerns and Information please contact our EU representative GPSR@taylorandfrancis.com
Taylor & Francis Verlag GmbH, Kaufingerstraße 24, 80331 München, Germany

www.ingramcontent.com/pod-product-compliance
Ingram Content Group UK Ltd.
Pitfield, Milton Keynes, MK11 3LW, UK
UKHW021016180425
457613UK00020B/955

* 9 7 8 1 1 3 8 1 9 9 3 4 7 *